PRACTICE OF THIN LAYER
CHROMATOGRAPHY

Practice of Thin Layer Chromatography

Second Edition

JOSEPH C. TOUCHSTONE
MURRELL F. DOBBINS
University of Pennsylvania
School of Medicine

A Wiley-Interscience Publication

JOHN WILEY & SONS INC.
New York • Chichester • Brisbane • Toronto • Singapore

Library of Congress Cataloging in Publication Data:

Touchstone, Joseph C., Dobbins, Murrell F.
  Practice of thin layer chromatography.

  "A Wiley-Interscience publication."
  Includes bibliographical references and index.
  1. Thin layer chromatography.   I. Title. [DNLM:
1. Chromatography, Thin layer.
II. Dobbins, Murrell F., joint author.   QD 79.C8 T631p]

QD79.C8T68 1982   543′.08956   82-13654
ISBN 0-471-09766-7

Printed in the United States of America

10 9 8 7 6 5 4 3 2

This book is dedicated to our wives
PHYLLIS and CLARE

Preface

There are several widely used texts on thin layer chromatography
(TLC) each of which is composed of two major sections: one section dealing with basics, the other dealing with specific compound
class applications. In most instances, the chapters dealing with
the basics of TLC do not cover their topics in enough depth and
detail to enable the reader to carry out the operation described
with confidence and success.

In this volume, we have attempted to describe in detail all
the basic operations necessary for successful TLC in such a way
that the reader may readily carry them out without hesitation.
Rather than presenting complete literature reviews in each of the
topic areas, we have drawn information from practical experience
and from the literature describing practical methods. For best
overall results, the beginning TLC practitioner should read this
book from beginning to end before beginning any chromatography in
order to familiarize himself or herself with the entire TLC process. This familiarity will allow each step to proceed more
smoothly than might otherwise be possible.

Using this information, the worker in TLC can readily perform
the desired separations in his or her own laboratory.

This volume would not have been possible without the generous
support of the Department of Obstetrics and Gynecology, School of
Medicine, the University of Pennsylvania. Our thanks are also extended to the firms and individuals who granted permission to use

graphs and illustrations of equipment and procedures. The inclu-
sion of equipment from a given supplier should not be construed as
a recommendation, nor should failure to cite any product or supplier be construed as a lack of recommendation.

  We also appreciate the devoted work of those who diligently
worked to assemble the data: Mr. Philip Blackwood, Mrs. Clare R.
Dobbins, Miss Cheryl Crowder. A very special thanks to Dr. Herman
Felton for constructive criticism of the manuscript.

Joseph C. Touchstone
Murrell F. Dobbins

Philadelphia, Pennsylvania
October 1982

Contents

Glossary of TLC Terms

absorbent

Substance that absorbs others, usually on the basis of wetting ability.

activation

The process of heating a TLC plate to drive off moisture resulting from layer preparation or adsorption from the atmosphere.

adsorbent

Substance adhering to another due to attraction between surface atoms of the two substances resulting from intermolecular forces such as hydrogen bonding, electrostatic forces, and charge-transfer forces.

adsorption chromatography

Process whereby a sample is separated by interaction between adsorptive forces of a medium (stationary phase) and a solvent (mobile phase).

alumina

Common adsorbent; Al_2O_3

argentation TLC

TLC employing silver nitrate impregnated in the layer material, usually
silica gel. This impregnation changes
the separation characteristics of the
silica gel.

ascending chromatography

Chromatography in which the mobile
phase moves upwards in the medium.

band

Chromatographic zone; region where the
separated substance is concentrated.

bed

A column or layer of porous material
containing the stationary phase, the
interstices being filled with mobile
phase.

CC

Column chromatography.

cellulose

Common medium for separation on a TLC
plate.

chamber

Tank, jar, or vessel in which chromatographic separation takes place.

chamber saturation

Equilibration of the chamber or tank
with mobile phase before the plate is
placed into it.

chromatogram

A series of separated bands or zones
in or on the stationary phase. The
end product of the chromatography
process.

chromatographic solvent

Solvent or mixture of solvents used as
the mobile phase.

chromatographic system

Combination of the solvent, the sorbent, and components of the sample
mixture. The interactions of the
system determine the selectivity of
the separation.

chromatography

A method of analysis in which the flow
of a mobile phase (gas or liquid) pro-
motes the separation of substances by
differential migration from a narrow
initial zone in a sorptive medium.

chromatoplate

A thin layer plate; a layer of sorbent coated on a solid support such as glass, aluminum, or plastic.

continuous development

Development occurring over a distance that is usually greater than one plate length. Development is often expressed as a function of time rather than distance.

deactivation

The process of making the chromatographic layer less active to decrease its separation capabilities. Usually done with water.

densitometry

Measurement of a zone on a layer with an instrument that determines the optical density of the zone.

descending chromatography

Chromatography in which the mobile phase moves downwards in the sorptive medium.

destructive detection

A detection process that changes the chemical nature of the substance being detected in an irreversible manner. Sulfuric acid charring is one example.

detection

The process of locating a separated substance on a chromatogram, whether by physical methods, chemical methods, or biological methods.

developing solvent

Mobile phase.

development

The movement of mobile phase in the chromatogram to effect separation.

diatomaceous earth

A naturally occurring fine white powder, formed from the skeletons of microscopic marine organisms. Also called kieselguhr.

distribution coefficient

$$k = \frac{\text{amount of solute per unit of stationary phase}}{\text{amount of solute per unit of mobile phase}}$$

eluent

Solvent that removes a sample from a medium.

eluotropic series

Series of solvents or solvent mixtures arranged in order of eluting power.

elution

Removal of a solute from a sorbent by passage of a suitable solvent.

flat-bed chromatography

Common term for thin layer or paper chromatography occurring in a single plane. Sometimes called planar chromatography.

front

The visible boundary at the junction of the mobile-phase wetted layer and the "dry" layer. If a trough chamber is used with equilibration for development, the "dry" layer can contain amounts of the mobile phase.

GC (GLC)

Gas (liquid) chromatography.

gradient elution

Development using a solvent system whose composition is continuously changing to effect separation; normally done to increase the strength of the eluent.

gradient TLC

Separation on a sorbent layer that has changing characteristics, that is, a gradient, from one portion of the layer through an adjoining portion of the layer.

HPLC

High pressure liquid chromatography or high performance liquid chromatography.

HPTLC

High performance thin layer chromatography.

hR_f

$100 \times R_f$

impregnation

Loading of the sorbent with a liquid or a solid to change the chromatographic behavior of the layer. An example is $NaNO_3$ impregnated silica gel.

in situ   Occurring in place, e.g., on the thin
                                 layer.

ion exchange   Process whereby ions of the same
                                 charge replace each other in a given
                                 phase. In chromatography, it usually
                                 refers to situations where the sta-
                                 tionary phase is made of an ionic
                                 polymer, which can be a synthetic res-
                                 in or a specially treated mineral.

IR   Infrared.

kieselguhr   See "diatomaceous earth."

migration   Travel of sample in the medium in the
                                 direction of the mobile phase.

mobile phase   The moving phase (solvent or gas) of
                                 a chromatographic system.

MS   Mass spectrometry.

multiple chromatography   Chromatography repeated a number of
                                 times using the same or different
                                 mobile phases.

nondestructive detection   Detection of a substance on a chro-
                                 matogram by a process that will not
                                 permanently change the chemical nature
                                 of the substance being detected.
                                 Visualization with iodine vapor is one
                                 example of a nondestructive method.

origin   Point where sample is applied.

partition   Divide or distribute between.

partition chromatography   Process in which sample is separated
                                 by partition between two liquid phases
                                 or between a gas and a liquid. One
                                 liquid is stationary while the other
                                 is mobile.

partition coefficient   Ratio of concentration of solute after
   or ratio (K_d) partition between two immiscible
                                 phases:

$K_d = C_s/C_m$, where C_s and C_m are the concentration in the stationary and mobile phases, respectively.

PC · · · · · Paper chromatography.

PLC · · · · · Preparative layer chromatography. Used for the separation of larger amounts of substance than are normally separated with regular, analytical TLC. Normally a thicker layer (500-2000 μ) or sorbent is employed than in TLC.

PMD · · · · · Programmed multiple development. The repeated development of a TLC plate with the same mobile phase in the same direction for gradually increasing distances.

polar · · · · · Highly charged, or with uneven electrical charges. Degrees of solubility 'in water can be used as a measure of polarity. In organic chemistry a polar molecule is one with a large dipole moment. In chromatography a polar molecule is one whoe distribution coefficient favors the polar phase. Affinity of substances for polar solvents depends on their dipole moments and their molecular volumes. It is clear that polarity in the strict sense is not always synonymous with solubility.

precoated plates or sheets · · · · · Commercially available thin layer plates or sheets ready for use in TLC.

resolution · · · · · The degree of separation between two substances expressed as.

$$Rs = \tfrac{1}{4}(\alpha - 1) \sqrt{N} \ (k'/k'+1),$$

where α is the separation factor or the ratio of the capacity factors between two solutes k_1/k_2; N is the number of theoretical plates in the sorbent bed; k' is the average of k_1 and k_2. The capacity factor k is the equilibrium ratio of total solute in the stationary phase to total solute in the mobile phase.

reversed-phase

Chromatography on a sorbent impregnated with a nonpolar and nonvolatile liquid as a stationary phase. Separation is effected by a polar mobile phase.

R_f value

A ratio: the distance from the origin to the center of the separated zone divided by the distance from the origin to the solvent front.

R_m value

$\text{Log } (\frac{1}{R_f} - 1)$

sandwich chamber

Developing chamber formed from the plate itself, a spacer, and another nonlayered cover plate that stands in a trough containing the mobile phase.

secondary front

An additional solvent (mobile phase) front, lower than the primary solvent front. Occurs because components of the mobile phase have demixed and migrated apart from the other components.

silica gel

Silicic acid. The most widely used sorbent for TLC. Also used in column chromatography.

solvent

Liquid used for mobile phase. Not identified *a priori* with mobile phase.

solvent front

The forwardmost point of the mobile phase during development.

sorbent

A generalized term for the chromatographic stationary phase in which the nature of the force (adsorption, ion exchange, or reversed phase) is not specified.

starting point line	Position on chromatogram where the sample is applied. Usually 10-20 mm from the bottom of the plate. Also called the origin.
stationary phase	The phase of the chromatographic system that is made up of the surface of an adsorbent or liquid held by the support of a partition or gel system.
stepwise elution	Development using an eluent whose composition is changed using discontinuous, stepped gradients, in contrast to *gradient elution*.
support	The sheet of glass, plastic, or aluminum upon which the TLC sorbent is coated. Gives physical strength to the layer.
tailing	Incomplete separation of zones, often resulting in elongation of a zone.
TLC	Thin layer chromatography (chromatogram).
TLG	Thin layer gel chromatography employing a gel, such as Sephadex, coated on a glass plate for the separation of molecules predominantly according to their size.
two-dimensional chromatography	Successive development of a chromatogram in directions orthogonal to each other with the same or different mobile phase.
two-dimensional development	A two-step development technique in which a plate is first developed in one mobile phase, then dried, turned through 90°, and developed in a second, different mobile phase.
UV	Ultraviolet light.

zone (also spot or
  band)
                        The distribution of the solute or
                        separated compound on the stationary
                        phase before, during, and after chro-
                        matography.

CHAPTER 1
Basics of Thin Layer Chromatography

1.1 INTRODUCTION

Thin layer chromatography (TLC) is one of the most popular and widely used separation techniques. The reasons for this are many, and include ease of use, wide application to a great number of different samples, high sensitivity, speed of separation, and relatively low cost. A variety of apparatus for TLC is commercially available.

TLC is used for the separation of substances in a wide variety of fields. Amino acids from food protein, hallucinogenic alkaloids from plants, steroids from the urine of a newborn infant, morphine in the blood of an overdose victim, and pesticides from soil may be separated by TLC, with sensitivities of one microgram or less. Many separations can be accomplished in less than an hour, at very reasonable cost. An initial investment of a couple of hundred dollars will purchase most of the equipment necessary to do TLC on a qualitative basis. Instrumentation for quantitative analysis on the TLC plate will cost additional thousands of dollars.

Chromatography is a method of separating a mixture into its various components. It makes use of heterogeneous equilibrium established during the flow of a solvent called the mobile phase through a fixed (stationary) phase to separate two or more components from material carried by the solvent. The stationary phase can be either solid or liquid. The mobile phase can be either liquid or gas. Thus, chromatography can be classified as (a) liquid-solid, (b) liquid-liquid, (c) gas-solid, or (d) gas-liquid.

1

At this point it is desirable to introduce a number of important concepts. The first of these is polarity. A polar compound (one with high chromatographic polarity) is one that is held by the stationary phase, whereas a nonpolar substance tends to move forward in the mobile phase and be held less by the stationary phase. This term is also used to describe the solvents used as the mobile phase. Methanol, for example, is a polar solvent, since it is extremely effective as a mobile phase in moving substances through the stationary phase. Cyclohexane is an example of a nonpolar solvent; it will not move many substances. This type of polarity should not be confused with the concept used in organic chemistry, in which polarity is expressed in terms of dipole moments. In a chromatographic sense, benzene is a more polar solvent than cyclohexane, yet neither has a dipole moment.

The most common form of chromatography is adsorption chromatography. Here the stationary phase is a solid such as alumina or silica gel. The sample substance to be separated is applied at one end of this stationary phase, and the mobile phase is allowed to flow through. In TLC, capillary action in the finely divided stationary phase causes the mobile phase to move. Separation occurs when one substance in a mixture is more strongly adsorbed by the stationary phase than the other components in the mixture. Since adsorption is essentially a surface phenomenon, the degree of separation is dependent on the surface area of the adsorbent available; hence the emphasis on small particle size of the adsorbent.

The main factor in any form of chromatography is the distribution coefficient of a substance between the two phases of the system in use, where

$$\text{Distribution coefficient (k)} = \frac{\text{amount of solute per unit of stationary phase}}{\text{amount of solute per unit of mobile phase}}$$

Usually in adsorption chromatography, k is dependent on the temperature and the concentration of the solute. At a given temperature, the relationship between the amount of solute in each phase can be expressed graphically as the adsorption distribution isotherm. The ideal isotherm is a 45° linear curve obtained from plotting the concentration of the solute in the stationary phase versus its concentration in the mobile phase.

A nonlinear isotherm in a given TLC system is exemplified by the shape of the separated zone (spot) on the plate. Spots have either a well-defined front half with a tail, appearing as an upside-down teardrop, or a well-defined back half, appearing as a normal teardrop, depending on whether the distribution isotherm is convex or concave, respectively.

It is usually possible with TLC to work at such concentrations that the distribution isotherms approach linearity. If two substances in a mixture have different isotherms, they can be separated. The movement of each substance in a given system is a characteristic of the particular substance and can be used for qualitative identification.

Partition chromatography is also used in the separation of many compounds. An equilibrium between two phases is involved; one of the phases is fixed, that is, it is held by an inert support such as cellulose or kieselguhr; the other phase is mobile. In TLC, layers usually contain water adsorbed onto the surface during manufacturing or exposure to the atmosphere. This water will behave as the stationary liquid phase, partitioning the sample components between it and the mobile phase. The support material in such a case is not directly involved with the separation. The equilibrium occurs between the stationary liquid and the liquid mobile phase. The separation of a compound takes place on the basis of nonbonding interactions in the phases. The chief advantage of partition over adsorption chromatography is that the distribution isotherms are reasonably linear over a fairly wide concentration range.

Thin layer chromatography can be used: (a) to simply check the purity of a substance, (b) to attempt to separate and identify the components in a mixture, or (c) to obtain a quantitative analysis of one or more of the components present.

From a purely experimental standpoint, chromatography can be carried out in columns or layers. In column chromatography the adsorbent or medium is packed into a tube. With modern columns it is possible to reproduce separations because of uniformity in particle size and characteristics. The sample is introduced into one end of the column, and the liquid or gas is allowed to flow through. In the case of thin layer separations, the sample is applied at one end of the layer, and this end is immersed in the mobile phase with the sample just above the liquid level. The tank used for this operation is usually rectangular. After the front of the mobile phase has reached to within a short distance from the top of the layer, the plate is removed and dried prior to detection procedures.

1.2 HISTORY

Thin layer chromatography was first referred to in 1938 by two Russian workers, Izmailov and Shraiber (1) in what they called drop chromatography on horizontal thin layers. Little notice was made of the method until 10 years later, when two American chemists described the separation of terpenes in essential oils by thin layer chromatography (2). Thin layer chromatography, as it

is presently known, began to attract attention through the work
of Kirchner and his associates, starting in 1951 (3-5). The procedure was not generally accepted in its early years because
available media and apparatus for coating the plates lacked uniformity.

It was not until 1958, when Stahl (6) described equipment
and efficient sorbents for the preparation of plates, that the
effectiveness of the technique for separation was shown.

The method is now one of the most frequently described separation techniques in qualitative as well as quantitative analysis. Stahl's book, *Thin Layer Chromatography* (7), first appeared in 1965. This work and Kirchner's *Thin Layer Chromatography* (8), among others, gave the real impetus to development in
the field. In its second edition, published in 1969, Stahl's
book (9) was greatly expanded.

1.3 TLC PROCEDURE

Thin layer chromatography is a separation method in which uniform
thin layers of sorbent or selected media are used as a carrier
medium. The sorbent is applied to a backing as a coating to obtain a stable layer of suitable size. The most common support is
a glass plate, but other supports such as plastic sheets and
aluminum foil are also used. The four sorbents most commonly
used are: silica gel, alumina, kieselguhr (diatomaceous earth),
and cellulose.

Silica gel (silicic acid) is the most popular layer material.
It is slightly acidic in nature. In order to hold the silica gel
firmly on the support, a binding agent such as plaster of Paris
(calcium sulfate hemihydrate) is commonly used. This binding
agent may be omitted if the silica gel employed has a very small
particle size. Fine particles will adhere well to the support
without a binder.

Two ultraviolet (UV) indicators, to aid in the location of
separated substances, are also incorporated, either singly or
together, in silica gel or other layer materials. Zinc
silicate fluoresces when exposed to ultraviolet light of 254 nm
(mμ) wavelength, so that substances absorbing this wavelength will
contrast sharply by appearing dark while quenching the greenish-
yellow fluorescing background. The sodium salts of hydroxy-
purene-sulfonic acids fluoresce at 366 nm and provide a con-
trasting background for substances that absorb at this frequency.

Alumina (aluminum oxide) is also widely used as a sorbent.
Chemically it is basic, and for a given layer thickness it will
not separate quantities of material as large as can be separated
on silica gel. Alumina is more chemically reactive than silica

gel, and care must be exercised with some compounds and compound classes to avoid decomposition or rearrangement of these substances during sample application, storage before development, or development.

Diatomaceous earth (kieselguhr) is a chemically neutral sorbent that does not separate or resolve as well as either alumina or silica gel. It is used mainly as the support for the stationary phase in partition chromatography.

Cellulose is used as a sorbent in TLC when it is convenient to perform a given paper chromatographic separation by TLC in order to decrease the amount of time necessary for the separation and increase the sensitivity of detection. Many separations achieved by paper chromatography (PC) can be directly transferred to TLC on cellulose. This is important when only a small amount of sample is available, as TLC does not generally require as much sample as PC. The primary separation mechanism is partition, where the cellulose becomes a support for a stationary phase of water adsorbed from the atmosphere. The mobile phase is that put into the tank and used for development. Because of its tenacious nature, cellulose is usually coated onto a plate without a binder. It is available with and without fluorescent indicator.

Other substances used as sorbents include a variety of ion-exchange cellulose powders, polyamide powder, and Florisil. Commercially available coated plates employing these sorbents are listed in Chapter 3. Modern sorbent materials are of uniform particle size and characteristics, and it is possible to reproduce the quality of a layer.

The term sorbent is used in a general sense to include layer materials that may be used for either adsorption or partition chromatography. Table 1.1 lists most of the common sorbents, the major form of separation that occurs when they are used, and typical compound applications for each. Some generalizations may be made about their use. Adsorption is used to separate nonpolar, hydrophobic (non-water-soluble) substances with nonpolar mobile-phase solvents. Partition may be used for polar, hydrophilic (water-soluble) substances, with polar mobile-phase solvents. See Table 1.2.

The standard size for TLC plates is 20×20 cm. For most separations, the mobile phase is allowed to travel on the layer for a distance of 15 cm. Other plate sizes used are 5×20 cm, 10×20 cm, and 20×40 cm. "Micro" plates have been made from microscope slides.

In practice, a sample to be separated is applied on the layer 1-2 cm from one end of the plate. The furthermost edge of the application is called the starting point, or origin. Separation is achieved by passing a solvent, the mobile phase, through the

TABLE 1.1   SORBENT MATERIALS AND MODE OF SEPARATION

Sorbent	Chromatographic Mechanism	Typical Application
Silica gel	adsorption	steroids, amino acids, alcohols, hydrocarbons, lipids, aflatoxins, bile acids, vitamins, alkaloids
Silica gel RP	reversed phase	fatty acids, vitamins, steroids, hormones, carotenoids
Cellulose, kieselguhr	partition	carbohydrates, sugars, alcohols, amino acids, carboxylic acids, fatty acids
Aluminum oxide	adsorption	amines, alcohols, steroids, lipids, aflatoxins, bile acids, vitamins, alkaloids
PEI cellulose	ion exchange	nucleic acids, nucleotides, nucleosides, purines, pyrimidines
Magnesium silicate	adsorption	steroids, pesticides, lipids, alkaloids

layer. The layer, with the sample zones at the bottom, is placed on a slight angle from the vertical into a closed tank containing a small amount of the mobile phase.

The nature and chemical composition of the mobile phase is determined by the type of substance to be separated and the type of sorbent to be used for the separation. The composition of a mobile phase can be as simple as a single, pure solvent (such as benzene used to separate dyes on alumina) or as complex as a three- or four-component mixture containing definite proportions of chemically different substances, such as a 1:1:1:1 solution of n-butanol, ethyl acetate, acetic acid, and water used to separate amino acids on silica gel.

Capillary action causes the mobile phase to travel through the medium in a process called *development*. Ascending development is the most common, but horizontal, descending, and centrifugal methods have also been reported. After the plate is dried, the

TABLE 1.2 COMPARISON OF TLC SEPARATION MECHANISMS

	Adsorption	Partition	Reversed-Phase Partition
Compound types separated	Hydrophobic (lipophilic) substances of low or medium polarity	Hydrophilic inorganics and hydrophilic polar organics	Closely related hydrophobic substances
Layer type used	Activated adsorbents	Sorbent containing water, buffer, or very polar organic liquid; no activation	Sorbent containing a very nonpolar liquid stationary phase; no activation
Mobile phase	Many organics	Usually organic liquids saturated with water or buffer	Polar liquids
Common layer materials	Silica gel, alumina	Cellulose, deactivated silica gel	Cellulose, silanized silica gel
Average time for 10 cm development	20-45 min	60-90 min	60-90 min

separated spots can be visualized (made visible) in a number of
ways, such as viewing under an ultraviolet light or spraying with
one of a wide variety of reagents. The entire TLC process is
summarized in Figure 1.1 and will be described in detail in succeeding chapters of this book.

The process presented in Figure 1.1 is the simplest form of
chromatography. In some respects it is the fastest. Ideally, it
also requires the least amount of equipment. Figure 1.2 shows a
TLC kit produced by one of the major TLC suppliers. Very little
is needed to perform a separation, and one does not need commercial supplies to become an experimenter in TLC. A screw-capped
jar can be substituted for the tank. Capillary pipets (e.g.,
melting point tubes) are readily obtained for applying samples in
qualitative work. One can produce plates by simply dipping microscope slides into a slurry. Ready-made commercial plates are
becoming less expensive, save time, and often yield more reproducible results.

As stated in the diagram in Figure 1.1, the initial decision
to be made by the experimenter, provided the basic chemical nature of the sample is known, is which sorbent to use. The more
information that is known about the sample, the easier it will be
to choose the mobile phase and sorbent that will most likely effect the desired separation. Silica gel can be used as the sorbent for most applications. However, before the whole process
can be performed, some attention must be paid to the characteristics of the sample to be examined. The nature of the sample determines the sorbent. But often the sample must be subjected to
some previous separation, such as extraction or filtration, to be
suitable for satisfactory TLC. When the sample is relatively
"clean" and uncontaminated, as in the case of pharmaceutical preparations containing only two or three components, then solutions
of the sample, with proper dilution, can be applied to the TLC
plate directly. Often, however, crude extracts of biological or
industrial material are too bulky to be applied directly. These
can be "prepurified" by a TLC or column separation.

The sample can be spotted or streaked on the plate. This
operation can be manual or automatic, using one of the many spotters or streakers now available. There are merits to both techniques. This subject is discussed in detail in Chapter 4. The
spots must be dried before the plate is inserted in the tank.

The chromatogram can be developed for a short distance (less
than half the length of the plate) or to the end of the plate; it
can be developed a number of times successively with the same or
different solvents, or multidimensionally (in more than one dimension). After the chromatogram is properly dried, the spots
can be located in a number of ways. Colored zones are readily
located. Visualization under ultraviolet light can often be used.

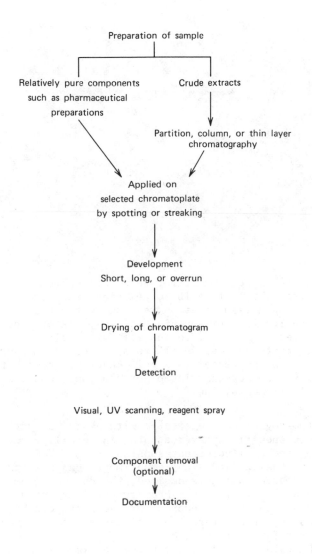

Figure 1.1. The process of thin layer chromatography.

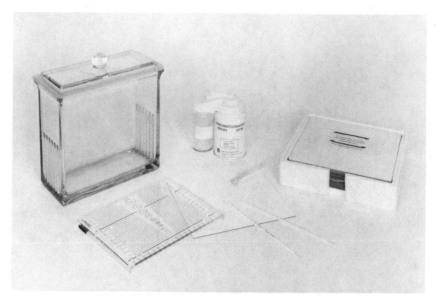

Figure 1.2. Basic equipment for thin layer chromatography. At
the rear, from the left, a TLC developing tank, propellant sprayer for applying a visualization reagent to the plate, vial of
spotting capillaries for applying samples, box of precoated commercial plates. In front, to the left, is a clear Plexiglas
spotting template used as an aid to spotting and measuring developed spots. In the center is a thin layer plate with the spotting capillaries and measuring pipets on top. Courtesy
Brinkmann Instruments, Inc.

Or the chromatogram can be treated with a wide selection of specific and nonspecific spray reagents, which will react to produce
visible colored or fluorescent substances.

After visualization, many experiments can be considered complete. If documentation is desired, the chromatogram can be
treated in a number of ways. Photography is relatively simple
and provides a permanent record. Or the plate can be covered
with a glass sheet and taped to seal the edges, and thus can be
preserved for future reference.

The R_f value is a convenient way to express the position of
a substance on a developed chromatogram. It is calculated as the
ratio:

$$R_f = \frac{\text{distance of compound from origin}}{\text{distance of solvent front front origin}}$$

R_f values are between 0 and .999, and without units. Distance is measured to the center of the sample zone or spot. Factors affecting R_f are discussed in Chapter 11.

If it is desired to express positions relative to the position of another substance x, the R_x can be calculated:

$$R_x = \frac{\text{distance of compound from origin}}{\text{distance of reference compound x from origin}}$$

It is possible for R_x values to be greater than 1. The R_x is thus the "relative retention value."

According to the types of compounds being investigated, the separations on the plate may then be evaluated by a number of instrumental techniques. If the substances àre radioactive or suspected to be so, the plate can be examined by a radioisotope scanner to locate them. Under proper conditions, it is also possible to measure both the area of a spot and its density, using a photodensitometer. These characteristics are related to the quantity of a given substance in the spot and can thereby be used to assess the amount of the substance in the sample. A quantitative calibration or reference curve should be established beforehand, using known amounts of the substance to be quantitated.

Elution methods may also be used for quantitation and measurement of radioactivity. In this technique, the substance to be measured in an area on the developed plate is scraped off the plate with a spatula or blade. It is dissolved out of the layer material (eluted) with a suitable solvent. The measurement technique is then applied to this solution.

The purpose of this book is to describe each one of these techniques in detail. The tools that can be usd will be itemized, and their relative merits discussed. The sections on sorbents and solvents and methods for their selection may be the most important. Visualization and the many detection reagents are also covered.

1.4   OTHER FORMS OF CHROMATOGRAPHY

The use of gas chromatography (GC), which developed almost concurrently with thin layer chromatography, is now widely appreciated. "High performance" liquid chromatography (HPLC) came into being with breakthroughs published in 1969. Thin layer chromatography is essentially the same as column chromatography and liquid chromatography, the only difference being that the medium or sorbent is in a layer and not in a column. These procedures are distinguished by details of equipment, material, technique, and theoretical basis. The term "liquid chromatography" (LC) refers to any chromatographic process in which the mobile phase is a liquid in contrast to the moving gas phase of gas chromatography.

A comparison of the relative merits of gas chromatography, liquid chromatography (HPLC), and thin layer chromatography illustrates some of the advantages of TLC. Gas chromatography has a high capability of separation and is used for the analysis of a wide variety of mixtures. GC can separate faster and sometimes better than other techniques. Moreover, automated equipment is now available for trouble-free unattended operation. GC has the advantage of readily reproducible results, since the parameters of operation such as temperature and gas flow can be controlled. The technique suffers in that most organic compounds cannot be handled due to their insufficient volatility or lack of thermal stability.

In contrast, TLC is not limited by the lack of volatility or thermal stability of the sample. Paper chromatography also enjoys this advantage but has the disadvantage that the times involved in achieving a separation are among the longest found in any form of chromatography.

The volatility or thermal stability of the sample is not a limitation with "modern" liquid chromatography. The term HPLC is used here to designate the more recent developments in instrumentation and packing materials that make high-pressure liquid chromatography possible. Separations by liquid chromatography can normally be done within short periods of time, although some may take hours.

One of the advantages of liquid chromatography over gas chromatography is in sample recovery. Generally, in paper chromatography, TLC, or LC, samples are readily recoverable. In column separations the mere collection of an effluent representing the desired fraction is all that is required. In paper and TLC the desired component must usually be "cut out" and eluted. The recovery of samples from GC is possible but is generally more difficult, less convenient, and not always quantitative.

In classical liquid chromatography, a column had to be packed for each separation; it could not be used repeatedly. In modern LC, columns can be used many times. In TLC, the chromatogram is generally not used again. This can represent a significant expense in material and time. However, the amount of sorbent used is small, and the average cost of a TLC plate is low, whereas packings in LC are expensive. But since packings and adsorbents available for chromatography today are very uniform, it is usually possible to reproduce the required chromatogram whether it be by TLC or HPLC.

Typical separations and analyses by classical LC were tedious and time consuming. HPLC or TLC can be done in relatively short times, often as little as 10 min, depending on the mobile phase or the sorbent. The time factor in this controversy becomes more diverse when detection methods are considered as well as the quantitative aspects. In HPLC, detection and quantitation are simple provided proper detecting instruments are available. It is possible to monitor separations other than those detected with instruments using 254 nm and 360 nm wavelengths of the light sources. Detection in TLC can follow a number of properties depending on the compound under consideration. Absorption in the ultraviolet or visible regions can be used, or the compounds can be rendered visible by chemical change before or after the development of the chromatogram.

There has, however, been some misconception about the quantitative aspects of TLC. Early workers and even contemporaries in the field eluted the compounds from the chromatograms and then carried out the appropriate quantitative analyses. Relatively few chromatographers using TLC realized that *in situ* densitometry provides a highly reproducible and sensitive methodology for quantitation of many compounds. Densitometric methodology is beyond the scope of this book; those interested should consult the recent literature on the subject (10-12). Many compounds can be determined at the picogram level with this technique, and determinations in the nanogram range are common.

1.5 ADVANTAGES OF TLC

TLC has a number of basic advantages over LC. While a methodology is being developed for a specific separation, TLC uses less solvent. The polarity of the solvent or the type of solvent mixture can be changed in a matter of minutes. Little equilibration is required, and only a small amount of solvent is needed for an exploratory chromatogram. Thus, because of short development time and easy change of the mobile phase, TLC is probably the easiest chromatographic method to set up for a specific compound. In fact, TLC is often used to develop solvent systems for LC.

TLC as well as LC can be performed using adsorption, ion-exchange, or partition media. The choice of any of these media as layers for TLC is limited only by the availability of the media in a suitable form for coating the support. Many commercial suppliers have a wide selection of these chromatographic media on precoated plates. Nearly any separation is possible if the proper combination of mobile phase and layer are used.

Probably one of the most advantageous features of TLC as opposed to other chromatographic methods is the number of samples that can be handled at one time. Gas or liquid chromatography are limited to the analysis of a single sample at one time. As many as 20 samples can be applied to a single 20×20 cm TLC plate for determination at one time. This is the main reason why TLC has been so popular for screening samples in drug abuse investigations.

The literature abounds in references to the separation of many classes of compounds. The serious practitioner of chromatography should have ready access to the basic texts and journals. Some theory of chromatography is covered in the excellent text of Giddings (13). The theory of LC is not much different from that of GC, and this book develops an understanding of the fundamentals. A comprehensive review of all books on chromatography through 1969 (14) is also valuable. Books by Marini-Bettolo (15), Bobbitt (16), Randerath (17), Macek (18), Hermann (19), among others, contain much of the early literature on TLC.

The newer technique of "high performance" thin layer chromatography (HPTLC) is discussed in two books devoted to this topic (20, 21) and is covered in Chapter 14.

Very useful is the two-volume *Handbook of Chromatography* by Zweig and Sherma (22). This handbook lists solvent systems and sorbents for the separation of many compounds. Also given are the various spray reagent formulations for detection of these compounds.

The abstracts found in the Journal of Chromatography (Elsevier, Amsterdam) are organized according to compound classes, and they often provide information about the separation and detection of the compound under examination. Literature is also published periodically by the companies that supply materials for TLC. For a list of such companies, refer to the Appendix at the end of this volume. Company newsletters often publish details of separations not yet available in conventional journals. Furthermore, these brochures may provide the detail necessary for the success of a separation, since journals do try to keep detail to a minimum in many instances.

REFERENCES

1. N. A. Izmailov and M. S. Shraiber, Farmatsiya, *3*, 1 (1938).
2. J. E. Meinhard and N. F. Hall, Anal. Chem., *21*, 185 (1949).
3. J. G. Kirchner, J. M. Miller, and G. J. Keller, Anal. Chem., *23*, 420 (1951).
4. J. M. Miller and J. G. Kirchner, Anal. Chem., *24*, 1480 (1952).
5. J. M. Miller and J. G. Kirchner, Anal. Chem., *26*, 2002 (1954).
6. E. Stahl, Chemiker Ztg., *82*, 323 (1958).
7. E. Stahl, Ed., *Thin Layer Chromatography*, Springer-Verlag, Berlin, 1965.
8. J. G. Kirchner, *Thin Layer Chromatography*, Wiley-Interscience, New York, 2nd Ed., 1978.

9. E. Stahl, Ed., *Thin Layer Chromatography*, 2 ed. Springer-Verlag, New York, 1969.
10. J. C. Touchstone, Ed., *Quantitative Thin Layer Chromatography*, Wiley-Interscience, New York, 1973.
11. J. C. Touchstone and J. Sherma, Eds., *Densitometry in Thin Layer Chromatography*, Wiley-Interscience, New York, 1979.
12. J. C. Touchstone and D. Rogers, Eds., *Thin Layer Chromatography: Quantitative Environmental and Clinical Applications*, Wiley-Interscience, New York, 1980.
13. J. C. Giddings, *Dynamics of Chromatography*, Marcel Dekker, New York, 1965.
14. Anonymous, J. Chromatogr. Sci., *8*, D2 (1970).
15. G. B. Marini-Bettolo, Ed., *Thin Layer Chromatography*, Elsevier, Amsterdam, 1964.
16. J. M. Bobbitt, *Thin Layer Chromatography*, Reinhold, New York, 1963.
17. K. Randerath, *Thin Layer Chromatography*, Academic Press, New York, 1963.
18. K. Macek, Ed., *Pharmaceutical Applications of Thin Layer and Paper Chromatography*, Elsevier, Amsterdam, 1972.
19. E. Heftmann, Ed., *Chromatography*, 3d ed., Van Nostrand Reinhold Co., New York, 1975.
20. A. Zlatkis and R. E. Kaiser, Eds., *HPTLC: High Performance Thin-Layer Chromatography*, Elsevier, Amsterdam, 1977.
21. W. Bertsch, S. Hara, R. E. Kaiser, and A. Zlatkis, Ed., *Instrumental HPTLC*, Springer-Verlag, Heidelberg, 1980.
22. G. Zweig and J. Sherma, Ed., *Handbook of Chromatography*, Chemical Rubber, Cleveland, 1972.

CHAPTER 2

Preparation of Thin
Layer Plates

2.1 BACKING SUPPORTS FOR SORBENT LAYERS

The primary prerequisites for a support material to hold the
chromatographic sorbent are mechanical strength, chemical resistance to solvents and visualization reagents, ability to with-
stand the temperatures needed for reaction on the plate, uniformity of thickness, and low cost.

Glass plates meet these requirements best and therefore are
the most widely used, both for "home-made" and commercial precoated layers. Borosilicate glass is preferred over lesser grades
for its chemical and thermal stability. Glass-backed plates must
be used when visualization is to be carried out with sulfuric
acid, for example. Aluminum and plastic supports are attacked
by this corrosive acid.

Flexible plastic sheets, commonly of polyethylene tere-
phthalate, are widely used as supports for commercial plates but
are seldom coated in the laboratory by the researcher. Two suppliers produce precoated plates using a glass fiber support impregnated with sorbent; this support is not available for home-
made plates, however. More is said on these in Chapter 3. Heavy
aluminum foil is also used for some commercial precoated plates.

The 20×20 cm plate has become the most widely accepted, and
most apparatus and accessories have been designed around this

size. Larger plates, 20×40 cm, are also available.

Using a normal hardware store glass cutter and a straight
edge, it is possible to cut a 20×20 plate into smaller plates such
as the widely used 5×20 cm and 10×20 cm. If a clean plate of
glass is used as a counter-top surface, it is possible to cut an
already coated plate, home-made or commercial, by laying it sorbent surface down on the glass and scribing the back of it with a
straight edge and glass cutter. It can then be picked up and
snapped into pieces of the desired size. Commercial plates are
available prescribed which can be broken into smaller convenient
sizes.

2.2 PREPARATION OF GLASS PLATES PRIOR TO COATING

It is important to employ good quality, uniform-thickness plates
(glass or aluminum) for the preparation of dependable home-made
thin layers. If glass plates are used, it is advisable to soak
them at least a day in saturated sodium carbonate solution. Rinse
the plates well with distilled water and dry them. The next step
is most important. Handling the plates by the edges, carefully
and completely wipe them well with a good lipid solvent such as
ether or petroleum ether, using cheesecloth or other lintless
cloth. If they are not ready to be coated immediately with sorbent, store them horizontally in a plate storage rack (Figure 2.1)
in a clean, fume-free, dust-free environment such as a cabinet or
desiccator, as shown in Figure 2.2. It is advisable to wipe them
with lipid solvent immediately before coating if they have been in
storage a while.

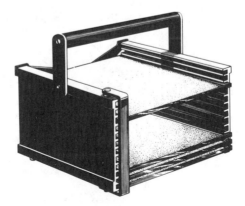

Figure 2.1. Storage rack for drying, storing, and transporting
thin layer plates. Courtesy of Kontes Glass Co.

Figure 2.2. Storage rack in metal desiccating cabinet for the
storage of thin layer plates. Courtesy of Brinkmann Instruments,
Inc.

2.3 AVAILABLE SORBENTS FOR PREPARING PLATES

2.3.1 Introduction

There are many experimentally proven sorbent materials suitable
for separating substances when they are coated in a thin layer
(usually 250 microns, 0.25 mm) on a support material. Most of
the well-proven adsorbents are commercially available from a
number of reliable firms, and it is neither necessary nor eco-
nomical to prepare them in the laboratory. The preparation, including extensive washing of the various sorbents presently available, is a very time-consuming process. Furthermore, it is difficult to exactly reproduce the steps from batch to batch. The
commercial availability of carefully controlled and prepared substrates precludes this effort on the part of the individual laboratory. The bulk of the primary supplies of many of the sorbents

are manufactured by only a few firms, and most TLC outlets use
these materials. These sorbents or commercially coated plates are
relatively uniform, differing only in method of coating and sometimes the binder used. Special-purpose sorbents, of course, may
involve further treatment in the final manufacturing process.

The factors to be considered when choosing a sorbent for TLC
are: type of compound to be separated; visualization technique
to be employed; thickness and stability of the layer desired; and
mobile phase characteristics. When these factors are considered
along with the wide variety of adsorbents available, it is usually possible to choose just the sorbent necessary for the desired separation. Also, as a minor consideration, the nature of
the visualization reagent is a factor. Some layers that contain
organic binders or use organic sorbents are not applicable when
a strong reagent such as sulfuric acid plus heat is required.
Each of these factors will be discussed in later chapters.

A number of commercial firms supply ready-to-use sorbents,
with and without binders, ultraviolet (UV) indicators, silver
nitrate, etc. The following pages will describe most of these
sorbents, including manufacturer and peculiar characteristics.

The specifications described in the following sections under
each supplier and sorbent were obtained from the most recent
catalogs and bulletins published by the suppliers at the time of
this writing. Table 2.1 lists the adsorbents available from a
number of suppliers.

Activation is a process whereby a thin layer plate, either
home-made or commercially prepared, is heated at a given temperature for a period of time to drive off excess water from the sorbent layer. Activated plates often produce sharper, better
separations than unactivated plates.

2.3.2 Binders

Sorbents are prepared with and without binding substances. A
binder changes properties of the sorbent so that it will adhere
to the support well enough not to readily flake off during handling, charomatography, visualization, or documentation. Binding
agents are not always necessary or desirable. Cellulose, a
widely used sorbent in TLC, does not usually require a binder; it
inherently has good properties of adhesion and strength. Binders
are undesirable when they interfere with development or visualization. Calcium sulfate, a commonly used binder, is slightly
soluble in water, and therefore it is not suggested that layers
employing it as a binder be used with aqueous solvent systems.
Layers employing starch or polymers as the binder (so-called organic binders) should not be used when very high temperature
(>150°) visualization techniques are to be used. They are not

TABLE 2.1   SUPPLIERS OF BULK TLC ADSORBENTS FOR COATING PLATES

	Silica Gel	Silica Gel with $AgNO_3$	Alumina	Cellulose	Polyamide	Cellulose Ion Exchangers	Sephadex	Kieselguhr	Other Available Sorbents
Analabs,	X								Magnesium oxide
Analtech	X			X					
Applied Science Laboratories	X	X	X	X		X		X	Silanized silica gel
Bio-Rad Laboratories	X								
Brinkmann Instruments, Inc.	X			X	X	X	X		Silanized silica gel
Camag	X			X	X			X	
J. T. Baker Chemical Co.	X			X	X	X		X	Other inorganics
MCB Reagents (E. Merck)	X			X	X	X		X	
Pharmacia Fine Chemicals, Inc.							X		
Schleicher and Schuell, Inc.	X			X		X			
Supelco, Inc.	X	X		X					
Universal Scientific, Inc. (Woelm)	X		X			X			
Whatman, Inc.	X			X		X			

suitable with universal reagents such as sulfuric acid spray. Starch is recommended only when it does not interfere and when tough, strong layers are needed.

Some commercially available precoated plates are made with a small amount of organic polymeric binder, which makes the layer very hard and resistant to abrasion. This binder will not dissolve in relatively strong polar solvents or in organic solvents, thus, it increases the applications for separation but precludes some detection procedures.

Calcium sulfate binder concentrations range from about 5% by weight in some silica gel preparations to 9% in aluminum oxide G sorbent, 13% in the widely used silica gel G, and up to 30% in some silica gel preparations for preparative (high capacity) TLC. The "G" in these names denotes the presence of calcium sulfate (gypsum).

Adsorbents without a binder are usually designated with an "H" or an "N" in the name, for example, silica gel H.

2.3.3 Ultraviolet Indicators

Most commercially available sorbents may be obtained with or without a chemically inert ultraviolet (UV) indicator for visualization of many compound types. The indicator, at a concentration of 2%, is usually manganese-activated zinc silicate with an activation peak at 254 nm of light. A thin layer prepared with the indicator will emit a light green light under a shortwave (254 nm) UV light. If a substance that absorbs light in this region is present in the separated zone, it will show up as a dark area against the green fluorescent background. A long-wavelength (366 nm) UV indicator, which has a blue fluorescence, is also available. The presence of one of these indicators in an adsorbent is designated with an "F," "F-254" or F-366." Normally when just an F is used it means the short-wavelength (254 nm) indicator is present. A sorbent with a UV indicator is useful in detection procedures only when the separated compounds absorb light in the region of the wavelength of maximum activation of the phosphor incorporated in the sorbent, i.e., at 254 or 366 nm.

2.4 SILICA GEL (SILICIC ACID)

Silica gel has become the most widely used sorbent for thin layer chromatography since being described by Kirchner et al. in 1951 (1). The book by Stahl (2) deals extensively with the properties and preparation of silica gels used in TLC.

Beesley (3) has been deeply involved with the technology of TLC sorbents. He elaborates on the concept that the separation process on silica is somehow a function of the surface hydroxyl content, which is responsible for selective adsorption onto the particle. Because of this, the hydroxyl content must be controlled. The higher the water uptake, the more active the sorbent.

The process of plate activation is used in an attempt to control the hydroxyl (i.e., water) content of the layer in order that reproducible separations will be obtained. Beesley notes that conditioning of an activated TLC plate to 40% relative humidity (6.67% water by weight) gives a more reproducible chromatographic system and seems to suggest the deactivation of very active sites and a more uniform hydrogen-bonding mechanism.

The standard deviation from R_f values was determined for three types of silica gel plates under these conditions by Beesley (3); silica gel G, 8.4%; polymer bound silica gel, 2.5%; and preadsorbent layer silica gel, 0.9%. It is therefore evident which plates are the preferable choice when highly reproducible separations are desired, according to Beesley.

Analabs, Inc., a division of Foxboro Analytical, offers its own grade of silica gel under the trade name Anasil. It is available with 13% calcium sulfate binder (G), with binder and fluorescent indicator (GF), without binder or indicator (H), and without binder but with UV indicator (HF). Available package quantities are 500 g, 1 kg, and 5 kg. Also offered is Merck silica gel G (Type 60) with and without (H) 13% calcium sulfate binder in package quantities of 500 g, 1 kg, and 5 kg.

Unique to the Anasil line are absorbents B and S, which are mixtures of silicon dioxide and magnesium oxide. Anasil B contains 10% calcium sulfate as binder; S does not. They are suggested by the manufacturer for the analysis of lipid classes including phospholipids where the "presence of binder would interfere with the normal migration of individual components."

Adsorbosil is the brand name of the silica gel adsorbents supplied by Applied Science Laboratories. Every batch manufactured is pretested to ensure that it will perform a specific separation. A quality check certificate supplied with each bottle gives all details of the test separation and directions for making the slurry. The manufacturer generally recommends mixing 50 g Adsorbosil with 60 mL distilled water, spreading the plates, then allowing them to air dry. Activation is recommended at 110° for 30 min before use.

The adsorbosil line represents a complete offering of silica gel adsorbents with all combinations of no binder, calcium sulfate, or magnesium silicate binders, and with or without fluorescent indicator.

Also available is another Adsorbosil that is impregnated with 20% silver nitrate, with or without 10% calcium sulfate binder. Chromatography on silica gel impregnated with silver nitrate allows compounds containing double bonds to be separated on the basis of their cis-trans configuration and degree of unsaturation. Cyclic olefins are readily separated from their straight-chain counterparts. Normally, the fewer the double bonds

present in the compound, the faster it migrates. Some applications
include separations of unsaturated fatty acids (4), mono- and
sesquiterpenes (5), resin acids (6), chlorophylls (7), and quantitation of pesticides (8). Visualization is often accomplished by
charring with potassium dichromate and sulfuric acid.

For convenience, Applied Science Laboratories also offers
silica gel with and without binder, and with and without UV indicator, prepared by E. Merck in Darmstadt, West Germany.

J. T. Baker Chemical Co. carries five forms of silica gel for
TLC. Silica gels G and GF are available; silica gel 7 is a neutral
form containing no binder; silica gel 7A contains calcium sulfate
as a binder. Silica gel 7GF contains this binder as well as long-
and short-wave UV indicators.

Brinkmann Instruments, Inc. is a supplier of TLC sorbents,
precoated plates, and apparatus. They are distributors for the EM
(E. Merck) and Brinkmann/MN (Macherey, Nagel) lines of sorbents and
plates and for the Desaga line of apparatus. In addition to the
usual forms of silica gel offered by EM and MN, Brinkmann also
offers silica gel specially prepared for making preparative (not
analytical) layers. Preparative layers are usually 500-20 μm
(0.5-2.0 mm) thick as compared with 100 or 250 μm thickness normally used for analytical layers. The thicker preparative layers
permit the application of larger sample amounts because of increased capacity. The primary function of preparative TLC is the
isolation of quantities of a substance by employing TLC techniques.
When sorbents that are intended primarily for analytical thin layer
layers are used to make thicker preparative layers, cracking or
pitting of the layer often results. The specially manufactured
"P" (preparative) sorbents were developed to eliminate some of
these problems. These "P" silica gels are available in four different preparations as the normal silica gel with and without calcium sulfate binder, and with or without fluorescent indicator.
Dual-wavelength (254+366 nm) indicator is also available with the
preparative silica gel.

Another large supplier of TLC apparatus, plates, and sorbents
is Applied Analytical Industries, Inc., which is the United States
distributor of Camag of Switzerland. They offer eight different
types of silica gel for layer preparation, including analytical
and preparative grades, binder or no binder, and indicator or no
indicator. Their binder is usually 5% calcium sulfate. The designation used for the analytical layer silica gel is "D"; D-0 is
without binder or indicator; D-5 contains binder only; DF-5 contains both; and DF-0 contains indicator only. The preparative
grade silica gel is designated "DS," and all forms follow the
same coding system as the analytical grades.

E. Merck adsorbents may be obtained directly from MCB Reagents. Their line of silica gel sorbents are prepared according

to Stahl (2), with extra pure (type HR) and silanized types available along with the usual types available from the other suppliers. They also have silica gel H, HG-254, and HG-254+366 grades; where the "H" designates that the binder is composed of fine particles of silicon dioxide or alumina. All preparative silica gel has an F-254 indicator, and a silanized type is also available for reversed-phase chromatography.

Supelco, Inc., offers eight types of silica gel for TLC. Their own product trademark is Supelcosil, which is available with and without UV indicator and with or without binder. They offer two unique specially prepared forms of silica gel: Supelcosil 12D, which uses silicon dioxide as a binder and contains 15% silver nitrate, and Aflasil, which has calcium sulfate binder and is used for the separation of glycolipids and aflatoxins. Whereas most silica gels have a pH of 7.0, Aflasil has a pH of 7.4-7.6.

2.4.1 Reversed Phase TLC

Silica gel and alumina retain solutes almost exclusively on the basis of polarity, and separations of close polarity are therefore difficult, even with different molecular weights. Selectivity can only be effected by the mobile phase in these cases, and changes in the surface of the layer are then required in order to change selectivity. One way to bring about this modification is to react the surface hydroxyls with organic silanes, particularly those containing octadecyl chains. This provides selectivity between solutes of not only very similar polarities but also different molecular weights (3).

A reversed phase is a sorbent that has been modified in this manner; most often this is a silica gel. Silica gels are designated in a way to indicate the number of carbon atoms involved in the silane: RP-2, RP-8, RP-18, or C_{18}. The RP stands for reversed phase.

Since the chromatographic mechanism in reversed phase separation is the retention of nonpolar compounds by the organic stationary phase, an increase in the polarity of the mobile phase will result in increased retention of the more nonpolar species. An increase in carbon loads therefore means that lower polarity solvents are required for retention, which results in faster developing systems of higher efficiency (3).

2.4.2 Preparation of Silica Gel Slurry

Universal Scientific, Inc. is the distributor of Woelm chromatographic products in the United States. These include adsorbents, precoated plates, and visualization sprays. They offer the four basic types of analytical layer TLC silica gel, with and without calcium sulfate binder, and with and without fluorescent indicator.

For batch spreading of five 20×20 cm plates (or the equivalent ten 10×20 cm or twenty 5×20 cm) with Woelm silica gel G, mix 25 g powder with 50 mL distilled water. For spreading one 20×20 cm plate by hand, mix 5 g silica gel G with 15 mL of a 2:1 (v/v) solution of ethanol:water.

The quantities are somewhat different when using silica gel without the binder. For the batch coating of five 20×20 cm plates using a spreader, mix 30 g powder in 45 mL water. For pouring one plate by hand without a spreader, mix 6 g powder in 15 mL of a 9:1 ethanol:water solution.

Woelm also manufactures a silica gel GPF-254 for coating preparative layers. Their formula for one 20×20 cm plate, 1 mm thick, is 20 g silica gel GPF mixed with 30-35 mL water. The plate can be activated at 130° before use.

Schleicher and Schuell, Inc., well known for filter and chromatography papers, supplies adsorbents in bulk as well as precoated TLC plates. Six different forms of silica gel are available: the four commonly supplied (with and without UV indicator and with or without 12% calcium sulfate binder) and two unusual ones, with starch (3%) as a binder, with and without UV indicator. The following procedures are offered for preparation of these silica gels.

For silica gel without binder, slurry 30 g powder in 60-65 mL water. Homogenize with an electric blender for 30 sec. Spread the slurry 250 μm thick on five 20×20 cm plates. Dry at ambient temperature, then activate at 110° for 30 min.

For silica gel with calcium sulfate binder, slurry 30 g powder in 65-70 mL water. Using an electric blender, homogenize for 30 sec and spread five 20×20 cm plates within 2 min. Dry, then activate at 110° for 30 min.

For silica gel with starch binder, slurry 30 g powder in 90 mL boiling water. Homogenize for 30 sec with an electric blender and spread while hot. This recipe will make five 20×20 cm plates. Dry and activate at 110° for 30 min.

Bio-Rad Laboratories also supplies a silica gel suitable for TLC. It is without binder and without indicator and is called Bio-Sil A for TLC.

Whatman, Inc. supplies a complete line of silica gel sorbents, with and without indicator.

For the preparation of home-made silica gel plates containing silver nitrate, Applied Science Laboratories (9) recommends making a slurry by stirring 20 g "good quality" TLC silica gel powder into a solution of 10 g silver nitrate dissolved in 50 mL water. Stir the slurry carefully to remove air bubbles; three 20×20 cm plates 250 μm thick may be coated. Allow the plates to

air dry for 2 hr in a protected area, and then activate them for
1 hr at 110° before using. Plates may be prepared in advance of
their use but should not be activated until ready for use. The
prepared plates may turn light gray upon activation, but will
fade after developing and visualization by charring with sulfuric
acid and heat. The coated plates should always be stored in the
dark and only for limited periods of time, as $AgNO_3$ layers will
gradually darken; the process is hastened by light.

2.5 ALUMINA

Aluminum oxide, commonly called alumina, is the second most
widely used TLC sorbent after silica gel. Many applications for
alumina in TLC have been adapted from similar applications in
column chromatography. As with silica gel, good TLC grade alumina is available from a number of commercial suppliers for making
plates in the laboratory. Precoated alumina plates are also
readily available; these are discussed in Chapter 3.

Alumina is manufactured with three pH ranges--acid, neutral,
and basic-- for separating different types of compounds. To obtain reproducible results with alumina, it is usually necessary
to control the amount of adsorbed water, which can block the
active separating sites on the alumina surface. To do this, it
is necessary to activate the coated TLC plate in an oven at specified temperatures between 75° and 110° for a specified period of
time before applying the samples to the plate. All the parameters
in the preparation of the plate must be defined in order to reproduce a given separation.

Applied Science Laboratories offers aluminum oxide GA, which
contains 10% calcium sulfate binder, and aluminum oxide HA, which
has no binder. Both have a pH range of 9.0-9.5 and a low iron
content (less than 0.01%). In the past, iron was often a troublesome contaminant, but with the specially prepared chromatographic
grades of alumina available now, this problem has been kept to a
minimum. The EM line of alumina sorbents is also carried by
Applied Science Laboratories.

A basic aluminum oxide containing an inorganic fluorescent
indicator that is activated at 254 or 366 nm is available from J.
T. Baker Chemical Co. as aluminum oxide 9F.

Brinkmann Instruments, Inc. supplies a complete line of
alumina sorbents. Included are acidic, basic, and neutral pH
forms with binder but without UV indicator. These are cataloged
as Type T aluminas; they are suggested for peptide and steroid
analyses, among other uses. Also available are aluminum oxide G
(binder), FG (binder and indicator), and H (no foreign binder);
all are suggested for analytical work.

Alumina suitable for the making of preparative layer plates is also supplied by Brinkmann Instruments, Inc. This alumina contains either short-wave (254 nm) or short-wave and long-wave (366 nm) ultraviolet indicators. Both indicator pigments are available separately and may be added to any sorbent before plate coating if desired.

Alumina without the calcium sulfate binder is recommended whenever there is a likelihood that calcium ions would interfere with the separation or detection. It is also suggested for charring visualization techniques. Without the binder, the layer is more resistant to the solubilizing effects of polar solvent systems used in development. Generally speaking, bulk alumina contains more impurities than other sorbents, making it difficult to obtain "clean" layers and reproducible results.

2.5.1 Preparation of Alumina Slurry

The following formulas are suitable for the preparation of alumina slurries. For analytical layers using aluminum oxide G with binder, mix well 30 g powder with 40 mL distilled water. This will coat five 20×20 cm plates or twenty 5×20 cm to a depth of 250 μm. Immediately after coating, allow the plates to dry for at least 30 min at room temperature before activation or use. Activation at 110° for 30 min immediately before use is suggested. Calcium sulfate binder sets very fast, and it is therefore imperative that the plates be spread within 2 min after preparation of the slurry.

To use aluminum oxide H (without binder) for analytical layers, mix 30 g with 80-90 mL distilled water. Since there is no fast-setting binder in this case, there is no urgency involved with coating, and the slurry may even be stored for future use. Also because of this, the plates must be allowed to dry for 3-4 hr before activation or use. They should be activated for 1 hr at 110°.

For preparative plates, it is suggested that the preparative quality aluminum oxide P be used. The slurry preparation is different. The following formula will coat two 20×40 cm preparative size plates, or four 20×20 cm plates, to a depth of 2 mm (2000 μm). A proportionate number of plates may be coated to 1 mm (1000 μm) or 0.5 mm (500 μm) if desired. Mix 320 g alumina with about 320 mL distilled water, shaking well for 1 min. An electric blender may also be used. Allow the slurry to stand for 15 min, then shake it again and spread it on the plates. The freshly coated plates must then be stored horizontally for 15 hr at room temperature. Activation and final drying are done at 110-120° for 3-4 hr.

Camag supplies aluminum oxide DS. The "D" designates that it is their own brand, and the "S" designates that it is made up of slightly larger particles than usual so that the separation is

faster than normal. Four types are available: aluminum oxide
DS-0 has neither binder nor indicator; DSF-5 has short-wave UV
indicator and 5% calcium sulfate binder; DS-5 has binder without
indicator; and DSF-0 has indicator without binder. Types DS-0 and
DSF-0 are recommended for preparative layers of 2 mm and over.

MCB Reagents offers a complete line of alumina sorbents. Two
basic types and quite a number of additional forms of each type
are offered. Type T is an aluminum oxide ignited at a higher
temperature than Type E. Brinkmann Instruments, Inc. uses the
same designations for its fundamental types. Acidic, basic, and
neutral forms of Type T are available for specific pH-dependent
separations. Type E aluminum oxide is offered with 10% calcium
sulfate binder as aluminum oxide G; with binder and indicator,
GF-254; with binder composed of fine particles of silicon dioxide
or aluminum oxide, H; with this binder and indicator, HF-254.

Preparative forms of both Type E and Type T are available,
but only with UV indicators, either short-wave (254 nm) or short-
wave combined with long-wave (254+366 nm).

Universal Scientific, Inc., distributes four different types
of Woelm alumina: alumina G with binder, and acidic, basic, and
neutral aluminas without binder. None of these is supplied with
UV indicator.

For coating plates with these sorbents, mix 35 g with 40 mL
distilled water. Using a spreader, spread the alumina G immediately because the binder is fast setting; the other three forms
do not have to be spread as quickly. This recipe is sufficient
for five 20×20 cm plates. Activate the plates for 30 min at 130°.

If a spreader is not available, the sorbents may be spread
by pouring, but this is not recommended on a routine basis because
of the nonuniform, unknown thickness of the layers produced. A
single plate may be coated with alumina G by mixing 6 g with 15
mL of 2:1 ethanol:water. If the other aluminas are used, mix 6 g
with 15 mL of 9:1 ethanol:water.

2.6 CELLULOSE

Silica gel and alumina are both inorganic substances that are
used chiefly to separate lipophilic ("fat-loving") compounds or
those soluble mainly in organic solvents. Cellulose, on the other hand, is organic and is mainly used to separate hydrophilic
("water-loving") compounds such as sugars, amino acids, inorganic
ions, and nucleic acid derivatives. These substances are soluble
in aqueous solutions. The behavior of cellulose coated in a thin
layer on a glass, plastic, or aluminum support is not much different from that of a paper strip in paper chromatography. The
developing solvents and visualization reagents employed in paper

chromatography are most readily used in cellulose thin layer chromatography. The separation process occurring on cellulose is mainly that of partition because the adsorbed water molecules cover the cellulose molecules, being joined to them by hydrogen bonds. The substances to be separated on a cellulose layer are actually partitioned between the developing solvent and the water coating the cellulose particles. Cellulose thus has many hydrophilic properties because of this adsorbed water, and very little adsorption of sample components actually occurs.

Two types of normal cellulose powder have been found suitable for TLC: native, fibrous cellulose and microcrystalline cellulose. The fibrous cellulose is produced primarily by Macherey, Nagel Co. (MN) under the name MN 300 cellulose. In the United States the distributor for MN products is Brinkmann Instruments, Inc. The major brand of microcrystalline cellulose is Avicel, made by FMC Corporation. They supply it to the chromatography companies for sale in bulk or for the manufacture of plates, known either as Avicel plates or microcrystalline cellulose plates, depending on the supplier.

Applications for the two celluloses are not always the same, one providing a better separation than the other for a particular set of compounds. Six forms of chemically treated cellulose are available commercially for TLC use. These forms and their suppliers are listed in Table 2.2 These are ion-exchange sorbents (except acetylated cellulose, which is used for reversed phase chromatography) and are widely used for separating larger molecules because the active substance is present mostly on the surface of the cellulose, where it can readily interact with the molecules to be separated. These larger molecules include proteins, peptides, steroids, and enzymes.

Three of these celluloses are derivatives of cellulose formed through chemical reaction. They include carboxymethyl (CM) cellulose, DEAE cellulose, and Ecteola cellulose. The other two forms, PEI and phosphorylated, are impregnated cellulose in which the cellulose behaves primarily as a support for the ion-exchanger. A brief explanation of each of these cellulose forms may be helpful.

Carboxymethyl (CM) cellulose, $R-O \cdot CH_2COOH$, is synthesized from monochloroacetic acid and alkaline cellulose. It is a weak cation exchanger.

Diethylaminoethyl (DEAE) cellulose, $R-O \cdot C_2H_4 \cdot N(C_2H_5)_2$, is the reaction product of 2-chloro-1-diethylaminoethyl hydrochloride and alkaline cellulose. It is a strong basic anion exchanger.

Ecteola cellulose is the product of the reaction of alkaline cellulose with triethanolamine and epichlorohydrin. It is a weak basic anion exchanger.

TABLE 2.2   SUPPLIERS OF CELLULOSES FOR THIN LAYER CHROMATOGRAPHY

	Cellulose, Fibrous "Cellulose 300"	Micro-crystalline*	DEAE†	Ecte-ola	PEI‡	Acetylated	Carboxy-methyl (CM)	Phosphory-lated	QAE§
Applied Science Laboratories		x	x	x	x				
Brinkmann Instruments, Inc.	x	x	x	x	x	x	x	x	
Camag	x	x							
J. T. Baker Chemical Co.		x	x	x		x	x	x	
MCB Reagents (E. Merck)		x	x	x		x	x	x	
Schleicher and Schuell, Inc.	x	x	x	x	x	x	x	x	x
Supelco, Inc.		x							

* Often called Avicel (trade name).
† Diethylaminoethyl.
‡ Polyethylenimine impregnated.
§ Quaternarized ion exchanger.

Polyethylenimine (PEI) cellulose is fibrous cellulose impregnated with polyethylenimine. It is a strong basic anion exchanger.

Phosphorylated (P) cellulose, $R-O \cdot PO_3H_2$, is a strong cation exchanger formed from alkaline cellulose and phosphorous oxychlorine or fibrous cellulose and urea phosphate.

Because cellulose has natural adhesive properties, the addition of a binder does not increase the stability of the layer. Most bulk cellulose sorbents, as well as precoated plates, are therefore offered without binder.

The relative popularity of the different forms of cellulose is shown by the degree of availability from the commercial suppliers.   Microcrystalline cellulose is the most widely used form and the most generally available.

Applied Science Laboratories supplies microcrystalline, DEAE, and Ecteola celluloses.

Microcrystalline cellulose, 10% acetylated cellulose, cellulose CM, DEAE, Ecteola, and PEI are available from J. T. Baker.

Brinkmann Instruments, Inc. stocks all nine forms of cellulose 300 to the phosphorylated form. All forms are normally supplied without UV indicator except for the normal fibrous cellulose 300, which is available with or without it. Brinkmann also has specially prepared cellulose 300-HR (MN), which is acid washed and fat-free, for use when a highly purified cellulose is desired. Four forms of acetylated cellulose 300 are available: 10%, 20%, 30%, and 40%. The percentage designation is the degree of acetylation of the cellulose.

Cellulose powder D is the Camag equivalent to cellulose 300. It is available with and without UV indicator. Microcrystalline cellulose is also available from Camag with and without UV indicator as cellulose powder DS.

EM Laboratories supplies all forms of cellulose except the 300 and phosphorylated types.

ICN Pharmaceuticals distributes all forms of cellulose for TLC application except the microcrystalline and acetylated types. They also offer a special form not readily found elsewhere, quaternarized ion exchanger (QAE). This is an anion exchanger used for the separation of proteins and other delicate biological substances as well as weakly acidic substances. It is also suitable for partition TLC in polar solvents.

Schleicher and Schuell, Inc. supplies all nine forms of cellulose for coating plates: normal acid-washed, microcrystalline, and 20% and 40% acetylated. Also available are DEAE, Ecteola, carboxymethyl cellulose (CM), PEI, QAE, and phosphorylated cellulose.

2.6.1 Preparation of Cellulose Slurry

Preparation of slurries is different for each of the celluloses. Each of the following formulations will coat five 20×20 cm plates or the equivalent in smaller plates: ten 10×20 cm or twenty 5×20 cm.

For normal cellulose (200 μm thick), dissolved 0.4 g starch in 10 mL distilled water and add to 90 mL boiling water. Boil for 1 min and add 20 g cellulose. Using an electric blender, homogenize for 30 sec; coat the plates immediately while the suspension is still hot. Dry in air or for 45 min at 110°. No activation is necessary if drying occurs at 110°.

To form a slurry of microcrystalline cellulose (without UV indicator, layer thickness 100 μm), mix 20 g in 60 mL distilled water. Homogenize for 30 sec using an electric blender. Let the plates dry in air or at 110° for 30 min. No activation is necessary if plates are dried at 110°. Blending time is critial; the degree of hydration increases with increased blending time. This changes the nature of the thin layer and affects the speed and quality of the separation. This variable should therefore be controlled for reproducible layers.

To prepare a slurry of microcrystalline cellulose with UV indicator, add 25 g to 40 mL methanol plus 20 mL distilled water. Homogenize for 30 sec in an electric blender. Dry the spread plates in air or at 110° for 30 min. No activation before use is necessary if drying occurs at 110°. The methanol in this formula, as in the silica gel with UV indicator formulation, is necessary to solubilize the fluorescent green indicator so that it will be uniformly distributed throughout the slurry.

A slurry of acetylated cellulose (for layers 130 μm thick) is prepared by mixing 30 g sorbent plus 4.5 g calcium sulfate $(CaSO_4 \cdot \frac{1}{2}H_2O)$ in 60 mL distilled water with 10 mL methanol. Blend for 30 sec in an electric blender. Air bubbles may be removed by carefully covering this suspension with 2-3 mL methanol and shaking. Spread the plates within 10 min, and dry them at room temperature or for 30 min at 110°. No activation is necessary if plates have been dried at 110°.

Randerath and Randerath (10), in an excellent paper on coating plastic sheets with PEI cellulose, state that "the coated sheets are allowed to dry overnight at room temperature. This will result in a layer with optimal ion-exchange properties. Under no circumstances should drying of the layer be accelerated by heating." This is the usual case, for most often cellulose is used for partition chromatography, where the atmospheric water retained by the cellulose in the layer is necessary for the proper separations to occur. It appears to be the exception, therefore, to recommend drying a cellulose plate with heat as stated in the previous paragraphs.

Brinkmann Instruments, Inc. suggests different formulas for the preparation of cellulose slurries using the Macherey, Nagel (MN) brand of celluloses. The following formulations will each coat five 20×20 cm plates or the equivalent number of smaller plates.

For MN cellulose 300, mix 15 g with 90-100 mL distilled water in an electric blender for 30-60 sec. Mixing by hand is not aggressive enough for good hydration, and an electric blender is mandatory for cellulose slurry preparation. (Other sorbents, e.g., silica gel, alumina, polyamide, and kieselguhr, are just as easily prepared by shaking in an Erlenmeyer flask.) The coated cellulose plates are air dried with no activation.

Slurry preparation of MN cellulose 300 with 254 UV indicator requires 15 g in 60 mL 95% methanol with blending for 30-60 sec in an electric blender.

Avicel cellulose (microcrystalline) must be prepared carefully. Mix 15-20 g with 100 mL distilled water and, using an electric blender, slurry for exactly 60 sec. Longer blending will result in an undesirable gel formation. Dry the plates 30 min at room temperature, then at 105° for 10 min to drive off excess water.

MN cellulose 300 acetate is prepared by first forming a stiff paste from 15 g powder and a small amount of 95% ethanol taken from a total volume of 60 mL. After the paste is formed, add the remaining alcohol and slurry in an electric blender for 30-60 sec. Air dry only after spreading the layers.

Carboxymethyl cellulose, from MN cellulose 300 CM, is prepared from 15 g powder and 60-70 mL distilled water. Blend for 30-60 sec in an electric blender. If dry layers made from this formulation crack, an alternative formulation may be tried. Mix 12 g with 3 g regular MN cellulose 300 in the same amount of water. Blend, spread, and dry in the same manner.

DEAE cellulose MN 300 slurry is prepared from 10 g powder and 75 mL distilled water. The formulation recommended if this one produces layers that crack is 8 g DEAE powder plus 2 g cellulose 300 powder in 75 mL water. Blend either formula for 30-60 sec with an electric blender. Air dry after spreading on the support.

MN cellulose 300 Ecteola slurry is prepared from 10 g cellulose and 50 mL distilled water. The alternative formulation used against cracking is 8 g Ecteola plus 2 g cellulose 300 with 50 mL water. Slurry for 30-60 sec in an electric blender. Allow the plates to air dry after spreading.

Phosphorylated MN cellulose 300 P is slurried from 15 g in 60-70 mL distilled water. An optional formulation to decrease the likelihood of cracking is made from 12 g cellulose P plus 3 g cellulose 300. Blend for 30-60 sec in an electric blender. Allow to air dry after coating on the support.

Predevelopment of cellulose plates in a tank containing distilled water is recommended to remove impurities normally found in most celluloses. This washing will concentrate the impurities along the upper edge of the plate out of the separation zone. Then apply the substances to be separated on the "clean" end of the plate, so that the impurities will not be redistributed over the plate by the developing solvent. Washed plates can be marked with an arrow or other notation using a scribe, e.g., syringe needle, on the upper or "dirty" edge.

2.7 POLYAMIDE

On polyamide thin layers, separation depends on the strength of the hydrogen bonds formed between the polyamide molecules and the

substances being separated.. This bond strength is a function of
the number of phenolic hydroxyl or carboxyl groups and their positions in the molecules to be separated. Compound types that can
be separated on polyamide therefore include phenols, carboxylic
acids, steroids, quinones, and aromatic nitro compounds. A mono-
carboxylic acid, having only one carboxyl group, is not as readily
bound to the polyamide molecule as would be a dicarboxylic acid or
an aromatic acid molecule, and will therefore migrate further on a
polyamide thin layer than either of the other two molecular types.
For elution, a solvent must be chosen that is able to break the
hydrogen bonds formed between the sample molecules and the polyamide molecules by displacement. Certain solvents are better at
doing this than others. Chapters 5 and 6 on solvent systems and
development explore this subject.
      Four different polyamides are currently being used for TLC:

      Polyamide 6 = Nylon 6 = aminopolycaprolactam
      Polyamide 6.6 = Nylon 6.6 = polyhexamethyldiaminoadipate
      Polyamide 11 = Nylon 11 = polyaminoundecanoic acid
      Acetylated polyamide = acetylated derivatives of the above

      Only two suppliers carry polyamide in bulk for coating
plates. Baker has polyamides 6 and 11. Brinkmann Instruments,
Inc. distributes the polyamides made by E. Merck (EM brand) and
Macherey, Nagel (MN brand).
      Brinkmann Instruments, Inc. supplies E. Merck polyamide pow-
der 11 without UV indicator, MN polyamide 6, MN polyamide 6 with
UV 254 indicator, and acetylated polyamide 6. Also available are
MN polyamide 6.6, MN polyamide 6.6 with UV 254 indicator, and
acetylated polyamide 6.6.

      2.7.1 Preparation of Polyamide Slurry

The polyamides commercially available for TLC are normally supplied without a binder, because it is not necessary. When preparing polyamide slurries an electric blender must be used for thorough mixing. Hand shaking in an Erlenmeyer flask will not adequately prepare a polyamide slurry.
      Macherey, Nagel (MN) brand polyamides, available from Brink-
mann Instruments, Inc., are prepared for coating according to the
following procedures. For a polyamide 6 slurry, add 15 g powder
to 65 mL distilled water. Mix well in an electric blender for
30-60 sec and spread the slurry on the support immediately before
any separation of liquid and solid occurs. Air dry after coating
the support. Polyamide 6.6 is prepared in exactly the same manner.

For a slurry of polyamide 11, blend 15 g powder with 60 mL methanol in an electric blender for 30-60 sec. Spread the plates immediately and allow them to air dry.

It is not necessary to activate polyamide layers before use. Occasionally, if a stable layer is necessary, 10% starch can be added as a binder.

Rosler et al. (11) developed a standardized procedure for the preparation of TLC quality polyamide.

2.8 KIESELGUHR

Kieselguhr, also known as diatomite or diatomaceous earth, is composed of the siliceous skeletal remains of microscopic marine plankton called diatoms. Massive accumulations of diatomite occur in many parts of the world. Because of its high porosity and large surface area, it is widely employed as a filter aid in the laboratory, in industry, and in swimming pool filtration. It is used in column chromatography and as a sorbent in TLC, where it is often coated with a liquid phase and used for partition chromatography.

When used in TLC, it normally has a binder of calcium sulfate and is known therefore as kieselguhr G, the "G" being the designation for the calcium binder, usually with a concentration of about 15%.

Brinkmann Instruments, Inc. and MCB Reagents are both suppliers of E. Merck kieselguhr G. Kieselguhr without binder is also offered with and without F-254 indicator, as well as mixed with silica gel. The only two sources of purified kieselguhr for TLC are E. Merck and Machery-Nagel.

A slurry is prepared from 30 g powder in 60 mL of distilled water. The plates should be coated within 2 min because the binder sets rapidly. The layers are allowed to dry at least 30 min at room temperature before activation. Activate for 30 min at 110°. Kieselguhr is not normally used for preparative TLC.

Kieselguhr G or N (no binder) is prepared from 20 g powder in 55 mL distilled water. Shake in an Erlenmeyer flask and spread the G sorbent on the support within 2 min. The kieselguhr G is allowed to dry at least 30 min, the kieselguhr N at least 4 hr, before activation at 110° for 30 min.

2.9 SEPHADEX

Sephadex is the trade name for a group of modified dextran gels for gel filtration that are produced by Pharmacia Fine Chemicals, Inc. Sephadex gels are hydrophilic, neutral stationary phases

that may be used in thin layers. The American distributor is
Pharmacia Fine Chemicals, Inc. in Piscataway, New Jersey.

Separation in gel filtration is achieved according to molecular size on a gel swollen in a carefully prepared solution. The
mechanism of separation is primarily one of partition governed by
steric hindrance. Determann (12) and Johansson and Rymo (13) were
the first to use the thin layer gel filtration technique (TLG).
It is basically the same as TLC but differs from it in that it is
carried out on a prewetted and wet layer and the movement of liquid is continuous. Solvent flow is potentiated by gravity. Development of Sephadex plates is covered in Chapter 6.

As in TLC, plates to be coated with Sephadex should be thoroughly cleaned by washing with warm detergent solution and distilled water. Washed plates may be stored in concentrated sodium
carbonate solution, then washed with distilled water and dried before coating.

Most of the Sephadex grades are available in a variety of
mesh sizes. For TLG (thin layer gel) chromatography, the "superfine grade" should be used. The consistency of the slurry is important. If it is too thick, the gel will not adhere. If it is
too thin, it will run on the plate. A binder is not normally used
with Sephadex because of its natural adhesive properties. Table
2.3 provides the information needed to prepare slurries of each of
the Sephadex types. It is necessary to swell the gel in the
eluent developing solvent to be used for at least the amount of
time shown in the table. Air bubbles may be removed from the
slurry by placing the container in a boiling water bath. The
slurry should be carefully stirred by hand to completely suspend
all the Sephadex and break up all clumps. Mechanical and magnetic stirrers should not be used, as they may cause destruction
of the particles.

TABLE 2.3   PREPARATION OF SEPHADEX SLURRIES FOR TLC

| Type of Sephadex Superfine | Gel wt. (g/100 mL) | Minimum Swelling Time | |
		At Room Temp.	On Boiling Water Bath
G-50	10.5	3 hr	1 hr
G-75	7.5	24 hr	2 hr
G-100	6.5	72 hr	3 hr
G-150	4.8	72 hr	3 hr
G-200	4.5	72 hr	3 hr

The gel slurry may be spread by hand or with any commercially available spreader. Pharmacia Fine Chemicals, Inc. offers their own spreader, which is simple in design and easy to use. It is recommended for spreading Sephadex. Its use is illustrated in Figure 2.3. This TLG spreader, as it is known, produces even layers of reproducible thickness, which include 0.4, 0.6, 0.8, and 1.0 mm. Table 2.4 gives the gel slurry volume for coating one plate. The thickness of 0.6 mm (600 μm) is recommended for most uses.

TABLE 2.4   GEL SLURRY VOLUME (mL) FOR COATING ONE PLATE

Layer Thickness	20×20 cm	20×40 cm
1.0 mm (1000 μm)	55	110
0.8 mm (800 μm)	45	90
0.6 mm (600 μm)	35	70
0.4 mm (400 μm)	25	50

2.10   COATING PROCEDURES

There are four methods that may be used to coat a plate with a thin layer of sorbent. These are: (a) pouring, (b) spreading, (c) immersing, and (d) spraying. Of these four, the spreading procedures are the most widely used, the others being occasionally used under particular circumstances. For instance, microslide plates are often prepared by the immersion technique because of the ease with which it is applied to small plates. Spreading procedures are preferred for a number of reasons, including good reproducibility, adjustment of layer thickness, and ease of operation. The other three techniques generally lack these virtues. Spraying techniques are messy and require large guns and high pressure, which are not ideal laboratory equipment. The spreading techniques, therefore, will be the only coating procedures covered in the present discussion.

The object in plate coating is to spread a smooth, uniform layer that dries without bubbles, pits, or cracks. All of these defects can cause poor migration of the substances to be separated due to "channeling" or trough effects. Three important steps should be taken in plate preparation to help prevent this. Although all have been mentioned previously, they are emphasized here. First, using a lint-free cloth (cheesecloth), wipe the plates well with a fat solvent such as ether or petroleum ether before coating. The sorbent will not adhere to a dirty, greasy

support. Careful slurry preparation is the second point to consider. Good quality solvents should be used to prepare the slurry, including deionized and distilled water.

The slurry should be stirred with care so that all lumps of
sorbent powder are broken up and no air bubbles remain. After
careful coating, the layer should be allowed to air dry at room
temperature in the horizontal position for at least the recommended time. Preparative layers normally require 15-18 hr drying
time at room temperature. Do not place freshly coated plates in
a hot oven to dry. This will result in drying that is too fast
and too severe and will invariably produce a cracked layer, often
with cracks not easily seen with the eye.

The instructions for slurry preparation for each sorbent
given in the previous sections should be followed, as these
"recipes" have proved to be reliable.

2.11 SPREADING THE LAYERS BY MANUAL PROCEDURES

The easiest way to coat a plate with a thin layer of sorbent is
to simply pour the slurry on top of the plate and spread it evenly
across the plate with a glass rod. This will not produce a layer
of uniform thickness nor one that can be reproduced unless some
provision is made to space the rod at a given height, for example,
0.25 mm (250 μm) above the plate surface so that the spreader layer
will be uniform. Spacers can be provided on the ends of the rod,
as is the case of the Pharmacia Fine Chemicals, Inc. TLG spreader
(Figure 2.3), or a spacer of the desired height may be placed on
either side of the plate to support the spreading rod. Often
spacers of this type are made with cellophane or masking tape.

The technique for coating a plate with the Pharmacia Fine
Chemicals, Inc. spreader rod is illustrated in Figure 2.3. After
preparing the necessary amount of gel (a) pour it onto the middle
of the precleaned plate and spread around with a spatula; (b)
rotate the spreader rod in the slurry until it is evenly wetted;
(c) place the rod behind the bulk of the slurry with the Pharmacia
Fine Chemicals, Inc. spreader (the number facing upwards on the
end collar corresponds to the layer thickness); and (d) using the
rod, draw the slurry slowly and evenly toward the end of the plate,
maintaining an excess in front of the rod. Finish with one continuous, even stroke, taking excess slurry off the plate. It is
convenient to place a clean glass plate at the end of the plate
being coated so that any excess slurry ends up here rather than
on the bench top. If the plates coated by this procedure are
Sephadex gel, they are then placed in the special developing
chamber used for gels to keep them wet until use. If the plates

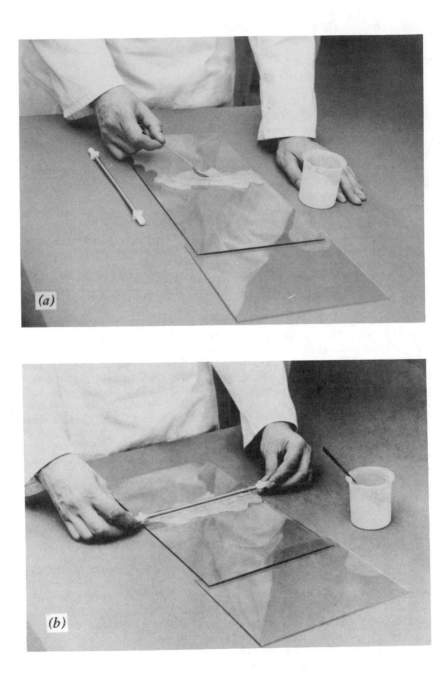

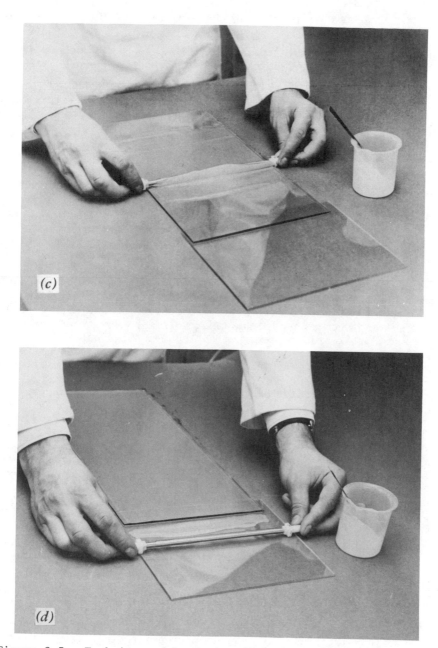

Figure 2.3. Technique of hand spreading a TLC plate using the rod apparatus. Courtesy of Pharmacia Fine Chemicals, Inc.

coated are of the normal type, then they would be allowed to dry
at room temperature for the minimum length of time before activation or use.

2.12 COMMERCIAL SPREADING APPARATUS

There are two basic types of commercially available thin layer
spreading equipment. One type has a fixed trough into which the
slurry is placed, and the plates to be coated are moved underneath. A manually operated form of this apparatus is pictured in
Figure 2.4; the motorized form is shown in Figure 2.5. The second type employs a movable trough containing the slurry, which is
pulled or driven over the top of the plates to be coated.
     The following points will be helpful when it is desired to
layer "home-made" plates. Set up the apparatus on top of a
chemically clean, dust-free, uncluttered, sturdy, and level bench
or table, preferably where there are no fumes that may be absorbed
into the finished plates while they are drying. Carefully wipe
the plates to be coated with a lint-free cloth soaked with a good
lipid solvent such as petroleum ether to remove any traces of
grease or dirt residue. Set them on the mounting board of the
spreader against the retaining edge, which should be closest to
the operator. Check the plates to make sure that they are all
the same thickness. Do not place plates of different thicknesses
on the board to be coated at the same time. Bumps of sorbent
would be produced, and the thinner plates would receive a thicker
layer of sorbent than the thicker plates because the trough is
set to deposit a given layer depth above the plate surface.
     Often when a movable trough spreader is used, narrow plates
are placed on either end of the plates to be coated to serve as
"start" and "finish" plates to support the trough and slurry so
that all the plates coated are uniform. An end plate is always
advisable to catch any leftover slurry so that it does not flow
over the mounting board or counter.
     Microscope slides may also be coated if they are placed
close together on a large 20×20 or 20×40 cm plate. If they move
about too readily when in contact with the trough, a drop of water under each slide in contact with the supporting glass plate
should help them adhere.
     The use of the fixed trough spreader is illustrated in Fig-
ure 2.4. In (a) the clean plates have been placed on the mounting
board and the first plate has been lined up with the edge of the
trough. The slurry has been evenly poured into the trough, all
ready to spread. In (b) the first plate has been slowly pushed
under the trough and has been coated with slurry. Photograph (c)

shows the completed plate left on the mounting board so that it
may be allowed to dry before the next plate is pushed through.
The length of the mounting board is often extended with plastic,
plywood, or glass, to support the coated plates as they come out
from under the trough and to allow room for the next plate. In
this way a number of plates may be coated in a continuous motion
without interruption or without disruption of the still wet
plates.

Operation of the movable trough applicator is not much different than for the fixed trough. All general steps are followed
except that the trough is pushed slowly and evenly across the top
surfaces of the plates.

Always refer to the instructions supplied with the commercial
applicator being used.

(a)

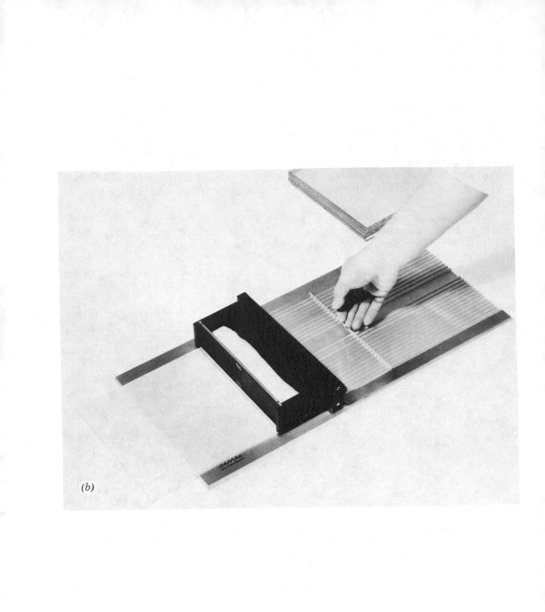

(b)

44

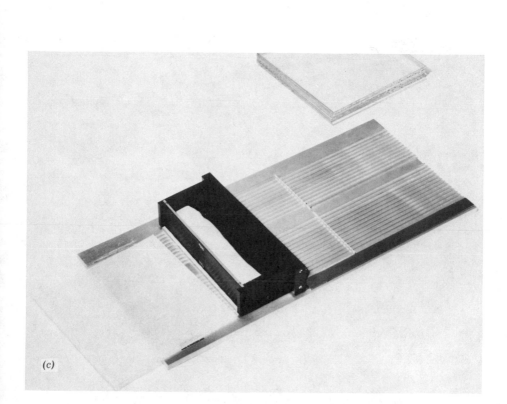

(c)

Figure 2.4.   Procedure for spreading thin layer plates using a
fixed trough apparatus.   (a) Clean plates to be coated are placed
on the mounting board, with the edge of the first plate lined up
with the edge of the trough.   Slurry is evenly poured into trough.
(b)   Plates are slowly pushed under trough at a constant rate of
speed.   Do not stop in the middle of a plate.   (c)   The finished
plate may be taken off the end of the mounting board before the
next is pushed through.   Courtesy of Camag, Inc.

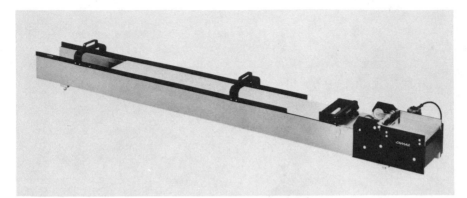

Figure 2.5. A motorized form of the fixed trough spreading apparatus. The finished plates are to the left. The plates to be coated are stacked on the right. A motor-driven wheel pushes the plates under the slurry trough to be coated. Courtesy of Camag, Inc.

REFERENCES

1. J. G. Kirchner, J. M. Miller, and G. J. Keller, Anal. Chem., 23, 420 (1951).
2. E. Stahl, Ed., in *Thin Layer Chromatography*, Springer Verlag, New York, 1969, p. 7.
3. T. E. Beesley, in *Thin Layer Chromatography, Quantitative Environmental and Clinical Applications*, J. C. Touchstone and D. Rogers, Eds., Wiley-Interscience, New York, 1980, p. 10.
4. D. S. Sgoutas and F. A. Kummerow, Biochemistry, 3, 406 (1964).
5. M. von Schantz, S. Juvonen, and R. Hemming, J. Chromatogr. 20, 618 (1965).
6. D. F. Zinkel and J. W. Rowe, J. Chromatogr., 13, 74 (1964).
7. J. W. Copius-Peereboom and H. W. Beekes, J. Chromatogr., 17, 99 (1965).
8. M. F. Dobbins and J. C. Touchstone, in *Quantitative Thin Layer Chromatography*, J. C. Touchstone, Ed., Wiley-Interscience, New York, 1973, p. 295.
9. Applied Science Laboratories, "Thin Layer Chromatography with Silver Nitrate," Technical Bulletin No. 4 (1964).
10. K. Randerath and E. Randerath, J. Chromatogr., 22, 110 (1966).
11. H. Rosler, W. Heinrich, and T. J. Mabry, J. Chromatogr., 87, 433 (1973).

12. H. Determann, Experientia, *18*, 430 (1962).
13. B. G. Johansson and L. Rymo, Acta Chem. Scand., *16*, 2067 (1962).

CHAPTER 3
Commercial Precoated Plates

3.1 INTRODUCTION

Commercially available precoated plates for thin layer chromatography have been available since 1961. They are supplied with all the basic sorbents on glass, plastic, or aluminum supports. Sorbents with and without binder and with and without UV indicator are available in a variety of layer thicknesses from 100 μm on most plastic plates up to 2000 μm for some preparative thin layer glass plates. The largest selection is glass, coated to a 250 μm thickness of "analytical layers."

Precoated plates have become quite popular for a number of justifiable reasons. First of all, they eliminate the expensive equipment necessary for the coating process and the trouble and mess that often go with the procedure. Directly associated with this is the saving of the time needed to prepare a series of plates, which are often only done five (20×20) at a time. This can be a very real saving when a large number of plates or a wide variety of sorbents are used, and when labor is in short supply and high priced. Five to six feet of clean, clear bench space is also required to coat a batch of plates, and equipment is necessary for storage. Precoated plates are usually supplied in divided boxes or individually wrapped for convenient shelf storage, which often eliminates the need for even a storage and carrying

rack. Convenience is the second major point in favor of the precoated plate.

   The most important considerations for their use are that the
layers are uniform in thickness and surface, and from package to
package the plates yield reproducible results. Both of these
features are difficult to duplicate in the laboratory and are of
prime concern in analytical work. High quality commercial precoated plates have been one reason why quantitative thin layer
chromatography has been successful in many applications (1).

   Precoated plates are available with any sorbent except
Sephadex. This allows the chromatographer complete freedom to use
the sorbent needed for a particular application and to conveniently stock as many different ones as desired right on the laboratory shelf.

   Precoated plates with a plastic support have an additional
feature not possible with glass-backed plates: they may be cut
with scissors. This is advantageous when smaller plates are desired but only 20×20 cm plates are on hand. When plastic-backed
plates are used, it is also easy to elute substances out of the
layer by cutting out the area to be eluted with a pair of scissors
and immersing the cut piece in a suitable solvent for extraction.
This is an important consideration when working with radioactive
substances. Counting the radioactivity of the substance is facilitated by simply immersing the piece of plate containing the
radioactive zone in the scintillation vial containing fluid. Because the layer is not scraped, contamination and losses caused
by flying dust are minimized.

   Considering only the cost of the sorbent and the plates, the
precoated plates are more expensive than home-made plates. However, in looking at the total expense involved, the cost of the
coating apparatus, the drying/storage racks, and the labor/time
factor to make the plates must be considered in addition to the
cost of the sorbents and plates when making home-made layers.
After this is done, it will be seen that the cost of the precoated
plates is indeed reasonable.

   3.2 SUPPLIERS

A number of commercial suppliers and manufacturers supply precoated plates, but they do not all carry a complete selection of sor-
bent/backing-plate combinations in the four major sizes: 5×20,
10×20, 20×20, and 20×40 cm. Some supply the 20×20 cm plates with
prescored lines on the back at 5, 10, and 15 cm, so that they may
be easily broken into the smaller plates. Not all suppliers provide preparative layer precoated plates because of their limited,
specialized use.

Four unusual forms of ready-made silica gel and silicic acid plates are manufactured by Gelman Instrument Co. as a part of their Instant Thin Layer Chromatography (ITLC) line. The sorbent is not actually coated on the support but is impregnated in a glass fiber sheet that serves as the support. These silica gel plates are produced by immersing very pure glass microfiber sheets into a supersaturated solution of potassium silicate in ammonium chloride. The precipitated silicic acid that is formed surrounds the glass fibers in a hydrated matrix. The ammonia by-product is removed by chromatography in distilled water and by heating at over 300° for a period of time. This silica gel is a weak, very porous stationary phase with a pH of 5 and is suitable for the separation of nonpolar compounds.

Silicic acid plates are prepared in much the same way as silica gel plates, but they contain a greater amount of silica in the form of polysilicic acid. This is denser than silica gel and produces a stronger attractive phase suitable for the separation of polar compounds. The pH of these plates is about 6. A list of the plates available is given in Table 3.15.

Whatman, Inc. plates incorporate a 3 cm wide strip as a spotting and preadsorbent sample purification area. Samples and standards are diffusely applied as dilute solutions to the preadsorbent strip in each line. As development occurs, the compounds travel with the solvent front to the preadsorbent-silica junction, where they are metered into the silica gel as a compact band for separation. Preadsorbent plates are also available from Analtech, Inc., E. Merck, Supelco, Inc., Analabs, Inc., etc.

One important factor should be considered when using commercially supplied precoated plates: plates with the same designation from two different manufacturers do not necessarily exhibit the same chromatographic behavior. Given a particular substance or extract to be separated, a precoated silica gel plate from manufacturer A will not always separate these substances in exactly the same manner as a silica gel plate from manufacturer B. All other factors remaining the same, even a given type of plate from the same supplier may exhibit variations (2). This is important to keep in mind when trying to repeat a given separation as described in the literature; always use plates of the brand specified in the article if the same results are desired. This is an extremely important consideration in quantitative analysis by *in situ* densitometry, because plates from different manufacturers have different layer characteristics. Even glass and plastic support thicknesses vary from manufacturer. Once a particular separation works well, it is important to "standardize" on the plate used; specify and use exactly the same type and brand when repeating the separation.

There is not much literature comparing plate layer characteristics and separations. To the benefit of the method, there are
only two or three major manufacturers for each sorbent type, and
therefore the amount of diverseness for each sorbent is at a minimum. However, manufacturers employ different coating techniques,
thereby changing the characteristics of the sorbent coated on the
plate. So even though two precoated plate companies buy their
sorbent from the same manufacturer, the resulting plates may not
have identical separation characteristics.

The hardness of a layer will vary according to the binder
used, and this in turn affects the speed of development and the
resulting R_f for a given substance. Factors such as activation
and developing-chamber saturation are also important influences in
this regard. Generalizations about silica gel layers are difficult to make.

Generalizations about cellulose layers are a little easier to
make. The two most popular forms of cellulose are fibrous and
microcrystalline. The major supplier of fibrous cellulose is
Macherey, Nagel and Co., whose main product is MN 300 cellulose.
The major supplier of microcrystalline cellulose is FMC Corporation, under the trade name Avicel. Microcrystalline cellulose
layers usually develop faster and produce zones that are lower in
R_f and more compact in size than fibrous cellulose layers.

In the following tables layer designations are those used by
the particular manufacturer. The organic binder descriptions are
not specific, because in most cases this information is of a pro-
prietary nature.

Precoated commercial plates employing organic binders do not
adsorb as much moisture from the atmosphere as those made with
Calcium Sulfate binder and therefore do not usually require activation prior to use.

The addresses of all suppliers are listed in the appendix at
the back of this book.

REFERENCES

1. J. C. Touchstone, Ed., *Quantitative Thin Layer Chromatography,*
    Wiley-Interscience, New York, 1973.
2. P. Turano and W. J. Turner, J. Chromatogr., *90,* 388 (1974).

TABLE 3.1   SORBENT SUFFIX DESIGNATIONS

F or F-254	UV indicator present, with activation peak 254 nm (short-wave).
F-254+366	Two fluorescent indicators present, with activation peaks of 254 and 366 nm.
G	Contains calcium sulfate ($CaSO_4$) binder.
H	Without binder.
HR	Pure sorbents that have been specially washed.
N	Without binder (Macherey, Nagel and Co. designation). Also designates sorbents containing silicon dioxide binder.
P	Special sorbent for making preparative (i.e., thicker) layers.
P+$CaSO_4$	Preparative sorbent containing calcium sulfate binder.
RP	Reversed-phase.
Type E/Type T	Aluminum oxides having different specific surfaces: Type E: 120-180 m^2/g Type T: 60-90 m^2/g (Type E is identical to aluminum oxide G.)
MN	Macherey, Nagel and Co. brand.
O	Hard layer (Analabs-Foxboro)

SIZES (cm)

a = 20×20
b = 10×20
c =   5×20
d = 15×20
e = 20×40
f =   5×10
g = 10×10
m = microscope slide, 2.5×8/5

TABLE 3.2 SILICA GEL PRECOATED PLATES, GLASS SUPPORT, ANALYTICAL AND PREPARATIVE

Manufacturer	Product	Layer Thickness (μm)	Binder	UV Indicator	Plate Size
Analabs, Inc.	Anasil G	250	13% CaSO$_4$	-	a,c
		500, 1000			a
	Anasil GF	250	13% CaSO$_4$	+	a,c
		500, 1000			a
	Anasil H	250	--	-	a,c
		500			a
	Anasil HF	250	--	+	a,c
		500			a
	Anasil 0	250	organic	-	a,b,c
	Anasil 0F	250	organic	+	a,b,c
	Preabsorbent & HPTLC plates also available.				
Analtech, Inc.	Silica gel G	250, 500, 1000, 1500, 2000	CaSO$_4$	-	a-e
	Silica gel GF	250, 500, 1000, 1500, 2000	CaSO$_4$	+	a-e
	Silica gel H	250	--	-	a,b,c
	Silica gel HF	250	--	-	a,b,c
	Silicar 7G	250	13% CaSO$_4$	-	a,b,c
	Silicar 7GF	250	13% CaSO$_4$	+	a,b,c
	Silica gel GHR	250	13% CaSO$_4$	-	a,b,c
	Silica gel GHL	250	Inorganic	-	a-e
	Silica gel GHLF	250	Inorganic	+	a-e
	Silica gel HL	250	organic	-	a-e
	Silica gel HLF	250	organic	+	a-e
	Whole line of HPTLC plates in GHL & HL also available.				

Adsorbent	Particle size	Binder		Notes
Silica gel GHR/F	250	13% $CaSO_4$	+	a,b,c
Silica gel G--Woelm	250	13% $CaSO_4$	-	a,b,c
Silica gel GF--Woelm	250	13% $CaSO_4$	+	a,b,c
Silica gel G--AgNO3	250	13% $CaSO_4$	-	a,b,c
Silica gel GF	250	13% $CaSO_4$	+	a,b,c
Applied Science Laboratories				
Silica gel G	250	13% $CaSO_4$	-	a,c
Silica gel GF	250	13% $CaSO_4$	+	a,c
Silica gel H	250	--	-	a,c
Silica gel HF	250	--	+	a,c
Adsorbosil-1	250	10% $CaSO_4$	-	a,c
Adsorbosil-1-P	250	10% $CaSO_4$	-	a,c
Adsorbosil-5	250	--	-	a,c
Adsorbosil-5-P	250	--	+	a,c
Silica gel 60	250	13% $CaSO_4$	-	a,b,c
(EM Brand)	2000			a
Silica gel 60 F-254	250	13% $CaSO_4$	+	a,b,c,f
	500, 2000			a
Silica gel 60/silanized RP-2	250		-	a,c
Silica gel 60/silanized RP-2 F-254	250		+	a,c
Silica gel 60/kieselguhr F-254	250	13% $CaSO_4$	+	a
Silica gel 60/with preconcen- tration zone	250	13% $CaSO_4$	-	a
Silica gel 60 F-254	250	13% $CaSO_4$	+	a
High Performance:				
Silica gel 60	200	13% $CaSO_4$	-	b,g
Silica gel 60 F-254	200	13% $CaSO_4$	+	b,g
Silica gel RP-2 F-254	200		+	g
Silica gel RP-8 F-254	200		+	g
Silica gel RP-18 F-254	200		+	g

TABLE 3.2 (continued)

Brinkmann Instruments, Inc.	Silica gel G	250	13% CaSO$_4$	−	a-d
		2000			a
	Silica gel G UV-254	250	13% CaSO$_4$	+	a-f
		500			a,e
		1000			a,e
		2000			a,e
	Silica gel GHR	250	13% CaSO$_4$	−	a
	Silica gel GHR/UV-254	250	13% CaSO$_4$	+	a
	Silica gel/kieselguhr/UV-254	250	13% CaSO$_4$	+	a
	Silica gel/cellulose/UV-254	250	--	+	a
Camag, Inc. (Applied Analytical)	Merck silica gel 60	250	13% CaSO$_4$	−	a
	Merck silica gel 60-F254	250	13% CaSO$_4$	+	a,b,c
	High Performance:				
	Silica gel 60/with or without preconcentration zone	200	+	−	b,g
	Silica gel 60 F-254/with or without preconcentration zone	200	+	+	b,g
	Silica gel RP-2 F-254	200		+	g
	Silica gel RP-8 F-254	200		+	g
	Silica gel RP-18 F-254	200		+	g
ICN Pharmaceuticals (Woelm)	Silica gel	250	organic	−	a,c
	Silica gel F-254	250, 500	organic	+	a,c
	Silica gel F-254/366	250	organic	+	a
J. T. Baker Chemical Co.	Silica gel Si 250/with and without preconcentration zone and channels	250	organic	−	a,b,c

Sorbent	Layer thickness (μm)	Binder		Codes
Silica gel Si 250 F/with and without preconcentration zone and channels	250	organic	+	a,b,c
Silica gel Si 500 F	500	organic	+	a
Silica gel Si-HPF/high performance	200	organic	+	b,g
Silica gel SiC$_{18}$/ reversed phase available with channels	200		-	a,c,g
Silica gel SiC$_{18}$F/reversed phase available with channels	200		+	a,c,g
MCB Reagents				
Silica gel 60	250	organic	-	a,b,c
	2000		-	a
Silica gel 60 F-254	250	organic	+	a,b,c,f
	500			a
	2000			a
Silica gel 60/silanized RP-2	250		-	a,c
Silica gel 60/silanized RP-2 f-254	250		+	a,c
Silica gel 60/kieselguhr F-254	250	13% CaSO$_4$	+	a
High Performance:				
Silica gel 60	200	13% CaSO$_4$	-	b,g
Silica gel 60 F-254	200	13% CaSO$_4$	+	b,g
Silica gel RP-2 F-254	200		+	g
Silica gel RP-8 F-254	200		+	g
Silica gel RP-18 F-254	200		+	g
Schleicher and Schuell, Inc.				
Silica gel	250	organic	-	a
	500			a
	1000			a
Silica gel/LS 254	250	organic	+	a
	500			a
	1000			a

TABLE 3.2 (continued)

Supelco, Inc.				
Redi-coat G/silica gel G	250, 500	$CaSO_4$	-	a,c
Redi-coat AG/silica gel G with 10% silver nitrate	250, 500	$CaSO_4$	-	a
Redi-coat H/silica gel H	250, 500	SiO_2	-	a
Redi-coat HF/silica gel H with phosphor	250, 500	SiO_2	+	a,c
Redi-coat HK/made with 0.3 m KH_2PO_4, for sugars	250	SiO_2	-	a
Redi-coat 2D/silica gel with 10% magnesium acetate for two-dimensional TLC of lipids	250, 500	--	-	a
L/S Redi-coat/silica gel with ammonium sulfate for lecithin/sphingomyelin ratio separation	250, 500	--	-	a
Aflasil (specifically prepared silica gel for aflatoxin separation)	250, 500	$CaSO_4$	-	a
Universal Scientific, Inc.				
Silica plates, Woelm	250	organic	-	a,b
Silica plates, Woelm F-254	250	organic	+	a,b
	500			a
Silica plates, Woelm F-254/366	250	organic	+	a
Silica rapid plates, Woelm	250	organic	-	a
Whatman, Inc.				
Silica gel K1	250	organic	-	a,b,c,f
Silica gel K1F	250	organic	+	a,b,c,f
Silica gel PK1F	500, 1000			a
Silica gel MK1F	200			m
Silica gel K4	250	$CaSO_4$	-	a,b,c

Silica gel PK4	500, 1000			a
Silica gel K4D	250			a,c
Silica gel K4F	250	CaSO$_4$	+	a,b,c
Silica gel PK4F	500, 1000			a
Silica gel K4DF	250			a,c
Silica gel K5	250	organic	−	a,b,c,f
Silica gel PK5	500, 1000			a
Silica gel K5F	250	organic	+	a,b,c,f
Silica gel PK5F	500, 1000			a
Silica gel K5DF				a,c
Silica gel K6	250	organic	−	a,b,c,f
Silica gel K6D				a,c
Silica gel K6F	250	organic	+	a,b,c,f
Silica gel PK6F	500, 1000			a
Silica gel MK6F	200			m
Silica gel K6DF	250			a,c
Dual-phase multi-K CS5	250			a
Preabsorbent silica gel LK5	250	organic	−	a,c
Preabsorbent silica gel PLK5	1000		−	a
Preabsorbent silica gel LK5D	250			a,c
Preabsorbent silica gel LK5F	250	organic	+	a,c
Preabsorbent silica gel PLK5F	1000			a
Preabsorbent silica gel LK5DF	250			a,c
Preabsorbent silica gel LK6	250	organic	−	a,c
Preabsorbent silica gel LK6D	250			a,c

TABLE 3.2 (continued)

High Performance:
| Silica gel LHP-K | 200 | organic | - | b,g |
| Silica gel LHP-KF | 200 | organic | + | b,g |

Legend for Whatman, Inc. products: Fast, hard, high polarity K6; moderately hard, medium polarity K5; high performance HP-K; preabsorbent layer, linear, LK; divided into 8 mm channels, D.

TABLE 3.3 ALUMINA PRECOATED PLATES, GLASS SUPPORT, ANALYTICAL AND PREPARATIVE

Manufacturer	Product	Layer Thickness (µm)	Binder	UV Indicator	Plate Size
Analabs, Inc.	Anasil AG	250	10% CaSO$_4$	-	a,c
	Anasil AGF	250	10% CaSO$_4$	+	a,c
Analtech, Inc.	Alumina G	250	CaSO$_4$	-	a-e,m
		500			a-e
	Alumina GF	250	CaSO$_4$	+	a-e,m
		500			a-e
		1000			a-e
	Alumina H	250	-	-	a,b,c
	Alumina HF	250	-	+	a,b,c
	Woelm alumina neutral	250	-	-	a,b,c
	Woelm alumina basic	250	-	-	a,b,c
	Woelm alumina acid	250	-	-	a,b,c

Supplier	Product				
Applied Science Laboratories	Aluminum oxide 60 F-254	250	–	+	a,c
	Aluminum oxide 150 F-254	250	–	+	a,c
Brinkmann Instruments, Inc.	Aluminum oxide	250	+	–	a,c
	Aluminum oxide/UV-254	250	+	+	a,c
		1000			a
(EM Brand)	Aluminum oxide F-254, Type 60	250	+	+	a,c
	Aluminum oxide F-254, Type 150	1500	+	+	a
Camag, Inc. (Applied Analytical Industries, Inc.)	Aluminum oxide Merck 60 F-254	250	+	+	a
MCB Reagents	Aluminum oxide 60 F-254	250	organic	+	a,c
	Aluminum oxide 150 F-254	250	+	+	a,c
		1500	+	+	a
Whatman, Inc.	Alumina K3	250	–	–	a,c
	Alumina K3F	250	–	+	a,c
	Alumina LK3F	250	–	+	a,c

TABLE 3.4 CELLULOSE PRECOATED PLATES, GLASS SUPPORT, ANALYTICAL AND PREPARATIVE

Manufacturer	Product	Layer Thickness (μm)	UV Indicator	Plate Size
Analabs, Inc.	Anasil C (Avicel)	250	-	a,c
	Anasil CF	250	+	a,c
Analtech, Inc.	Avicel*	250	-	a-e,m
		500		a-e
		1000		a,b,d,e
	Avicel F	250	+	a-e,m
		500		a-e
		1000		a,b,d,e
	MN300	250	-	a-e,m
		500		a-e
	MN300F	250	+	a-e,m
		500		a-e
	Cellulose DEAE	250	-	a,b,c
	Cellulose Ecteola	250	-	a,b,c
	PEI Avicel	250	-	a,b,c
	10% Acetylated cellulose	250	-	a,b,c
	20% Acetylated cellulose	250	-	a,b,c
Applied Science Laboratories	Cellulose	100	-	a,b,c,g
	Cellulose F	100	+	a,b,c
	Cellulose F	100	+	a
Brinkmann Instruments, Inc.	Cellulose MN300	100	-	a
	Cellulose MN300/UV254	100	+	a

62

	Cellulose	100	−	a,c
	Cellulose F	100	+	a,b
	PEI cellulose F	100	+	a
Camag, Inc. (Applied Analytical Industries, Inc.)	Cellulose, Merck	100	−	a
	Cellulose, Merck F-254	100	+	a
	HPTLC Cellulose, Merck	100	−	b,g
MCB Reagents	Cellulose	100	−	a,b,c,g
	Cellulose F-254	100	+	a,b,c
	PEI cellulose F	100	+	a
Schleicher and Schuell, Inc.	Cellulose	100	−	a
	Cellulose/LS 254	100	+	a
	Cellulose, acetylated 20-25%, starch binder, 4%	100	−	a
	Cellulose, acetylated 40-45%, starch binder, 4%	100	−	a
	Cellulose with Dowex 2-X8, 5% Dowex	100	−	a
	Cellulose with Dowex 2-X8, 10% Dowex	100	−	a
	Cellulose with Dowex 50W-X8, 5% Dowex	100	−	a
	Cellulose with Dowex 50W-X8, 10% Dowex	100	−	a
	PEI cellulose	100	−	a
	PEI cellulose/LS 254	100	+	a
	Cellulose, Fibrous	500	−	a
Whatman, Inc.	Microcrystalline cellulose K2	250	−	a,b,c
	Microcrystalline cellulose K2F	250	+	a,b,c
	Microcrystalline cellulose PK2F	500, 1000	+	a

* Anicel is a trademark of the FMC Corporation for their brand of microcrystalline cellulose.

63

TABLE 3.5 KIESELGUHR PRECOATED PLATES, GLASS SUPPORT

Manufacturer	Product	Layer Thickness (µm)	UV Indicator	Plate Size
Analtech, Inc.	Kieselguhr G (CaSO$_4$ binder)	250	–	a,b,c
	Kieselguhr GF	250	+	a,b,c
Applied Science Laboratories	Kieselguhr F-254	250	+	a
Brinkmann Instruments, Inc.	Kieselguhr UV-254	250	+	a
MCB Reagents	Kieselguhr F-254	250	+	a

TABLE 3.6 SILICA GEL PRECOATED PLATES, PLASTIC SUPPORT

Manufacturer	Product	Layer Thickness (µm)	Binder	UV Indicator	Plate Size
Applied Science Laboratories	Silica gel 60	200	+	–	a
	Silica gel 60 F-254	200	+	+	a
J. T. Baker Chemical Co.	Silica gel IB	200	+	–	a,c,e,m
	Silica gel IB-F	200	+	+	a,c,e,m
	Silica gel IB2	200	+	–	a,c
	Silica gel IB2-F	200	+	+	a,c

Supplier	Product				
Brinkmann Instruments, Inc.	Silica gel G	250	+	–	a,c,e,h
	Silica gel G/VU-254	250	+	+	a,c,e,h
	Silica gel G, hydrophobic	250	+	–	a
	Silica gel N-HR (MN)	200	+	+	a,c,e
	Silica gel N-HR/UV-254	200	+	+	a,c
	Silica gel 60	250	+	–	a
	Silica gel 60 F-254	250	+	+	a
Camag, Inc. (Applied Analytical Industries, Inc.)	Silica gel Merck 60 F-254	200	+	+	a
Eastman Kodak Co.	Chromagram silica gel	100	polyvinyl alcohol	–	a
	Silica gel with indicator	100	polyvinyl alcohol	+	a
MCB Reagents	Silica gel 60	250	+	–	a
	Silica gel 60 F-254	250	+	+	a,c
		200	+	+	a, 20×500 roll
Schleicher and Schuell, Inc.	Silica gel	200	organic	–	a
	Silica gel with indicator	200	organic	+	a

* All J. T. Baker Chemical Co. products are called Baker-flex.

TABLE 3.7 ALUMINA PRECOATED PLATES, PLASTIC SUPPORT

Manufacturer	Product	Layer Thickness (µm)	Binder	UV Indicator	Plate Size
Applied Science Laboratories	Aluminum oxide 60 F-254 (Neutral)	200	+	+	a
Brinkmann Instruments, Inc.	Aluminum oxide N, MN brand	200	+	-	a,c,e
	Aluminum oxide N, MN brand, UV-254	200	+	+	a,c,e,h
	Aluminum oxide 60 F-254	250	+	+	a
Eastman Kodak Co.	Alumina	100	polyvinyl alcohol	-	a
	Alumina with indicator	100	polyvinyl alcohol	+	a
J. T. Baker Chemical Co.*	Aluminum oxide IB	200	+ inert	-	a,c,e,m
	Aluminum oxide IB-F	200	+ inert	+	a,c,e,m

* All J. T. Baker Chemical Co. products are called Baker-flex.

TABLE 3.8 CELLULOSE PRECOATED PLATES, PLASTIC SUPPORT, ALL BINDER-FREE

Manufacturer	Product	Layer Thickness (µm)	UV Indicator	Plate Size
Applied Science Laboratories	Cellulose	100	-	a, roll
	Cellulose F-254	100	+	a
	PEI cellulose	100	-	a
	20% acetylated cellulose (w/4% starch binder)	100	-	a

Manufacturer	Product			
Brinkmann Instruments, Inc.†	40% acetylated cellulose (w/4% starch binder)	100	−	a
	Cellulose MN300	100	−	a,c,e,h
	Cellulose MN300/UV-254	100	+	a,c,e
	Cellulose MN400, Avicel Microcrystalline	100	−	a,c,e
	Cellulose MN 400/UV-254	100	+	a
	10% acetylated cellulose, MN300	100	−	a
	30% acetylated cellulose, MN300	100	−	a
	CM cellulose, MN300	100	−	a
	DEAE cellulose, MN300	100	−	a,e
	Ecteola cellulose, MN300	100	−	a
	PEI cellulose, MN300	100	−	a,e
	PEI cellulose/UV-254	100	+	a,e
	Cellulose	100	−	a,e
	Cellulose F	100	+	a
	PEI cellulose F	100	+	a
Eastman Kodak Co.	Cellulose	160	−	a
	Cellulose, fluorescent indicator	160	+	a
J. T. Baker Chemical Co.	Cellulose	100	−	a,c,e,m
	Cellulose F	100	+	a,c,e,m
	Cellulose, microcrystalline	100	−	a,c
	Cellulose, F microcrystalline	100	+	a,c
	Cellulose AC10 (10% acetylated)	100	+	a,c
	Cellulose CM	100	−	a,c
	Cellulose DEAE	100	−	a,c
	Cellulose Ecteola	100	−	a,c
	Cellulose PEI	100	−	a,c
	Cellulose PEI-F	100	−	a,c

TABLE 3.8 (continued)

		Layer Thickness (μm)	UV Indicator	Plate Size
MCB Reagents	Cellulose	100	-	20×500 roll
	Cellulose F-254	100	-	a
	PEI cellulose F	100	+	a
Schleicher and Schuell, Inc.	Cellulose, microcrystalline	100	-	a
	Cellulose, microcrystalline LS254	100	+	a
	PEI cellulose	100	-	a
	PEI cellulose LS254	100	+	a

* All J. T. Baker Chemical Co. products called Baker-flex.

† All plates coated with MN brand cellulose.

TABLE 3.9 ION-EXCHANGE PRECOATED PLATES, * PLASTIC SUPPORT

Manufacturer	Product	Layer Thickness (μm)	UV Indicator	Plate Size
Brinkmann Instruments, Inc.	Ionex-25 SA-Na, strong acid, Na$^+$ active	250	-	a
	Ionex-25 SB-Ac, strong basic, CH$_3$COO$^-$ active	250	-	a
	Ionex-25 SB-Ac/UV 254, strong basic, CH$_3$COO$^-$ active	250	+	a

* All are manufactured by Macherey, Nagel and Co. and distributed by Brinkmann Instruments, Inc.

TABLE 3.10 POLYAMIDE PRECOATED PLATES, PLASTIC SUPPORT, NO BINDER

Manufacturer	Product	Layer Thickness (µm)	UV Indicator	Plate Size
Brinkmann Instruments, Inc.	Polyamide 6	100	−	a
	Polyamide 6/UV-254	250	+	a
	Polyamide 11	100	−	a
J. T. Baker Chemical Co.*	Polyamide 6	100	−	a,c
	Polyamide 6F	100	+	a,c
Schleicher and Schuell, Inc.	Micropolyamide, double-faced	25 both sides	−	
	Polyamide (w/4% starch binder)	120	−	a
	Polyamide LS 254 (w/4% starch binder)	120	+	a
	Polyamide/acetylated cellulose	100	−	a

* All J. T. Baker Chemical Co. products are called Baker-flex.

TABLE 3.11   SILICA GEL PRECOATED PLATES, ALUMINUM SUPPORT

Manufacturer	Product	Layer Thickness (μm)	Binder	UV Indicator	Plate Size
Applied Science Laboratories	Silica gel 60	200	+	–	a
	Silica gel 60 F-254	200	+	+	c
	Silica gel 60/kieselguhr F-254	200	+	+	a
Brinkmann Instruments, Inc.	Silica gel 60	250	+	–	a
	Silica gel 60 F-254	200	+	+	c
	Silica gel 60/kieselguhr F-254	200	+	+	a
Camag, Inc. (Applied Analytical Industries, Inc.)	Silica gel 60	200	+	–	a
	Silica gel 60 F-254	200	+	+	a
MCB Reagents	Silica gel 60	200	+	–	a
	Silica gel 60 F-254	250	+	+	c
Universal Scientific, Inc.	Silica sheets, Woelm	200	+	–	a
	Silica sheets, Woelm F-254/366	200	+	+	a

TABLE 3.12 ALUMINA PRECOATED PLATES, ALUMINUM SUPPORT

Manufacturer	Product	Layer Thickness (μm)	Binder	UV Indicator	Plate Size
Applied Science Laboratories	Aluminum oxide 60 F-254	200	+	+	a
	Aluminum oxide 150 F-254	200	+	+	a
Brinkmann Instruments, Inc.	Aluminum oxide 60 F-254	200	+	+	a
	Aluminum oxide 150 F-254	200	+	+	a
MCB Reagents	Aluminum oxide 60 F-254	200	+	+	a
	Aluminum oxide 150 F-254	200	+	+	a

TABLE 3.13 CELLULOSE PRECOATED PLATES, ALUMINUM SUPPORT

Manufacturer	Product	Layer Thickness (μm)	UV Indicator	Plate Size
Applied Science Laboratories	Cellulose	100	-	a
	Cellulose F-254	100	+	a
Brinkmann Instruments, Inc.	Cellulose	100	-	a
	Cellulose F-254	100	+	a
MCB Reagents	Cellulose	100	-	a,roll
	Cellulose	100	+	a
	PEI cellulose F	100	+	a

TABLE 3.14 KIESELGUHR PRECOATED PLATES, ALUMINUM SUPPORT

Manufacturer	Product	Layer Thickness (μm)	UV Indicator	Plate Size
Applied Science Laboratories	Kieselguhr F-254	200	+	a
Brinkmann Instruments, Inc.	Kieselguhr F-254	200	+	a

TABLE 3.15 SILICA GEL PLATES, IMPREGNATED GLASS FIBER

Manufacturer	Product	UV Indicator	Plate Size
Gelman Sciences, Inc.	Silica gel SG ITLC-SG*	–	a,c
	Silicic acid SA ITLC-SA	–	a,c
	Silicic acid SAF ITLC-SAF	+	a,c
		(254 and 350 nm)	a,c
	SAF-D drug identification media	+	a

* ITLC is a trademark of the Gelman Sciences, Inc. and is an abbreviation for instant thin layer chromatography.

CHAPTER 4
Preparation and Application of the Sample

4.1 INTRODUCTION

The field of TLC has increased considerably in recent years. However, in spite of the numerous reports on methodology, there has been no real treatise on the art of preparing a sample and applying it to a thin layer. The technique appears simple enough, so much so that one can tend to become careless and then wonder why the results are not satisfactory. TLC can generally be done with less preparation and attention than any other separation method. But, for reproducibility and for quantitative determinations as well as mass screening, some attention to detail is required.

The room used for TLC should be of constant temperature and humidity. It must be clean and free of dust, particularly in industrial areas, and well ventilated; open tanks release many solvent fumes. Dust particles interfere with fluorometric evaluation of TLC because many dust particles are fluorescent, some react with the reagents used for visualization, and lint particles from clothing are often fluorescent. Since TLC is essentially a microtechnique, extreme cleanliness should be in order, particularly when quantitation is to be performed.

The environment should also be free of chemical fumes that may alter the sample or be adsorbed into the plate while the sample is being applied.

   With the sample in hand and a micropipet of some sort and the
proper plate before you, what more simple technique for separation
is there anywhere? When you dip the pipet into the sample, it is
easy to see the microcapillaries fill. Microsyringes need some
manipulation to be sure they are filled. Touch the sides of the
pipet to the container to remove excess solution.
   Now carefully lower the tip of the delivery device to the
layer. Capillary action removes the solution from the glass capillary. Plunger action drives it from syringes. The amount of
solution to be transferred to the layer determines the procedure
from here. It is better to apply small proportionate amounts repeatedly. The aim is to form the smallest spot amenable with the
solvent that the sample is dissolved in.
   Now that the plate has a spot or streak or a series of them,
it can be put in the tank.
   All appears very simple. However, there is more to it than
just this. This chapter examines aids to spotting, pitfalls of
sample application, the different types of delivery systems (both
manual and automatic), selection of solvents for application, and
the reasons certain spots appear the way they do on finished chromatographic plates.

   4.2   SOLUTE SOLVENT SELECTION

The first consideration before any chromatography can be done is
to decide which solvent should be used to dissolve the sample.
Often the analyst receives a sample whose physical characteristics and solubility are not known. If the general nature of the
compound is known, experience will tell which solvent to use as
a starting parameter. Reference books such as *The Merck Index*
and *The Handbook of Chemistry and Physics* are extremely useful in
this regard and should be in the library of every chromatographer.
However, when the sample is a derivative, the solubility characteristics can be drastically changed, or there may not be enough
sample to perform the general solubility tests. If the material
is crystalline, its solubility can be determined readily with the
aid of a microscope. Place a few crystals of the sample on a microscope slide and mount the slide in the field of the microscope.
While viewing the crystals, use a micropipet to place a drop of
solvent on the crystals. If the crystals do not dissolve, evaporate the solvent and select another for testing. Judicious testing of a number of solvents will expedite selection of the optimal
one for the experimental analysis.
   The process of application of the sample to the plate may be
one of the most important steps for the success of the separation.
Proper preparation of the sample itself is prerequisite to application of the sample to the layer. The nature of the sample, or

the components in the sample that is to be analyzed, determines
the actual extraction or preparation of the compound in question.
This chapter will cover the handling of the sample before and during its application to the chromatogram. Extraction of the compounds of interest from a complex sample and dissolution after-
wards in proper vehicles (solvents), depend on the compound class
under consideration. Polar compounds do not dissolve well in nonpolar solvents. Nonpolar compounds will not dissolve in polar
solvents. Usually ionic compounds dissolve best in a polar solvent to which the opposing ion has been added. Generally speaking, however, for good application of samples to the layers the
solvent used should be highly volatile and as nonpolar as possible. The reason for this is that nonvolatile solvents spread
through the plate layer matrix during their period of contact,
taking the sample with them.

Likewise, a polar solvent will spread its solute through the
matrix because of high solvent power as well as low volatility.
Samples can, however, be changed to derivatives prior to chromatography to facilitate solution in solvents of lower polarity and
higher volatility. This also can aid in detection, identification,
or analysis. Many polar compounds are difficult to chromatograph
because of their poor solubility and high polarity. Cases have
been known where a nonpolar solvent had been used to handle a polar solute and the operator wound up wondering why no spots appeared on the plate in spite of following directions described in
the literature. Basically, a knowledge of the compound to be separated aids in selection of the sorbent, the developing solvent
(mobile phase), and the solvent to be used for solutions and during application of the sample.

The terms polar and nonpolar deserve comment. A polar solvent is one of high ion potential. Conversely, a nonpolar solvent
is one with very low ion potential. Water and acids or bases are
highly polar, whereas hydrocarbons are solvents of the nonpolar
class. The eluotropic series described in detail in Chapter 5
lists solvents in order of polarity as well as power of elution
from a sorbent. This type of table is sometimes useful in selecting a solvent for the most favorable sample application conditions.

4.3 SAMPLE PREPARATION

A variety of sample preparation methods are used to make a sample
ready for chromatographic analysis, including dissolving of samples, extraction, column chromatography, centrifugation, and evaporation. Often it is necessary to use a number of them together,
and when working with biological samples, many times all of the

methods are used in sequence in order to sufficiently purify a sample so that it is suitable for chromatography.

J. C. Touchstone has logically emphasized that the amount of sample preparation necessary is dependent upon the ratio of the substance to be determined (the analyte) to the amount of un-wanted substance (the junk). The higher the proportion of "junk," the greater the amount of sample preparation that may be necessary.

4.3.1 Direct Application of Samples

A raw sample can be applied directly to the TLC plate. When the analyte (or analytes) is present in sufficient concentration or purity, it produces a detectable zone after development of a suit-ably sized aliquot. The impurities present must not be so con-centrated that they obscure or modify the analyte zone if they mi-grate, nor retain it at the origin.

The availability of preabsorbent layer plates has made it possible to apply many different types of samples directly to the layer without any prepurification. There are many advantages to direct application, including saving time and reagents, prevent-ing loss of analyte through reduced handling, and being able to work with smaller initial quantities of sample. For high sensi-tivity work, it has been found that predevelopment for at least 3 hr in a 1:1 solution of methanol:chloroform reduces the back-ground on some preadsorbent layers.

Touchstone et al. (1) directly applied 2-5 μL aliquots of human gall bladder bile which had been diluted 1:100 with methanol. Reversed phase plates were used to separate conjugated bile acids.

Using preadsorbent layers with silica gel, it has been possible to separate and quantitate phospholipids, including lecithin and sphingomyelin, from the direct application of 50 μL of human amni-otic fluid (20) or 10 μL of a 1:1 dilution of human seminal fluid (3).

Caffeine can be determined in various beverages by direct ap-plication of 5-10 μL of the liquid such as cola, coffee, or tea to a silica gel layer. The absence of an appreciable amount of caf-feine in a decaffeinated beverages can be shown in the same manner.

4.3.2 Application of Sample Solutions or Extracts

When dealing with solid samples in which the substance to be de-termined is a major component, simply dissolving the sample in a solvent suitable for the analyte may be all the sample preparation that is necessary. An appropriate sized aliquot or series of ali-quots would then be applied to the TLC plate with comparative standards (knowns) if available. This mode of preparation is used extensively in analytical method development when working with known substances.

In biological and medical work, it is often necessary to determine very small amounts of substance in a substrate which cannot be directly applied to the plate, or which is in too dilute a
concentration to be determined in a reasonably sized aliquot. Under these circumstances, it is necessary to extract the substance
(or substances) of interest, concentrate the extract, and apply an
aliquot to the plate. Examples of applications where this is the
usual sample preparation method include pesticides in foods or
tissue, and drugs in body fuilds. The least polar solvent that
will dissolve the substances of interest should be used in order
to minimize the extraction of more polar impurities.

As an example, the drugs phenobarbital and p-hydroxyphenobar-
bital were extracted from 5 mL of urine in a 50 mL glass stoppered
centrifuge tube to which had been added 2 mL of saturated NaH_2PO_4
solution (4). This gave a final pH of 4.5. This was extracted
twice with 5 volumes of methylene chloride. The extracts were
concentrated by evaporation, and aliquots were applied to the TLC
plate.

4.3.3 Purification of Extracts

Extracts of samples may have to undergo further purification steps
before being totally suitable for chromatography. Solvent partitioning and column chromatography are the two methods that are
most frequently used to purify extracts.

Solvent partitioning is used to leave impurities behind in
one solvent while extracting the desired substance into another
solvent. Control by pH is often utilized here. Organic acids are
converted into salts and are soluble in aqueous solutions at high
pH. At low pH the acids are un-ionized and extractable into organic solvents. At high pH, basic compounds are extracted into
organic solvents and into water as their salt forms at low pH.

It is very important that the highest quality solvents, such
as glass distilled analytical reagent, be used for sample preparation. Thin layer chromatography is a sensitive procedure and
impurities from solvents can often be detected, particularly if
they become concentrated in the solvent through evaporation of
extracts.

As an example of solvent partitioning, Dobbins (5) partitioned organochlorine pesticides from vegetables between benzene-
isopropanol. The isopropanol was then extracted into water, leaving a benzene solution of pesticides. Some colored plant pigments
were also dissolved in the benzene and were removed by column
chromatography on neutral alumina. The benzene solution was evaporated and subjected to chromatographic analysis.

A number of commercial suppliers offer specially packed plastic tubes for the purification of sample extracts. Usually the

tubes are packed with silica gel or C_{18} reversed phase, and some
are made to fit onto a syringe so that the barrel becomes a column and the plunger can be used to force the sample through the
packing.

The silica gel packing eliminates low polarity substances by
eluting with a nonpolar solvent, while the analyte is retained.
The packing is then eluted with a more polar solvent to remove the
analyte (or analytes). The least polar solvent which will do this
should be used so that very polar impurities will still remain on
the packing.

The reversed phase packing retains nonpolar impurities but
elutes polar analytes.

The principle of liquid-liquid partitioning is used with another type of sample preparation tube, the Extube. It is a disposable, plastic syringe barrel type tube packing with an inert
fibrous matrix of large surface area. The sample and extraction
solvent are poured onto the tube where liquid-liquid partitioning
occurs. Analyte is extracted into the organic phase, which passes
through the matrix while impurities and other polar components,
including water, are retained. The procedure is simple and quick.
The organic effluent is concentrated by evaporation before TLC.

As an example of this sample preparation technique, Touchstone
and Dobbins (6) extracted cortisol from human plasma in an unbuff-
ered Extube. Plasma (2 mL) plus 4 volumes of water were vortexed
and radioactive cortisol was added to minitor recovery. Methylene
chloride (5 mL) was poured into the tube, followed immediately by
by the plasma solution. Four 10 mL portions of methylene chloride
were passed through the tube to extract and elute the cortisol.
The eluate was collected and evaporated prior to TLC. Recovery
of cortisol was 79%.

4.3.4 Evaporation of Sample Solutions

Evaporation of the sample preparation to concentrate the analyte
is usually the last step before applying the sample to the plate.
The evaporation must be carefully controlled so that loss or degradation of the sample does not occur.

A rotary evaporator is a necessary piece of apparatus wher-
ever large volume solutions of samples are prepared. A round bottom flask containing the solution to be evaporated is rotated in a
water bath while under reduced pressure. The water bath can be
heated, if necessary, to an appropriate temperature, and this combined with the reduced pressure will bring about evaporation.
When the solution is reduced to a small volume, the flask is
rinsed with a little solvent and the total transferred to a small
conical tube for the final evaporation under nitrogen. It is convenient to keep a selection of 24/40, 19/38 and 14/20 flasks with

adapters on hand for varied evaporation flexibility. Disposable
Pasteur pipets are also a must for sample handling, as are a selection of different size conical centrifuge tubes for evaporation.

Nitrogen is preferred for evaporation under a gas stream, for
unlike air, it will not promote decomposition of the sample or
carry moisture to it. Commercial evaporation manifolds complete
with water bath are available, or a home-made one can be fabricated.

Sample solutions are usually carefully evaporated just to
dryness so that the residue can be redissolved in a known amount
of solvent. Knowing this, aliquots can be taken for application
to the TLC plate. The least polar solvent in which the sample is
soluble should be used. If some highly polar impurities should be
in the residue, possibly they may be left behind. The solvent
must be chosen carefully because those with too high a polarity or
too high a boiling point are not easily removed from a sorbent
layer during sample application. It is imperative that sample
zones on a layer be totally dry before the plate is placed in the
mobile phase for development. If they are not dry, adverse affects may occur, including zone spreading or a change in R_f value.

Techniques for microsample preparation are covered in the
book by Dünges (7).

4.4 SAMPLE APPLICATION

Samples should be applied as a 0.01-1.00% solution in the least
polar solvent in which they are soluble. Normally 1-5 µg/µL solution concentrations are practical. Solvents with too high a
boiling point or a high polarity are difficult to remove from the
sorbent during application. If a small amount of solvent is retained in the sorbent after application of the sample, it will ad-
versely affect the separation, due to spreading of the sample in
the matrix. Be sure the samples are dry before placing the plate
in the developing chamber. Some knowledge of the relative concentration of the components under examination is usually helpful in
this respect. This will determine the smallest amount of sample
or extract that can be applied to the plate. Dilute solutions can
be applied to the layer with sorbent drying between successive applications, or they can be concentrated. Overspotting is a useful
technique, but care must be exercised since too rapid an application of sample results in "rings"; that is, the solvent washes the
sample to the outer edges of the application area. This sometimes
results in spurious separation of two zones where there should be
only a single spot, particularly if the ring formed at the starting line has a wide diameter. Sometimes drying between applications with a hair dryer or a stream of filtered air will facili-

tate solvent evaporation, but only if the sample components to be
separated are nonvolatile; otherwise they will be lost. The zone
of application should be as narrow as possible, or the spots
should be as small as is compatible with the concentration. Spots
that are too highly concentrated will give poor separation, since
the mobile phase solvent tends to flow through the point of least
resistance and will travel around the spot. This will result,
after development, in separated spots that are unsymmetrical.
Ideally, if the spots are properly applied the separated zones are
symmetrical and compact. The spots tend to diffuse and spread
somewhat as they migrate further up the plate (at increasing R_f).
Width of separated zones is a function of retention in the system.
Applying the sample as a narrow streak has an advantage over spot
application; a streak usually produces a sharper separation, which
in turn gives improved reproducibility. The mobile phase is less
likely to flow around a streak than a spot during development.

Samples should be applied near one side of the layer, usually
about 1.5-2 cm from the edge, so that the layer makes contact with
the mobile phase but the sample zone itself is not immersed in the
liquid. This avoids dissolution of the sample in the developing
phase. In practice a sheet of white cardboard can be used as a
background while the sample is applied to the TLC plate. A
dark line can be drawn on this background so that when the plate
is positioned on it the line will be seen through the layer 2 cm
(or the predetermined distance) from and parallel to the edge.
This serves to provide a uniform marker across the layer to guide
the sample application.

The sample may be applied as a spot or as a series of spots
to form a streak. The application of a continuous streak is made
possible with modern instrumentation. Streaking of the sample
along the baseline is not easily performed manually, but it has
been made easier by the use of the trough device of Achaval and
Effelson (8) or the automatic applicator described by Coleman (9).
Automatic devices for streaking, however, overcome the difficulties; they are described later in this chapter.

The band width or spot diameter must be as small as possible,
since as the chromatogram develops, the areas occupied by the separated zones tend to increase and the band or spot widens due to
diffusion. Nybom (10) has observed that there is no direct relationship between the amount per spot and the spot size variations,
being due to characteristics of the separated substances and to
layer thickness within limits. The load applied is critical and
affects the shape and position of the zone after development.
Overloaded sample applications produce cometlike vertical streaks.
If the adsorption isotherm is convex, the R_f measured from the
midpoint of the streak will be greater than normal. If the adsorption isotherm is concave, the R_f value will be less than normal.

The ends of the streak will overlap other spots on the chromatogram, rendering isolation and resolution poor. This is particularly true when one compound is present in larger quantity than
another with a similar R_f value. This is illustrated in Figure
4.1. Herein lies one of the most common faults of the practice
of thin layer chromatography. It should be remembered that this
is a sensitive technique, and the smallest sample size that can
provide visualization of all desired components is the one to be
applied. This is one of the prerequisites for reproducible R_f
values. This general rule also holds true for sample application
in quantitative work. Too concentrated a zone will tend to give
nonlinear calibration curves.

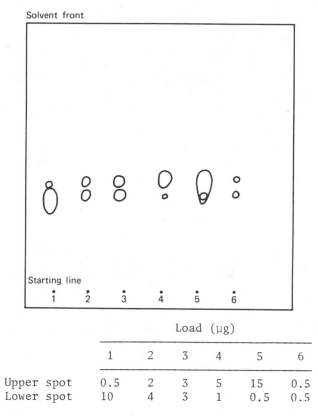

	Load (µg)					
	1	2	3	4	5	6
Upper spot	0.5	2	3	5	15	0.5
Lower spot	10	4	3	1	0.5	0.5

Solvent system: isooctane:acetone:n-butanol
(58:33.6:8.4); silica gel G layers.

Figure 4.1 The effect of loan on the resolution of a mixture of
reserpine and rescinnamine.

The technique of applying microliter amounts of solutions is not easy. The accuracy of methods of delivering small volumes has already been discussed. In practice, depending on the application instrument used, small drops do not leave the tip of a needle in contact with a thin layer, and care must be used to avoid disturbing the layer. A hole in the layer may cause distorted triangular or crescent spots in the final chromatogram (see Figure 4.2). Truter (11) describes this type of sampling effect.

Techniques for applying samples to TLC plates under specialized conditions such as from a gas chromatograph are discussed in Chapter 16.

4.5 SAMPLE APPLICATION ON CONCENTRATION ZONE LAYERS

Concentration zone, or "preadsorbent" zone, layers are commercially available; they are precoated glass plates that contain a zone at the bottom of the plate which is intended to aid the separation of large volumes of sample. This zone is usually three centimeters wide and is composed of a specially treated kieselguhr. At the present time, only silica gel, alumina, and reversed phase plates are available with the concentration zone.

Figure 4.2 Distorted shapes of spots, due to excessive pricking of the surface of the adsorbent.

This concept was first developed in 1970 by Quantum Industries, which has since become Whatman, Inc. Their plates with the preadsorbent zone or layer were called "Linear K" plates, and they are now designated with an "L," e.g., LK5, LK6F. High performance (HP) plates are also available with preadsorbent zones.

Sample application onto a plate containing a preadsorbent zone can be quite different than onto a regular plate because of what the zone is intended to do. Preadsorbent plates, as they are often called, are available either scored into lanes or unscored.

Unscored plates can be scored by the user for application of many individual samples, or they can be used as is for preparative work.

Often as much as 50-100 µL of biological sample can be applied to a single 0.8-1 cm wide preadsorbent lane. Samples can be organic or aqueous, crude or prepurified. One of the major advantages of using a preadsorbent layer plate is that samples often do not have to be purified before they are applied to the layer.

Sample components undergo processes in the preadsorbent area such that they are extracted, cleaned and concentrated. Using either the regular mobile phase or a special "pre-developing" phase, the solvent front carries the essentially purified and concentrated samples to the analytical layer. They are at this point virtually ideally applied samples because they are in the form of a concentrated band of purified material.

Because the preadsorbent-layer interaction serves to idealize the sample application, the sample does not have to be applied with the usual care and can be streaked onto the preadsorbent layer. Use of these layers eliminates variations due to nonideal sample application techniques.

The sample is normally applied to the preadsorbent layer at least 5-10 mm below the interface of the preadsorbent layer with the analytical layer. This same depth should be allowed for at the bottom of the layer because of the mobile phase in the bottom of the development chamber.

Measurement of R_f values is from the layer/preadsorbent-area interface.

Preadsorbent area plates have some beneficial feature; they reduce sample application time; accept crude, dilute, organic, or aqueous samples, often eliminating the need for prepurification of complex or biological samples; and they will accept higher sample loading per linear distance than a conventional layer, often providing better resolution and sensitivity with the larger sample than would be available on a nonpreadsorbent layer.

4.6 MANUAL SAMPLE APPLICATORS

It is unfortunate that in many reports of quantitative TLC methodology the preparations of the solutions, the solvent used, and the pipets used for applying the sample itself are not described in detail. Some expression of this methodology sometimes is necessary for the success of the separations as well as for the qualitative and quantitative results.

For much general work where exact sample size is not the consideration, capillaries fabricated in the laboratory are often used. These are readily made as they are needed. However, they are not suitable for quantitative work or when duplication of the size of the zones on the starting line is desired. There are many types of capillaries in a variety of forms. It is even possible, if a concentrated sample is available, to transfer it as a drop from a glass rod. However, these miscellaneous means of transfer should be considered only for those who do not have to reproducibly apply samples.

The sample can also be applied with micropipets or melting point capillaries. A number of automatic spotters of varying design are available commercially. Calibrated disposable capillaries as well as ordinary capillaries are generally available. These are preferable for most routine work when few samples are involved, the advantages being the saving of time as well as avoidance of sample contamination. Since chromatographic methods used today utilize only small test samples, 1 μL or even 0.1 μL sample solutions will suffice for a complete qualitative or quantitative analysis. The analyst is faced with the problem of delivery precision of the various micropipets. This poses a source of error that may be greater than the combined errors in the chromatographic process itself. The repetitive spotting of microliter volumes is a tedious operation, particularly if a large number of samples must be analyzed. Automatic spotters have become increasingly useful in this regard.

In spite of the availability of a number of sample delivery systems, little attention seems to have been given to reproducibility or determination of whether the measuring instruments actually do deliver replicates of constant volume. Volumes of 1-5 μL are frequently used, and therefore errors in repetitive delivery would invalidate any attempts to reproduce identical conditions from day to day. A certain amount of experience is necessary to obtain maximum accuracy. Fairbairn and Relph (12) report details of the results of repetitive measurement of volumes by a number of workers, each using a favorite syringe or micropipet. These results are given in Table 4.1. The participants used radioactive solutes to facilitate the determination. Each delivered 10-16 replicates (a) directly into glass vials for counting or (b) onto thin layer plates or paper chromatograms. Areas of the paper or sorbent containing the zone (from b) were transferred to vials for counting. The results as seen in the table indicate that pipetting errors ranged from ± 6 to ± 20% for individual mea-

TABLE 4.1 ERRORS DUE TO MEASUREMENT OF SMALL VOLUMES OF SOLUTION BY EXPERIENCED WORKERS, EXPRESSED AS COEFFICIENT OF VARIATION (S.D. OF INDIVIDUAL RESULTS CALCULATED AS A PERCENTAGE OF THE MEAN)

Worker	Solute	Volume of Solution Measured (μL)	Coefficient of Variation	
			Direct into Phosphor (%)	Via Adsorbent (%)
A	Morphine-2-T	10 (Agla)	7.8	7.3 (paper)
B	Morphine-2-T	10 (Agla)	5.8	10.1 (paper)
C	Morphine-2-T	2 (micropipet)	6.6	3.3 (paper)
C	d-Glucose-^{14}C(U)	2 (micropipet)	4.7	
		5 (micropipet)	2.5	
D	d-Glucose-^{14}C(U)	1 (Hamilton)	11.1	8.2 (paper)
			10.4	7.8 (paper)
		1 (Hamilton)		8.8 (TLC)
				9.8 (TLC)
B	d-Glucose-^{14}C(U)	5 (Agla)	5.5	
E	d-Glucose-^{14}C(U)	5 (Agla)	3.4	3.2 (paper)
				6.0 (TLC)
F	1-Tyrosine-^{14}C(U)	5 (Agla)	4.5	9.1 (paper)
		10 (Agla)	10.9	11.9 (paper)
B	1-Tyrosine-^{14}C(U)	5 (Agla)	7.9	
G	1-Tyrosine-^{14}C(U)	5 (Ag-a)	4.7	

surements. This is indeed a large source of error in quantitative
work as well as in repetition of sample analysis.

Emanuel (13) more recently reported results of a survey of
reproducibility in delivery by a number of micropipets. It appears from this report that simple glass capillary pipets may be
the most accurate microdelivery system. Since they are disposable
and relatively inexpensive, they are useful in applying samples to
thin layer plates. Table 4.2 shows the results of this study.
These experiments evaluated reproducibility of delivery by the
pipets, not the ease of delivering the sample to the layer.

A number of companies produce capillary pipets for general
use. Disposable uncalibrated glass capillaries are widely used
for applying samples to layers. The "Microcaps" produced by
Drummond Scientific Co. can be obtained both calibrated and uncalibrated. These pipets are simple to use. They fill readily, and
once touched to the layer at a *right angle*, the capillary action
of the layer draws out the liquid completely. Since they are disposable, the time-saving factor is considerable and the problem of
contamination between samples is eliminated.

An interesting apparatus using glass capillaries was described by Flinn (14). The apparatus was essentially designed for
use with samples from tablets or capsules of pharmaceutical formulations. It is a technique that provides maximum avoidance of
sample contamination or decomposition. The apparatus will handle
28 samples simultaneously in a single operation. The powdered
sample is loaded into an end of a 10 µL micropipet by plunging the
pipet into the powdered sample material. The operation is repeated as often as necessary to load sufficient sample in the pipet.
The same end of the pipet is then repeatedly plunged into a 2 mm
layer of sorbent. Three parts of sorbent to one part of sample
was found to be enough to filter the sample. The pipets are then
positioned in the holder lying on its side. The holder is then
placed upside down over a tank of solvent; all the micropipets are
filled with the solvent simultaneously by placing the empty ends
in the solvent long enough for them to fill by capillary action.
Then the holder is placed over the TLC plate, and the loaded pipets are pressed against the plate so that the solvent flows down
through the sample, thus extracting the active ingredient and de-
positing it on the layer. The plate is then developed in the desired manner. This may not necessarily be a quantitative method,
but it does treat each of the samples in the same manner and gives
comparison under the same relative conditions.

Since most samples for separation are small, the containers
that hold the sample as it is being applied to the layer must be
discussed. The capillary pipets discussed so far are of limited
length, usually 5-10 cm. This would in some cases limit the size
of the container used to handle the sample. Very convenient sam-

TABLE 4.2 ERRORS DUE TO MEASUREMENT OF SMALL VOLUMES OF SOLUTION
USING THE MACHINE REFERRED TO IN THE TEXT

Worker	Solute	Volume of Solution Measured (μL)	Coefficient of Variation	
			Direct into Phosphor (%)	Via Adsorbent (%)
B	Morphine-2-%	2	2.4	1.8 (paper)
B	Morphine 2-T	5	2.2	2.5 (paper)
B	Morphine 2-$\overline{1}$T	9.6	1.6	2.1 (paper)
B	d-Glucose-^{14}C(U)	5	2.0	2.7 (paper)
B	l-Tyrosine-^{14}C(U)	5	1.5	1.5 (paper)

ple holders are glass-stoppered microcentrifuge tubes of 0.5, 3,
and 5 mL volume. Most solvents used for transfer of the sample
should be highly volatile (by definition) for ideal spot size.
This means that in quantitative TLC it is necessary to have on
hand containers that are easy to open and close to prevent undue
evaporation of the solvent while samples are being applied.

For easy handling of the glass capillary during sample transfer, a holder can easily be made, or those commercially available
should be used. A set of capillaries supplied by Drummond Scientific is accompanied by one of these holders as pictured in Figure
4.3. A piece of 8 mm glass tubing can be cut to the length desired, and the medicine dropper bulb with a small hole in the top
is placed on one end. A small rubber stopper, cut to fit the base
at the other end (such as from a Microcap holder), is pierced to
form a hole through which the capillary pippt fits tightly. The
apparatus so formed facilitates handling of the sample and is useful in withdrawing sample from larger test tubes. The bulb facilitates sample uptake when the capillaries are long. The device is
particularly useful in forcing out the total sample as well as in
controlling the rate of dispersal.

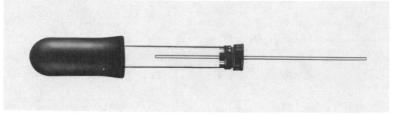

Figure 4.3   Glass capillary micropipet in holder. Courtesy of
A. H. Thomas Co.

The capillary pipets available from Drummond Scientific can
also be used in a transparent barrel graduated in 0.1 µL divisions. The disposable bore is the precision glass capillary,
which takes a stainless steel plunger. The plunger tip travels
the length of the bore and is used as the reference point for
readings. The bores (capillaries) and plunger are inexpensive
enough to justify discarding after a single use to avoid contamination of samples and the time-consuming cleanup. The finger
loop handle of the plungers allows one-hand operation (Figure 4.4).
The capillary passes through a Teflon collar at both ends of
the plastic barrel and is held by a collet chuck. The wire plunger is held in place by a springlock clamp of the handle. These
micropipets are supplied by Drummond under the name Microtol and

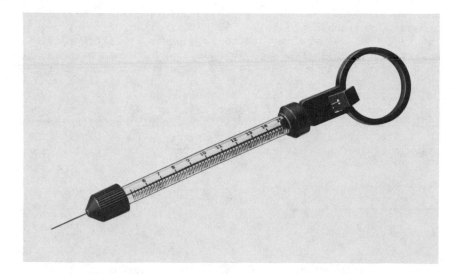

Figure 4.4 Disposable glass barrell microsyringe. Courtesy of
A. H. Thomas Co.

are available in sizes of 15 μL and 20 μL with graduations of 0.1
μL. If larger sizes are required, the syringe type of adaptation
is available. The reusable stainless steel plunger has a Teflon
top and finger loop for one-hand operation. The disposable bores
are precision glass capillaries held in place in a glass barrel by
threaded plastic end fittings. The glass barrel is graduated in
1 μL divisions for the 100 μL capillaries.

Glass capillaries are stocked by most chemical supply firms.
Since they are inexpensive, they are the pipets of choice when
large numbers of samples are to be applied, particularly in qualitative work. They are widely used in drug screening work where
TLC is the most commonly used analytical procedure.

Disposable capillaries are ideal when radioactivity is involved. Their use avoids cross contamination, the bane of those
engaged in metabolic studies. The conventional use of labeled
internal standards as well as reference substances added for assessment of recovery is facilitated when these capillaries are
used for sample transfer. As pointed out earlier, they are among
the most accurate transfer vehicles for use in TLC.

Microsyringes such as those manufactured by Hamilton Co.,
Unimetrics, and Labcrest, among others, have seen wide acceptance
and can be used for delivery of calibrated volumes of solution.
They usually are made with glass barrels and a stainless steel or

tungsten plunger. Some have a replaceable needle; others have
permanently attached needles. They are available in volumes of
1 μL to 100 μL and are calibrated to deliver as little as 0.05 μL.
They were originally designed for use in gas chromatography and
are equally adaptable to use in TLC. These syringes require some
practice in use for reproducibility of sample delivery in quantitative work. Figure 4.5 shows an example of this type of syringe.
With care and proper preparation of standard solutions it is possible to obtain reproducibility in the separations with a precision better than ±1% with the same syringe. These pipets must be
meticulously cleaned between the application of repetitive samples.

Figure 4.5 Microsyringe with a capacity of 1 μL. Courtesy of
A. H. Thomas Co.

Smooth delivery of the samples is hindered by bores and
plungers that are not clean. Spurious zones on chromatograms can
be due to solutes carried over from sample to sample in improperly
cleaned pipets. An uneven and sporadic travel of the plunger
through the bore prevents smooth application of the sample. With
extraneous dirt buildup in the bore, the plunger sometimes becomes
frozen.
Microsyringes are readily cleaned. Draw up fresh solvent
with the plunger and dispel it, repeating this a number of times.
The preferred method is to remove the plunger and wipe it with a
lint-free tissue saturated with solvent. Then, using a water as-
pirator with a rubber hose, draw a volume of clean solvent through
the needle and barrel. This method is used with nondisposable
glass micropipets.

4.7 AUTOMATIC SAMPLE APPLICATORS

Any treatise on sample application would not be complete without a
discussion of automatic sample applicators. A number of these are
commercially available. These instruments can be divided into
four types:

1. Holders of x number of disposable capillaries.
2. Syringe-type sample streak applicators.

3.   Large capacity tube fitted with a capillary delivery system to deliver a single sample.

4.   Multiple syringe holders with provision for automatic depression of the plunger.

A holder with 19 individual capillaries has already been described.   This device is now available commercially (A. H. Thomas Co.) and is shown in Figure 4.6.   A description from their catalog follows:

"U-shaped base A of applicator accommodates sample trough B or micro tube rack C (and subsequently the coated plate G) beneath perforated bar for self-filling capillary spotting pipets D.   Bar H is supported by coil springs on guide rods at each end.   Bar H is pressed down to lower the pipets for filling and for delivering samples on plate 22 mm from edge."

"Micro tube rack takes 19 tubes approximately 6 mm o.d. Suitable tubes can be improvised by sealing one end of short length of glass tubing.   Length 8 inches; holes are 1/2 in × 3/16 in deep.   Nickeled brass."

"Sample trough is V-shaped, 7-5/8 in long, width at top 3/16 in, depth 1/2 in.   Stainless steel."

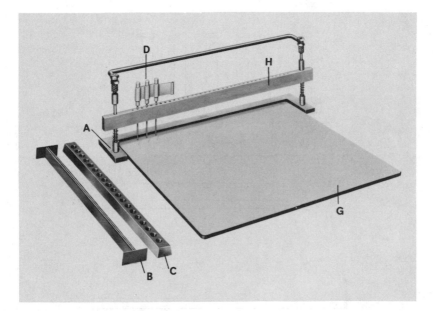

Figure 4.6.   Thomas-Morgan sample spotter.   For description see text.   Courtesy of A. H. Thomas Co.

The individual samples can be added to the perforated bar and all samples taken up at once, or the pipets may be used to withdraw samples from the original containers.

Streakers are not applicable when multiple screening of many samples is required. They are limited to applying relatively large samples as a streak across the origin of the plate. In general they work well, and resolved areas on developed chromatograms consist of well-defined linear zones across the plate.

The sample applicator by Kontes Instruments Division, shown in Figure 4.7, essentially consists of large-bore glass tubes fitted with needles. The needle is positioned just above the layer to deliver the solute gradually to the layer by capillary action. The essential feature is that the needle is surrounded by a template with orifices that permit compressed gas flow over the layer to evaporate the solvent as the solution percolates into the layer. The gas (air, nitrogen, etc.) can be cooled or heated. Because of this, regardless of the volume of solvent, a small spot will result. This type of sample applicator can be very useful when dilute samples must be applied as a small zone to the layer.

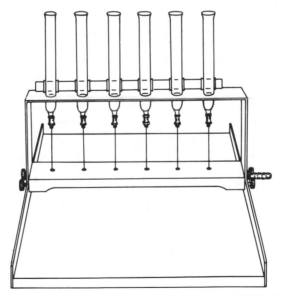

Figure 4.7 Kontes Chromaflex spotter with manifold for drying samples under air or nitrogen. Courtesy of Kontes Glass Co.

The automatic spotter made by Analytical Instrument Specialties permits application of up to 19 samples simultaneously. The
amount of the sample delivered is determined by the position of
the plunger. The syringes are clamped horizontally in a holder.
A bar is placed across the needles to depress them to a position
in which the tips are just above the layer. A motorized bar is
placed behind the plungers to drive each plunger through the syringe at a predetermined rate. The plungers of the syringes with
the larger volumes will be removed the farthest from the bore and
will be depressed first by the driving bar. The plungers of the
syringes with the least volume to be delivered will be depressed
last by the driver. This spotter is equipped with a heater, and
the layer can be warmed during sample application. Compressed gas
can also be directed over the spots. The whole operation results
in reproducible spot size. Temperature, gas flow, and rate of
travel of the depressing bar can be easily controlled. Uniform
sample application can be obtained from day to day with this type
of controlled spotting.

Figure 4.8 shows one of the automatic sample spotters made by
Analytical Instrument Specialties. These allow application of up
to 10 or 19 samples simultaneously depending on the model chosen.
Glass syringes can be chosen to accommodate 1, 5, 50 or 100 μL and
have blunt tipped needles coated with Teflon to prevent solvent
creep-back and solute deposition on the tip of the needle during
delivery.

The contents of a 100 μL syringe can be delivered over a 2-30
min period. Unattended use of the instrument will result in the
formation of circular spots on the layer. "Streaking" may be
achieved by slightly shifting the plate several times during sample application such that the tips of the syringes deliver the
sample to several spots across the origin in each lane.

The Linomat III Sample Applicator made by Camag is shown in
Figure 4.9. The sample to be dispensed is loaded into a Hamilton
syringe. The TLC plate is secured onto the plate table with a
magnetic clamp. The syringe is placed on the dosage turret and
the nitrogen flow is turned on. Lower actuating lever by quick
motor drive to make contact with the plunger head. Move plate
table by quick motor drive into desired position for sample application. Select length of band (between 0 = spot and 199 mm)
and sample volume by digital thumb wheel. Press start button.
The electronic system determines whether the nitrogen flow has
been turned on. If so, then the plate table and syringe plunger
start their movements, both controlled by stepping motors at synchronized speeds.

As the plate table moves back and forth according to the
preselected length of stroke, the sample is sprayed on, and the
volume delivered is continuously displayed in digital form. When
the preselected volume has been delivered, the syringe plunger

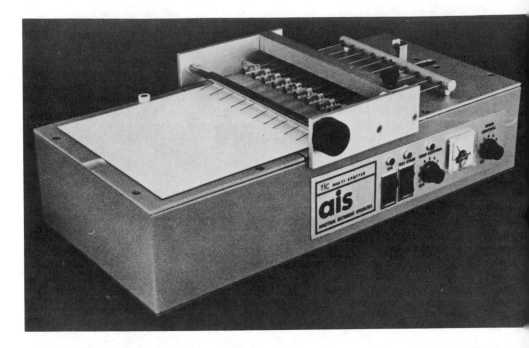

Figure 4.8 Multi-syringe sample applicator. Courtesy of Analytical Instrument Specialties.

movement stops. The final value is displayed. The plate table returns to its original position. Then the stop signal lights up, indicating that the next sample may be applied.

The Linomat has been developed to perform high precision narrow band application of samples for precise qualitative and quantitative analysis. The spray-on technique allows a large volume to be concentrated into a narrow band. Sample application in the form of narrow bands usually results in the best possible resolution of the particular sample in a given chromatographic system. The Linomat works well with plates having a concentration zone.

J. T. Baker Chemical Co. is now supplying an "Automated TLC Spotter." This apparatus will simultaneously apply up to 16 samples to a TLC plate. In use, each sample is first transferred to a disposable porous wick. The wicks are then inserted into a row of numbered holes on the wick bar which is placed over a reservoir containing the spotting solvent. A TLC plate is positioned over the wicks and is depressed, making contact with the wick heads. This contact starts the flow of solvent through the wick, trans-

Figure 4.9   The Linomat III sample applicator.   Courtesy of Camag, Inc.

ferring the sample to the plate.   A row of small uniform spots
result.   The solvent is evaporated from the plate with the aid of
a built-in manifold connected to a clean air or nitrogen supply.

4.8   AIDS AND GUIDES FOR MANUAL SAMPLE APPLICATION

Application of samples to a TLC plate, especially when multiple
samples are involved, becomes a tedious operation.   Herein may lie
one of the difficulties included with reproducible and ideal separation.   A number of aids to spotting are available to help overcome some of these difficulties.   The simplest of these is the
spotting template available from most suppliers of TLC equipment.
The plate is positioned in the apparatus so that one edge is
matched below a notched guide along a line (usually 1 cm from the

edge) designated as the starting line. There are notches evenly
spaced along this edge of the guide. The samples are applied in
the notches to form evenly spaced replicate applications along the
starting line. One of these is illustrated in Figure 4.10.

Also available from Brinkmann Instruments, Inc. is a calibrated multipurpose TLC template. Made of acrylic plastic, it can
be used as a guide to sample application points, and after chromatogram development it can be used for measurement of the R_f values. It has a sliding hairline for easy reading. It may also be
used as a storage tray for a plate.

More recently introduced by Sindco Corporation is a unique
apparatus known as a "spot-dry sampler" (Figure 4.11). This instrument is a template with provision for gas flow (cold or warm)
to the point of application. The apparatus is designed as a hand
rest for chromatographic spotting and features a drying system
that allows the gas flow to be easily stopped while applying the

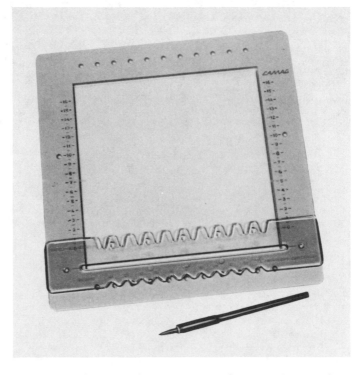

Figure 4.10. Plastic spotting template for evenly spacing sample
application on a TLC plate. Courtesy of Camag, Inc.

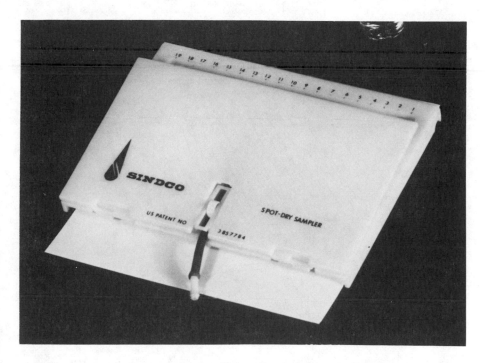

Figure 4.11 Spotting template with gas drying system. Details in
text. Courtesy of Sindco Corp.

sample. When the hand is raised between sample applications, the
gas flows over the zone to quickly evaporate the solvent. In this
way the application zone (starting spot) is kept as small as possible, a prerequisite for good resolution as well as uniform final
zones. This instrument is particularly valuable when application
using high boiling solvents is required. It also assures that the
application zone is dry before the plate is placed in the development chamber. This apparatus is inexpensive and simple to use.
    In some applications there are those who advocate scoring the
layer into individual lanes. Plate scorers are available for this
purpose, if it is not desired to score the lanes manually using a
straight edge and blade, which is difficult to do reproducibly.
The one pictured in Figure 4.12 is supplied by Schoeffel Instrument Co. and divides the 20 cm layer into 20 individual lanes.
Applying the samples in the lanes provides a means for uniformly
spacing the samples across the starting line. Cross contamination
between samples is avoided in close sample application since the
sample will not cross the scored lines, which are devoid of sor-

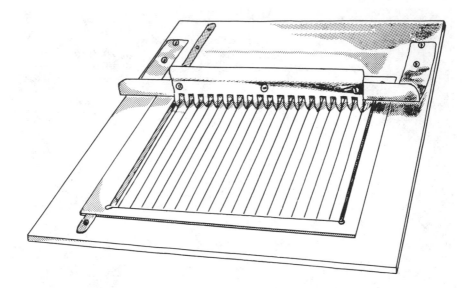

Figure 4.12. Layer plate scoring device for producing sample lanes on a TLC plate. Courtesy of Schoeffel Instrument Corp.

bent. Precoated plates that are prescored into 0.8 or 0.9 cm lanes are available from the major manufacturers and their suppliers.

4.9 SOURCES OF ERROR IN VOLUME MEASUREMENTS

Fairbairn (15) has discussed at length some of the errors in the use of pipets and syringes in sample application. Among these sources is "creep back" of the solvent as the drop forms on the end of the syringe needle. The following discussion, taken from his work, should be of aid to the chromatographer.

"The delivery of a number of drops from the needle of an Agla syringe by free fall was observed with a lens and it was noted that from time to time an accumulating drop suddenly slipped up slightly from the point and when the drop finally fell on to the paper a definite proportion remained on the stem. This 'creep back' effect was cumulative and sometimes a sizeable volume remained on the stem for some time; then quite unpredictably it would disappear with a succeeding drop. The creep back effect varied with the solvent used and was particularly noticeable with methanol; occasionally after the delivery of many drops of a methanolic solution of morphine, crystals of the latter substance were seen to have formed a tide mark on the stem of the needle. At times as exceptionally high tide of creep back would reach the crystals, redissolve them, and wash them into a succeeding drop. Accordingly some of the needles used in the experiments recorded in [Table 4.1] were washed after each series of radioactive solutions had been delivered. In several cases the 'remainder' on the needle represented a significant proportion of the total radioactive substance measured, e.g., for morphine-2-T (Worker B) after delivering 16 volumes the remainder represented 9.6% of the total delivered. Worker E, on the other hand, avoided creep back by using a hooked needle previously dipped in silicone; there was no detectable remainder after his series of deliveries."

"A second source of error arises from the fact that the measured drop does not always fall freely from the end of the needle as its weight may be insufficient to overcome surface tension effects. To assist this most workers touch the drop on to the surface of the paper or adsorbent or against the glass vial. We found that this process withdrew fluid from the lumen of the needle by capillarity and sometimes it required the delivery of 0.6-0.8 µL of solution from the barrel of the syringe into the needle before the succeeding drop made its appearance. The amount withdrawn varied according to the bore of the needle, time of contact and state of absorbency of the absorbent material, which decreases as more liquid is added to it."

"It is interesting to note that with a micropipet the degree of creep back is greatly reduced since only one delivery is made and capillation is probably taken into account when calibrating. These facts may explain why the errors of Worker C [(Table 4.1)] were somewhat lower than those of the workers using syringes. As the latter are obviously preferred for replicates or for measuring varying volumes an attempt was made to overcome the two sources of error by (a) siliconing the needles, (b) using different bore thicknesses, and (c) filing the tapered ends till they were at right angles to the long axis, but none of these methods were entirely successful. Success was finally obtained by means of a machine...which automatically delivers small volumes by rapid ejection or throwing. Creep back was prevented and since even quite small drops could be forced on to the paper or adsorbent without touching the surface, capillation was also prevented. This twofold advantage was demonstrated by producing a series of drops from the machine by forcible ejection in the normal way; the coefficient of variation was found to be ±2.5%. A second series was produced from the machine in identical circumstances except that the needle was brought sufficiently near to the paper for the drops to be drawn off by capillation, that is, to fall rather than be thrown. In these conditions the coefficient of variation rose to ±7%. Creep back had also occurred as the remainder left on the

needle after this series represented 3.27% of the total quantity
delivered. When the normal throwing procedure was used the remainder was only 0.32%."

"In [Table 4.2] some results obtained with the new machine
are shown and it is obvious that the errors are considerably less
than those for hand delivered drops [(Table 4.1)]. The results
are also more consistent, tedious hand spotting with likelihood of
fatigue is avoided and the resulting spots are small and of constant area since equal volumes (0.2-0.4 µL) are ejected each time
and the circle of solution allowed to dry before the next drop is
ejected."

During the operation of syringes for the delivery of samples,
some experience by the operator must be acquired before reproducible solvent delivery will be attained. In context with the two
previous examples of error the contact of the needle tip must be
the same each time. The edges should be clean at all times. Possibly more important is the uptake of the solvent into the barrel
of the syringe, since it is not possible to see the liquid in the
steel needles usually found on syringes. The operator must be sure
that there are no air bubbles in the needle. One must be sure that
the same type of delivery is used each time. Some touch the tip
to the layer during delivery of the sample while pushing down the
syringe plunger. Some collect a drop of the solvent on the tip and
touch this to the layer. Others hold the tip just above the layer,
and the capillary action of the layer draws the liquid while the
plunger is driven down. In any of these operations it is necessary
to repeat the particular method while avoiding disturbance of the
layer. Odd shaping of final spots results when holes or scratches
are made in the layer during these operations.

The syringes or delivery devices must be kept clean. Oil
film or build up of solute in the bore or on the outer surface will
affect solvent delivery. In this respect disposable pipets have an
advantage. The washing of the pipets used in automatic delivery
systems must be thorough and done in the same manner each time.
Multiple rinsing between deliveries is imperative. The syringe
must be dried between applications, especially in systems which
are all metal, because it is not possible to see the amount of
washing solvent remaining in the needle.

4.10 MANUAL VS. AUTOMATIC APPLICATION

Touchstone and Levin (16) have investigated manual versus automatic sample application when dealing with larger sample volumes.

Working with a solution of cholesterol in a concentration of
10 ng/µL, they found that with manual (i.e., spotting) application
up to 15 µL of sample could be applied to give a linear relation-

ship between the amount of cholesterol in the sample and the recorder response. Above 15 µL of sample, the spectrodensitometric
scans showed "double peaking," indicating that the sample spots
had migrated as rings. That is, the solute had diffused to the
periphery of the spots and migrated as two bands. This is illustrated in Figure 4.13.

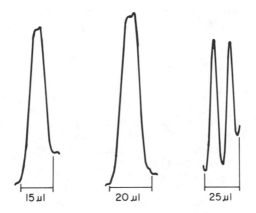

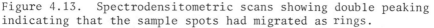

Figure 4.13. Spectrodensitometric scans showing double peaking
indicating that the sample spots had migrated as rings.

     In contrast, up to 250 µL of cholesterol standard could be
safely applied using a commercial automatic spotter, with only a
single peak resulting on a densitometric scan.
     Another important consideration contributing to the success
of a particular separation in TLC, particularly HPTLC, is the configuration of the applied sample at the origin. How is the sample
applied? Generally speaking, the application of the sample as a
"spot" in the middle of a lane decreases resolution of the sample
components as compared to the same amount of sample applied as a
series of small spots or "streak" across the sample lane. This is
illustrated in Figure 4.14, which shows the separation of a five
component dye mixture after it was applied to a plate by both
"spotting" and "streaking." Streaking of the sample results in a
sharper, less diffuse separation with narrower component zones.
The resolution for the separations resulting from the samples that
were streaked was as much as 32% greater than that from the samples that were applied as single spots.

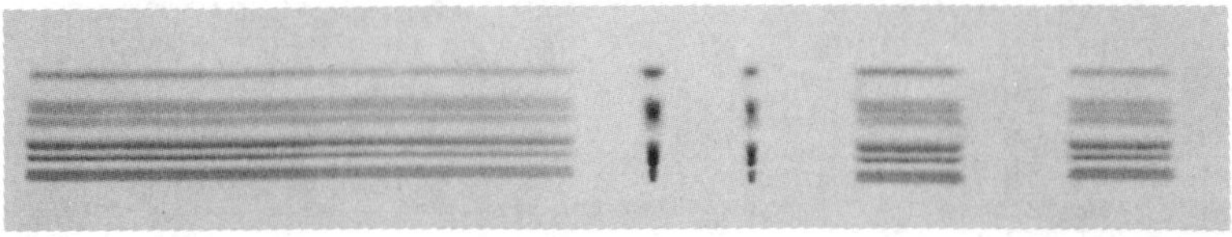

Figure 4.14. Sample spotting vs. streaking for resolution of components.

REFERENCES

1. J. C. Touchstone, R. E. Levitt, S. S. Levin, and R. D. Soloway, Lipids, *15*, 386 (1980).
2. J. C. Touchstone, S. S. Levin, and M. F. Dobbins, in press.
3. J. C. Touchstone, S. S. Levin, M. F. Dobbins, N. Goldman, and L. Blasco, N. B. Goldman, and P. J. Chang, "The Phospholipids of Human Seminal Fluid," 72nd Annual Conference, American Society Biological Chemist,, St. Louis, Mo., May 31-June 4, 1981.
4. S. S. Levin, M. F. Schwartz, D. Y. Cooper, and J. C. Touchstone, J. Chromatogr., *154*, 349 (1978).
5. M. F. Dobbins, Chemistry, *38*, (11) 27, (1965).
6. J. C. Touchstone and M. F. Dobbins, in *Densitometry in Thin Layer Chromatography*, J. C. Touchstone and J. Sherma, Eds., Wiley Interscience, New York, 1979, p. 654.
7. W. Dünges, *Pre-Chromatographic Micromethods*, Heuthig Publishers, New York, 1982.
8. A. Achaval and R. D. Effelson, J. Lipid Res., *7*, 329 (1966).
9. M. H. Coleman, Lab. Pract., *13*, 1200 (1964).
10. N. Nyborn, J. Chromatogr., *28*, 447 (1967).
11. E. V. Truter, *Thin Film Chromatography*, Cleaver-Hume Press Ltd., London, 1963.
12. J. W. Fairbairn and S. J. Relph, J. Chromatogr., *33*, 494 (1968).
13. C. F. Emanuel, Anal. Chem., *45*, 1568 (1973).
14. P. E. Flinn, J. Chromatogr., *82*, 117 (1973).
15. J. W. Fairbairn, in *Quantitative Paper and Thin Layer Chromatography*, E. J. Shellard, Ed., Academic Press, London, 1968.
16. J. C. Touchstone and S. S. Levin, J. Liq. Chromatogr., *3*, 1853 (1980).

CHAPTER 5
The Mobile Phase

5.1 INTRODUCTION

Success in the technique of thin layer chromatography depends to a great extent upon selecting the mobile phase that will give the desired separation. Much of what is known about selection of a satisfactory mobile phase has been derived from the experiences of many TLC investigators working by trial and error. Work in column chromatography has also been helpful. There has been little work on the theoretical aspects of selecting a sorbent for particular analyses, most of the focus being on mobile phase selection. The nature of the compounds to be separated dictates the sorbent to be used and perhaps even more the selection of the mobile phase.

Few reports describing separations carried out by TLC indicate the rationale behind the choice of mobile phase. It is hoped that the discussions in this chapter will aid the TLC practitioner toward a more organized selection procedure. This should not only save time but also result in better separations. This chapter is divided into sections according to the type of separation, adsorption, ion-exchange, and partition chromatography. In TLC there are other types of sorbents such as layers impregnated with silver nitrate or modified cellulose layers. Although these are not as widely used, the discussions of the mobile phases herein will give some general guidelines to selecting the mobile phase for use with them

also. Selection should not be confused with selectivity. Selectivity is the property of a solvent that causes separation, whereas selection refers to acts of the analyst in use of a particular
mobile phase.

5.2 THE SOLVENT PROPERTIES

Before any discussion of choice of a mobile phase can be meaning-
ful, a brief treatment of the problems related to the properties
of the solvents used is appropriate. If the solvents used for the
mobile phase are not pure further complications may result. A
number of general rules follow:

1. Chloroform or ether contain traces of ethanol as a preservative. It may not be possible to repeat a chromatogram if
this is not removed routinely by distillation or replaced after
distillation to duplicate the original concentration.
2. Water content of various solvents at different ambient
conditions can vary widely. Herein is one reason that reproducibility of conditions reported in the literature is poor. Many
solvents are hygroscopic and can be affected by humidity. It is
often beneficial to store such solvents over drying agents.
3. The conditions and duration of storage of the solvent can
affect reproducibility, since solvents can deteriorate. For example, solvents can become contaminated by autooxidation. Storage
in poorly sealed containers can result in contamination by absorption of atmospheric pollutants and water.
4. Interaction between mobile phase and solute irrespective
of the sorbent can change chromatogram behavior. Advantage can
sometimes be taken of this to give better separation. Impurities
in solvents can also enter such interactions.
5. Solvents in mixed mobile phases may interact. If solvents are not pure or vary in composition from batch to batch, the
nature of the mixture will vary even if its preparation is similar
from day to day.
6. Solvent demixing is sometimes seen during development.
This can happen when a small concentration of a strong solvent is
used with a weaker solvent. At the beginning there is a tendency
for preferential adsorption of the stronger solvent, and the front
can consist of 100% weaker solvent. As the sorbent becomes saturated with the stronger solvent, the second front consisting of
the mixture appears. Thus, two solvent fronts are seen on the
layer.

Solvent mixtures should not be used repeatedly. The mobile
phase should be used only once, when it contains a high proportion

of a volatile component such as ammonium hydroxide or ether. Single-solvent mobile phases can be used repeatedly until they become contaminated. The following give the reasons for this:

1. Evaporation of the lower boiling component of the solvent mixture can be caused simply by opening the chamber at the end of the development. In the case of a benzene-methanol mixture the methanol concentration gradually lowers on repeated use of the mobile phase.
2. Humidity can change the composition of the solvent mixture when the chamber is opened.
3. Preferential adsorption of one solvent by the sorbent in the layer is often seen. The solvent of highest strength is adsorbed more strongly, and if its proportion is low at the beginning the composition will eventually be affected.
4. The individual components of the mobile phase may react with each other and cause further changes in the real composition.
5. Continued elution of extraneous material from the sorbent can cause changes in the mobile phase composition and produce spurious zones on the layer. This effect is exaggerated the more often a mobile phase is used.

5.3 NATURE OF PHASE INTERACTIONS

There are a number of chemical and physical characteristics of compounds that determine the degree to which a pair can interact with each other. This is true for mobile phase-solute, solute-sorbent and mobile phase-sorbent interactions. The important ones are listed belog; they should be kept in mind since the mobile phase-solute interaction controls selectivity and thus separation, as discussed later.

1. Intramolecular forces, which hold neutral molecules together in the liquid or solid state, are well known. These are physical in character and do not involve formation of chemical bonds. These forces are physical and they are characterized by low equilibrium and result in good chromatographic separation,
2. Inductive forces, which exist when a chemical bond has a permanent electrical field associated with it (for example, C-Cl or C-NO$_2$ groups), affect interactions. Under influence of this field, the electrons of an adjacent atom, group, or molecule are polarized so as to give an induced dipole moment. This is a major contributing factor in the total adsorptive energy on alumina. Thus, the interaction is strong in these cases.
3. Hydrogen bonding makes a strong contribution in adsorptive energies between solutes or solvents having a proton-donor

group and a nucleophilic polar surface such as that of alumina or
silica gel. These are covered by hydroxyl groups (OH). The hydroxyl groups of silica can also react weakly with electrophilic
groups such as ethers, nitriles, or aromatic hydrocarbons to form
strong linkages (interactions).

4. Charge transfers between components of the mobile phase
and the sorbent can also take place to form a complex of the type
S^+A^- (where S = solvent or solute, and A = surface site of sorbent). This is prominent in ion-exchange chromatography.

5. Covalent bonds can be formed between solute and/or the
mobile phase and the sorbent. These are strong forces and result
in poor chromatographic separation.

It should be kept in mind that these factors are present
whether mobile phase-solute, mobile phase-sorbent, or sorbent-sol-
ute interactions are considered, and furthermore each one of these
interactions is independently variable.

5.4 NATURE OF THE SOLUTE

The choice of mobile phase depends on the nature of the compound
to be separated. The interaction between the solute-mobile phase
or solute-sorbent will *a priori* be determined by the number and
nature of the functional groups in the solute. A carboxyl group
confers high interactivity to an organic molecule. At the other
extreme, the series hydrocarbons with no functional group represent substances of low interactivity. The old rule of thumb ap-
plies well here: "like dissolves like." A polar compound will
require a mobile phase of high interactivity if it is to migrate on
the TLC plate. Saturated hydrocarbons are the least interactive of
the organic compounds, and a mobile phase of lower strength can
be used. Unsaturated hydrocarbons are more strongly adsorbed;
the more double bonds and conjugation, the greater the adsorption.
Introduction of functional groups into the molecule can increase
interactivity and thus the adsorptivity or interaction with the
sorbent and mobile phase. Functional groups affect the interactivity in increasing degree in the order: RH; $ROCH_3$, $RN-(CH_3)_2$;
RCO_2CH_3; RNH_2; ROH; $RCONH_2$; RCO_2H. Thus knowledge of the nature
of the component to be separated provides a starting point for
choosing a mobile phase. The above sequence also holds for components of the mobile phase. These properties provided some of the
earliest data relating solute parameters to separation efficiency
in chromatography. Usually some knowledge of the solubility of
the solute is available due to experience with the substance as
well as some knowledge of the class into which the compound fits.
The degree to which the solute interacts with the mobile phase as

opposed to the sorbent will determine its distribution between the two phases. The mobile phase is usually less interactive than the sorbent. If the solute has stronger interaction with the mobile phase, it will prefer the mobile phase and separation will be poor. Conversely, if the solute has a higher interaction with the sorbent, separation will be greater. For some intermediary interactions of the solute with the mobile phase and the sorbent, the distribution will be equal in both phases. These two interplays affect separation efficiency.

Sorbents are usually described by their chemical composition; that is, they are identified as alumina, ion-exchange, silica gel, or polyamide. There are significant differences in separation selectivity from one sorbent to another. However, there has been no real attempt to deliberately select a particular sorbent for any specific separation. Rather, the composition of the mobile phase is usually altered to produce changes in separation factors. The activity of alumina, for instance, can be changed by varying its water content. This is also true of silica gel. In practice, however, silica gel is usually heated to have high adsorptivity, and separation can be varied by changing the water content of the mobile phase.

5.5   THE SOLVENT STRENGTH PARAMETER

The solvent strength parameter ϵ° is defined by Snyder (1) as the adsorption energy per unit of standard sorbent. The physical and chemical forces listed in Section 5.3 are involved in determining the level of interaction of the sorbent and mobile phase. The higher the mobile phase strength (greater interaction with the sorbent), the greater will be the R_f of the solute in simple liquid-adsorption TLC. Usually one tries to select a mobile phase such as to obtain R_f values of 0.3-0.7, particularly if well-separated zones are desired. In practice, however, single-solvent mobile phases usually do not give separation although they might give proper mobility. Decreasing solvent strength can increase resolution, but in difficult separations changing mobile phase composition to change interactions may be required. However, at the same time decreasing solvent strength will decrease the R_f and the required separation may not be achieved. Subtle changes in selectivity are the rule. (Selectivity is discussed in later sections.) Mobile phase strength in liquid-solid chromatography is independent of the mobile phase-solute interaction. Strength as defined by Snyder is a function of free energy of adsorption on the solid surface (sorbent). Thus, when a mobile phase has proper strength to give the desired R_f it may not achieve the desired separation.

A starting point for the beginner can be found in some reported mobile phases (based on class of compound) as given in Table 5.1. These mobile phases have been developed for specific separations. A more precise approach to separations based on these mobile phases and how to modify them can be obtained by understanding the theories of mobile phase strength and selectivity. Varying the selectivity (changing interactions) can affect the separation. Solvent strength is dependent on its reactivity with the sorbent. With a large number of sorbents available, one has to resort to some trial and error to find the best mobile phase for separating the components of interest. Activity of the sorbent, particularly in the case of silica gel and alumina, can be controlled by controlling its water content. This is probably the main reason why it has been so difficult to duplicate many separations reported in the literature. There is at present no way to state sorbent activity other than to indicate the proportion of water content. The surface activity of the sorbent is important and can be related to surface area. Modern sorbents for chromatography are more carefully controlled so separations can be more reproducible. Stronger mobile phases give greater interaction with the solute and will decrease adsorption, thus causing faster migrations. Conversely, a weak mobile phase will give a low R_f.

An expression of solvent strength can be obtained in the water solubility of a variety of solvents. Table 5.2 gives the water solubility of a number of solvents. The sequence of solvents based on increasing or decreasing solubility results in an eluotropic series. Modern eluotropic series are based on solvent strength parameter ($\varepsilon°$), as will be discussed in following sections.

5.6 THE ELUOTROPIC SERIES OF SOLVENTS

In the expression of mobile phase strength the eluotropic series has been widely used. As early as 1940, Trappe (2) described a series of solvents in order of eluting power. Steiger and Reichstein used the principle in 1938 in their early work on the chromatography of adrenal steroids (3). Jacques and Mathieu (4) have stated that eluting power of solvents was proportional to their dielectric constant ε. They showed that the solvents function by being themselves adsorbed. In a mobile phase, the solvent with the greatest strength (highest ε) is the most strongly adsorbed. These early reports stimulated attempts to systematize the choosing of mobile phases in chromatography.

A "mixotropic" series of solvents was described by Hecker in 1954 (5). He noted that the farther apart two solvents were from each other in a series, the less miscible they were with each other.

TABLE 5.1   SELECTED SOLVENT-SORBENT SYSTEMS FOR THE TLC
SEPARATION OF VARIOUS COMPOUNDS

1. *Chloroplast Pigments*
   a. Isooctane-acetone-ether (3:1:1) on silica gel or polyamide
   b. Petroleum ether-benzene-CHCl$_3$-acetone-isopropanol (50:35:
       10:5:0.17) on cellulose
   c. Petroleum ether-*n*-propanol (99.2:0.8) followed by 20%
       CHCl$_3$ in petroleum ether (two-dimensional) on sucrose
   d. Petroleum ether-acetone (4:6) on alumina
   e. Petroleum ether-*n*-propanol (199:1) on starch
2. *2,4-Dinitrophenylhydrazones of Aldehydes and Ketones*
   a. Hexane-ethyl acetate (4:1 or 3:2) on silica gel
   b. Benzene or CHCl$_3$or ether or benzene-hexane (1:1) on alu-
       mina
3. *Alkaloids*
   a. Benzene-ethanol (9:1) or CHCl$_3$-acetone-diethylamine (5:4:
       1) on silica gel
   b. CHCl$_3$ or ethanol or cyclohexane-CHCl$_3$ (3:7) plus 0.05%
       diethylamine on alumina
   c. Benzene-heptane-CHCl$_3$-diethylamine (6:5:1:0.02) on cellu-
       lose impreg. with formamide
4. *Amines*
   a. Ethanol (95%)-NH$_3$ (25%) (4:1) on silica gel
   b. Acetone-heptane (1:1) on alumina
   c. Acetone-H$_2$O (99:1) on kieselguhr G
5. *Sugars*
   a. Benzene-acetic acid-methanol (1:1:3) on silica gel buffer-
       ed with boric acid
   b. *n*-Propanol-conc. NH$_3$-H$_2$O (6:2:1) on silica gel G
   c. Butanol-pyridine-H$_2$O (6:4:3) or ethyl acetate-pyridine-
       H$_2$O (2:1:2) on cellulose
   d. Ethyl acetate-isopropanol-H$_2$O (65:24:12 or 5:2:0.5) on
       kieselguhr G buffered with 0.02N sodium acetate
   e. Ethyl acetate-benzene (3:7)(for sugar acetates) on starch-
       bound silicic acid
6. *Carboxylic Acids*
   a. Benzene-methanol-acetic acid (45:8:8) on silica gel
   b. Methanol or ethanol or ether on polyamide
   c. Isopropyl ether-formic acid-H$_2$O (90:7:3) on kieselguhr
       G-polyethylene glycol (M-1000) (2:1)
7. *Sulfonamides*
   a. CHCl$_3$-ethanol-heptane (1:1:1) on silica gel G
8. *Food Dyes*
   a. Methyl ethyl ketone-acetic acid-methanol (40:5:5) on
       silica gel G
   b. Butanol-ethanol-H$_2$O (9:1:1,8:2:1,7:3:3,6:4:4 or 5:5:5) on
       alumina

TABLE 5.1   (Continued)

c.   Aq. sodium citrate (2.5%)-NH$_3$ (25%) (4:1) on cellulose

9.   *Essential Oils*
a.   Hexane on starch-bound silicic acid
b.   Benzene-CHCl$_3$ (1:1) on silica gel G

10. *Flavonoids and Coumarins*
a.   Ethyl acetate-Skellysolve B on starch-bound silicic acid
b.   Methanol-H$_2$0 (8:2 or 6:4) on polyamide
c.   Toluene-ethyl formate-formic acid (5:4:1) on silica gel G
     + sodium acetate
d.   Petroleum ether-ethyl acetate (2:1) on silica gel G

11. *Metal Ions*
a.   Dilute HCl on starch-bound alumina or Celite
b.   Acetone-conc. Hcl-2,5-hexanedione (100:1:0.5) on silica
     gel G
c.   1M aq. NaNO$_3$ on Dowex 1 + cellulose
d.   Methanol on alumina

12. *Insecticides*
a.   Cyclohexane-hexane (1:1) or CCl$_4$-ethyl acetate (8:2) on
     silica gel G
b.   Hexane on alumina
c.   Heptane saturated with acetic acid on starch-bound silic-
     ic acid
d.   Chloroform on silica gel G + oxalic acid

13. *Lipids*
a.   Petroleum ether-diethyl ether-acetic acid (90:10:1 or
     70:20:4) on silica gel G
b.   Petroleum ether-diethyl ether (95:5) on alumina
c.   CHCl$_3$-methanol-H$_2$0 (80:25:3) on silicic acid

14. *Fatty Acids*
a.   Petroleum ether-diethyl ether-acetic acid (70:30:1 or 2)
     on silica gel G
b.   Acetic acid-CH$_3$CN (1:1) on dieselguhr impreg. with
     undecane
c.   Benzene-diethyl ether (75:25 or 1:1) on starch-bound si-
     licic acid
d.   CH$_3$CN-acetic acid-H$_2$0 (70:10:25) on silica gel G impreg.
     with silicone oil

15. *Glycerides*
a.   ChCl$_3$-acetic acid (99.5:0.5) on silica gel G impreg. with
     AgNO$_3$
b.   CHCl$_3$-benzene (7:3) on silica gel G
c.   ChCl$_3$-methanol-H$_2$0 (5:15:1) on silica gel G impreg. with
     undecane

16. *Glycolipids*
a.   Propanol-12% NH$_3$ (4:1) on silica gel G

TABLE 5.1   (Continued)

17. *Phospholipids*
    a. $CHCl_3$-methanol-H_2O (60:35:8 or 65:25:4) on silica gel G
18. *Nucleotides*
    a. 0.15M NaCl or 0.01-0.06N CHl on Ecteola cellulose
    b. Sat. aq. $(NH_4)_2SO_4$-1M sodium acetate-2-propanol (80:18:2) on cellulose
    c. 0.02-0.04N aq. HCl on DEAE cellulose
    d. Gradient elution: Start with 1N formic acid and add 10N formic acid that is 2M in ammonium formate DEAE Sephadex A-25
    e. 1.0-1.6M LiCl on cellulose PEI
19. *Phenols*
    a. Xylene, $CHCl_3$ or xylene-$CHCl_3$ (1:3,3:1,1:3) on starch-bound silicic acid or silicic acid-kieselguhr (1:1)
    b. Benzene on alumina plus acetic acid
    c. Benzene-1,4-dioxane-acetic acid (90:25:4) on silica gel G
    d. Diethyl ether on alumina
    e. Hexane-ethyl acetate (4:1 or 3:2) on silica gel plus oxalid acid
    f. Hexane or cyclohexane or benzene on δ-polycaprolactam
    g. Ethanol-H_2O (8:3) containing 4% boric acid and 2% sodium acetate on silica gel G + boric acid
    h. CCl_4-acetic acid (9:1) or cyclohexane-acetic acid (93:7) on polyamide 6
20. *Amino Acids*
    a. Butanol-acetic acid-H_2O (3 or 4:1:1) or phenol-H_2O (75: 25) or propanol-34% NH_3 (67:33) on silica gel G
    b. Butanol-acetic acid-H_2O (4:1:1) on cellulose
    c. Butanol-acetic acid-H_2O (3:1:1) or pyridine-H_2O (1:1 or 80:54) on alumina
    d. Ethanol-NH_3 (conc.)-H_2O (7:1:2) on silica gel G buffered with equal portions of 0.2M KH_2PO_4 and 0.2M Na_2HPO_4
    e. n-Butanol-acetone-NH_3-H_2O (10:10:5:2) followed by isopropanol-formic acid-H_2O (20:1:5) (two-dimensional) on cellulose
21. *Polypeptides and Proteins*
    a. $CHCl_3$-methanol or acetone (9:1) on silica gel G
    b. H_2O or 0.05M NH_3 on Sephadex G-25
    c. Phosphate buffers on DEAE Sephadex A-25
22. *Steroids and Sterols*
    a. Benzene or benzene-ethyl acetate (9:1 or 2:1) on silica gel G
    b. $CHCl_3$-ethanol (96:4) on alumina
23. *Terpenoids*
    a. Hexane or hexane-ethyl acetate (85:15) on starch-bound silicic acid

TABLE 5.1   (Continued)

   b.   Benzene or benzene-petroleum ether or -ethanol mixtures
        on alumina
   c.   Isopropyl ether or isopropyl ether-acetone (5:2 or 19:1)
        on ailica gel G
24. *Vitamins*
   a.   Methanol, CCl_4,xylene, $CHCl_3$ or petroleum ether on alumi-
        na
   b.   Methanol, propanol, or $CHCl_3$ on silica gel G
   c.   Acetone-paraffin (H_2O sat.) (9:1) on silica gel G impreg.
        with paraffin
25. *Barbiturates*
   a. $CHCl_3$-n-butanol-25% NH_3 (70:40:5) on silica gel
26. *Digitalis Compounds*
   a. $CHCl_3$-pyridine (6:1) on silica gel
27. *Polycyclic Hydrocarbons*
   a. CCl_4 on alumina
28. *Purines*
   a.   Acetone-$CHCl_3$-n-butanol-25% NH_3 (3:3:4:1) on silica gel

NOTE: From G. Zweig and J. Sherma, Eds., *Handbook of Chromatogra-phy*, C.R.C. Press, Cleveland, 1972, by permission of the Editors.

TABLE 5.2   ELUOTROPIC SERIES BASED ON WATER SOLUBILITY

Solvent	Solubility in water (%)
Ethanol	∞
Ethyl acetate	8.60
Butanol	7.90
Ether	7.50
Methyl chloride	2.0
Methyl isobutyl carbinol	1.8
Isopropyl chloride	0.34
Isopropyl ether	0.20
Benzene	0.08
Toluene	0.05
Hexane	0.014

In his work with steroids, Neher (6) applied the principles
of mobile phase selectivity and strength to problems of separation
by TLC. His report may be one of the first real attempts to use
the interplay of mobile phase-sorbent and mobile phase-solute interactions for the separation of solutes difficult to resolve.

Figure 5.1 shows how a series of mobile phases with different
strengths were used to separate a number of steroids (6). It illustrates the effect of interaction on the separation. Figure 5.2
shows the series of "equieluotropic" solvents that was used by
Neher to select mobile phases of the same strength but of different composition that resulted in changes of interactions and hence
separation.

In the first figure there were only six solvents. However,
there are hundreds of possible and useful mixtures for all types
of steroids or other compounds. This is why the type of "nomo-
gram" illustrated in Figure 5.2 evolved. It was derived from the
mean R_f values for TLC and 20 steroids and a number of dyes. On
the left side of the mixing line is pure solvent of the weaker
strength, and on the right side is the pure solvent of higher
strength. The proportions in between are given on a logarithmic
scale. Pure methanol (high strength) is somewhere out of the picture to the right. Using this figure, it is possible by trial and
error to formulate a mobile phase that will separate the steroid
or other compound under consideration.

In Figure 5.2, the pure solvents and their binary mixtures,
that is, the mixing lines, are placed, empirically under each
other in such a way that the solvents of equal average eluting
power for the tested steroids are to be found in the vertical dimension. Thus, the ability to move solutes increases from left to
right, whereas it remains practically constant along the vertical
lines. The numbered circles 1-6 from left to right in Figure 5.2
refer to the composition of the mobile phases of increasing
strength shown in Figure 5.1. On the other hand, the vertical
line x has been drawn through the binary mixing lines in an arbi-
trary position. If, for example, estradiol-17β and cholesterol
were developed together in the mobile phase whose composition is
given by this line x, the R_f values of each steroid would remain
more or less constant in the various mobile phases, but because of
mobile phase-solute interaction different mobilities can result.

The results in practice are shown in Figure 5.3. At the top
are listed the mobile phases composed according to line x of Figure 5.2. On the lines between start and front, the experimental
R_f values of (a) testosterone, (2) estradiol-17β, and (c) cholesterol have been marked. The R_f values fluctuate about a certain
mean value.

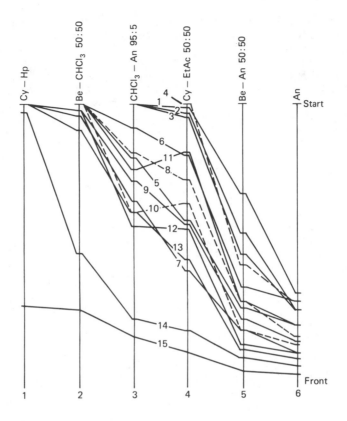

Figure 5.1. Migration distances of sterols and steroids of different polarities in mobile phases with different eluotropic properties. The numbers 1 to 6 (down) correspond to the numbered
circles in Figure 5.2. The R_f values in system No. 1 represent
the mean values of the very similar values in cyclohexane (Cy) and
heptane (Hp); Be = benzene, An = acetone, EtAc = ethyl actate.
The numbers 1-15 referring to joint R_f lines represent the steroid
as given in the text. This diagram gives a guide to the choice of
solvents over a wide range of polarity. Here, the R_f values of a
number of sterols and steroids, i.e., from tetrahydrocortisol (No.
1) down to cholestane (No. 15), in six pure and mixed solvent systems of increasing eluotropic power, are indicated graphically.
The intermediary substances are: (2) cortisol, (3) cortisone,
(4) corticosterone, (5) estradiol-17β, (c) 5β-pregnane-3α, 20α-
diol, (7) estrone, (8) testosterone, (9) 3β-hydroxy-pregn-5-en-
20-one, (10) androst-4-ene-3,17-dione, (11) cortexone, (12) pro-
gesterone, (13) cholesterol, and (14) cholesterol acetate.

114

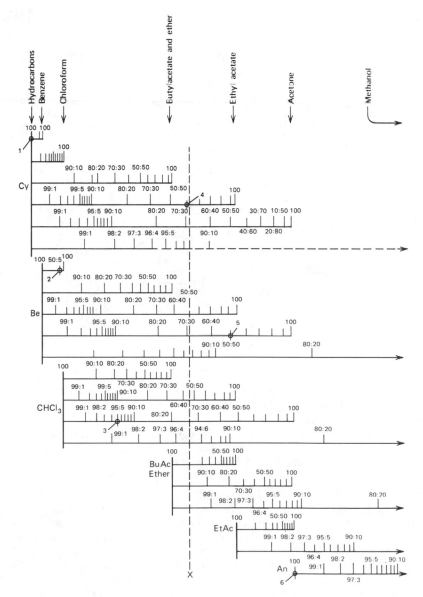

Figure 5.2.   Solvents and binary solvent mixtures arranged in the
"equieluotropic series" on the basis of average R_f values for 20
steroids.   Each vertical line connects solvent mixtures of the
same average eluotropic properties; pure methanol is womewhere
out of the figure to the right.   The first line section from the
left or right in the "mixing line" joining two solvents always represents a mixture of 90% and 99%, respectively, of the nearest
pure solvent, and 10% or 1% of the other, etc.

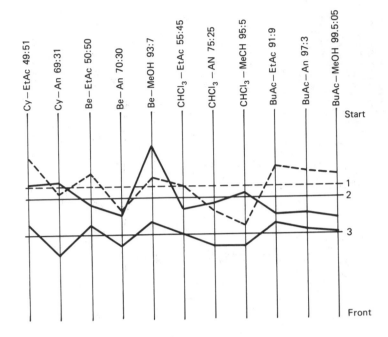

Figure 5.3. R_f values of (1) testosterone, (2) estradiol-17β, and (3) cholesterol in the binary solvent mixtures given by the vertical line x in Figure 5.2. The R_f values of each steroid vary within a certain "equieluotropic" range about the mean value (straight lines). An - acetone; Be - benzene; BuAc = ny-butyl acetate; Cy = cyclohexane; EtAc = ethyl acetate; MeOH = methanol.

As far as "strength" is concerned, not many mobile phases are needed to cover the whole range of steroid chromatography; but to separate all the mixtures that are encountered, it is advisable to use several mobile phases of approximately equivalent strength, as may be seen from the intersecting zig-zag lines in Figure 5.3. This means that if a steroid mixture cannot be separated in one mobile phase of suitable strength, a mobile phase of the same strength but of completely different composition may be required. This system of equieluotropic solvent mixtures is, of course, also applicable to column chromatography. The principle is based on the fact that different solutes interact differently with the different solvent components in the mobile phase. If two solutes do not separate in a given system, changing to a mobile phase of the same strength that interacts differently with the solutes will often increase the resolution.

Mobile phase strength affects mobility. As a general rule decreasing strength increases resolution. With a large number of solvents available there is an infinite number of mobile phase combinations possible. Therefore, it is appropriate to consider selectivity or interactions between the solute and mobile phase. This is discussed in detail in the following sections.

5.7 SOLVENT SELECTIVITY

The organization of mobile phase selection parameters is based on the work of Snyder, which is summarized in his widely used monograph (1). The expression of resolution resulting from this and other work can prove very useful in choosing an appropriate mobile phase in both column and thin layer liquid chromatography. Snyder's mathematical expressions of adsorption chromatography have made it possible to systematize choice of mobile phase. Further advances in the theory were described in his report published in 1971 (7). The basic books on liquid chromatography (8-10) contain complete discussions of the theories of adsorption and interactions between the mobile and stationary phases and the solute.

Solvent selectivity refers to the ability of a solvent to create differences in the relative migration of two solutes. The general formula for separation power or resolution (Rs) is expressed in the equation:

$$Rs = \frac{1}{4} (\alpha - 1) \sqrt{N} \left(\frac{k'}{k' + 1} \right)$$

The separation factor α is the ratio of capacity factors k_1/k_2 for two solutes. N is the number of theoretical plates in the adsorbent bed through which the mobile phase flows and through which the solute passes. k' is the average of k_1 and k_2 (capacity factor k is the equilibrium ratio of total solute in the stationary phase, to total solute in the mobile phase). Thus, resolution, a separation power, is dependent on three factors: N is largely a function of the nature of the sorbent whether in a column or in a thin layer. k' reflects the average migration speed of the two solutes and is determined by mobile phase strength $\varepsilon°$. k' and $\varepsilon°$ must be held between narrow limits for optimum separation to be possible; α is determined by the sorbent and the composition of the mobile phase. In thin layer chromatography little attempt has been made to evaluate N; consequently, emphasis has been placed on choice of mobile phase. Selectivity is largely based on differences in interaction between the mobile phase and solute. Different solutes will interact differently with the mobile phase. Therefore, if two solutes cannot be separated with a mobile phase of predetermined strength, they can sometimes be separated by

using a different mobile phase of the same strength. Although
mobility is not greatly changed the resolution will be improved.

In practice the application of Snyder's theories require the
assembly of a number of sample, sorbent, and mobile phase parameters. Saunders (11) has condensed some of the basic concepts of
adsorption activity, group adsorption strength, and mobile phase
strength into two graphic forms. The use of the two charts presented here in Figures 5.1 and 5.2 is very simple and rapid and
gives a first approximation to a mobile phase appropriate for a
given sample. The results are approximate and in a few cases may
be in error. However, the technique has been found to be a very
favorable alternative to tedious calculations and the trial-and-
error approach. Since the purpose of this text is to emphasize
practice rather than theory, and Saunders' method is applicable to
TLC, we shall describe that method.

The sample partition ratio k' is defined as the ratio of the
amount of solute in the stationary phase to the amount in the mobile phase (see previous formula from Snyder). This value can
also be calculated from the equation:

$$k' = \frac{t_r - t_0}{t_0}$$

where t_r is the sample retention time and t_0 is the retention time
of an unadsorbed component.

Since R_f or mobility factors in TLC also depend on the ratio
of solute in the mobile phase to that in the stationary phase, a
valid practical use can be made of the expressions that follow:
For a given sample and sorbent, log k' varies linearly with $\varepsilon°$.
Saunders in his example defines the E_3 value of a sample as the
solvent strength required to give k' $\stackrel{3}{=}$ 3, as shown in Figure 5.4.

For good results the k' value should be less than 10. With
k' of 10 the solute becomes diffused in the sorbent; the separation is poor and because of a long retention time the R_f is low.
A careful evaluation of the equation as related to retention times
and the limitation of k' to between 1 and 10 indicates that these
limits correspond relatively closely to the limits of R_f 0.3-0.7
proposed in this book and by other authors as the region of useful
separation in TLC. A value for k' of 1.0 in columns corresponds
roughly to an R_f of 0.1 in TLC. Thus, a k' value of 3 is a good
working point corresponding to an R_f between 0.3 and 0.5. The
essential point is to keep the R_f above 0.1 and less than 0.7.
Thus, the following is applicable to TLC as well as columns.

Figure 5.4 shows the E_3 values for a variety of compounds on
a typical silica. The following describes how to approximately
determine the solvent strength for a very broad variety of mono-

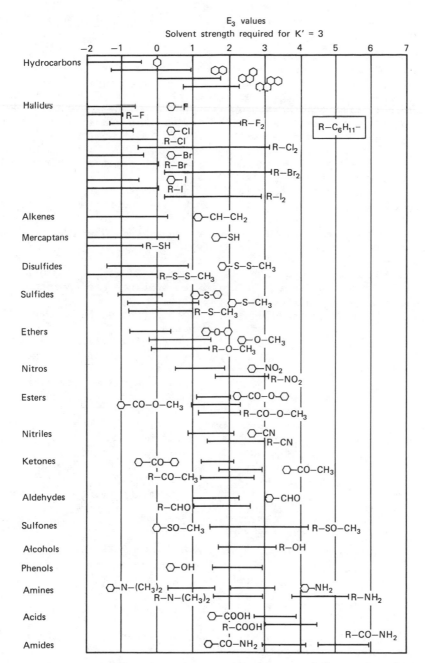

Figure 5.4 E_3 values of various compounds on silica gel. See text for nature of E_3.

and polyfunctional compounds and how to obtain several alternative mobile phases for any solvent strength using only the two simple graphs. Consider naphthalene as a hydrocarbon (second bar from top) in Figure 5.4. The range of E_3 values extends from 0.09 for fully activated adsorbent (equilibrated with dry solvent) down to -0.13 for highly deactivated adsorbent. This means that with an active silica and a solvent strength of 0.09, naphthalene will separate with k' = 3. With deactivated silica, weaker solvents will be needed to give k' = 3 down to a lower limit of about -0.13 for highly deactivated silica.

By grouping various parameters and comparing their relative effects on E_3, it is possible to formulate the following approximate rules for estimating E_3 in a broad variety of polyfunctional compounds. In the examples accompanying these rules, E_3 values cited are for fully activated silica.

1. Saturated hydrocarbons and simple olefins are eluted with k' = 3 with all solvents. Fluorinated solvents are an exception because of lower solvent strength.

2. Alkyl- and halogen-substituted aromatics have E_3 values within ±0.05 unit of the parent compound.

3. Aliphatic compounds with 2-20 carbons will have E_3 within ±0.05 unit of the model compound (R = C_6H_{13}) used in Figure 5.4.

4. Difunctional compounds will have E_3 values larger than either of the monosubstituted compounds. For example, for benzaldehyde E_3 = 0.23 and for methoxybenzene E_3 = 0.15, but the E_3 value for methoxybenzaldehyde will be larger than 0.23 by an increment that depends upon the difference between E_3 values (ΔE_3) of the monosubstituted compounds as follows:

ΔE_3	Increment
0-0.1	0.15
0.1-0.2	0.07
0.2	0.00

In our example E_3 = 0.23 - 0.15 = 0.08; thus for methoxybenzaldehyde, the increment is 0.15 and $E_3 \simeq$ 0.23 + 0.15 = 0.38. Similarly, for diaminohexane, ΔE_3 = 0.0, $E_3 \simeq$ 0.54 + 0.15 = 0.69.

5. Additional functional groups will add successively smaller increments to E_3, but the size of these increments cannot be easily predicted.

6. Addition of an aliphatic halogen or mercaptan moiety will increase E_3 by an increment 50-100% larger than the value indicated in 4 and 5 above. For example, for chlorohexyl ethyl ether $E_3 \simeq$ 0.15 + 2(0.07) = 0.29.

7. Compounds containing polynuclear aromatic moieties should
be considered benzene compounds substituted with additional aromatic rings. For example, nitrophenanthrene should be considered
a nitrobenzene (E_3 = 0.19) substituted with two additional rings,
that is, naphthalene (E_3 = 0.09). Therefore, $\nabla E_3 \simeq 0.1$ and the
corresponding increment is 0.15; thus, E_3 = 0.34. Once the E_3
values have been estimated from Figure 5.4 and the preceding rules,
we can choose an appropriate solvent mixture from Figure 5.5. The
six solvents used in Figure 5.5 are among the most useful for adsorption chromatography. Viscosities are low, allowing maximum
resolution per unit time, and the whole range of solvent strengths
is covered.

In Figure 5.5, solvent strength ($\varepsilon°$) is plotted across the
top versus various binary solvent compositions, in a manner similar to that of Neher in Figure 5.2. Each horizontal line corresponds to a range (0-100% volume) of binary solvent mixtures. The
first five lines represent mixtures of pentane with other solvents.
The top line of this series corresponds to mixtures of
pentane and isopropyl chloride. For any solvent strength intermediate between pentane and isorpopyl chloride ($\varepsilon°$ = 0.22), the
required solvent composition is found by dropping a vertical line
from the $\varepsilon°$ scale. Thus, for $\varepsilon°$ = 0.10 the mixture is 26 vol-%
isopropyl chloride in pentane, and for $\varepsilon°$ = 0.20 it is 80 vol-%
isopropyl chloride in pentane. The next four lines in this series
correspond to mixtures of pentane with methylene chloride ($\varepsilon°$ =
0.32), ethyl ether ($\varepsilon°$ = 0.38), acetonitrile ($\varepsilon°$ = 0.50), and
methanol ($\varepsilon°$ = 0.73). The second series of lines correspond to
binary mixtures of isopropyl chloride and a solvent of higher $\varepsilon°$.
For any given solvent strength, Figure 5.5 indicates several binary mixtures. Consider solvent mixtures with $\varepsilon°$ = 0.30. One of
these mixtures is 76 vol-% methylene chloride in pentane. Similarly, 49 vol-% ethyl ether in pentane or 37 vol-% ethyl ether in
isopropyl chloride, etc., would also give $\varepsilon°$ = 0.30. Any of the
indicated solvent mixtures would be an appropriate first choice
for a sample with E_3 = 0.30.

Because of the assumptions and approximations used in the
derivation of Figures 5.4 and 5.5 and Rules 1-7, and the limitations of the basic theory, the initial choice of solvent is seldom exactly correct. However, it is usually close enough to allow a rapid adjustment to the optimum. It has been found that the
following rules are helpful.

1. An increase in solvent strength ($\varepsilon°$) will decrease separation and increase the R_f.
2. A decrease in $\varepsilon°$ will increase separation and decrease
R_f.

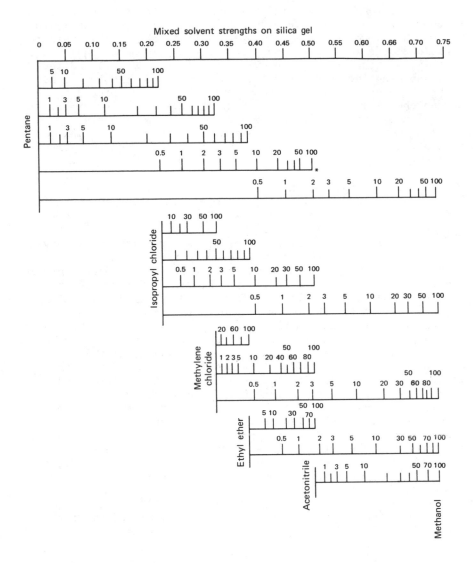

Figure 5.5. Mixed solvent strengths on silica gel. See text for details.

3. Selectivity will be greatest if the concentration of the stronger component of the mobile phase is < 5 vol-% or > 50 vol-%. For example, at ε° = 0.30, either 76 vol-% methylene chloride in pentane or 1.7 vol-% acetonitrile in isopropyl chloride would be expected to give maximum selectivity due to more favorable solute-mobile phase interactions.

4. Substitution of ethyl ether or methanol for one of the other strong components of the mobile phase can often improve selectivity due to the formation of hydrogen bonds (9), a potential of these solvents.

5. If a dry solvent results in tailed spots, a more highly deactivated sorbent should be used to decrease the solute-sorbent interactions. This is usually done by adding water to the sorbent.

For the extension of Figure 5.4 to more complex molecules, Rule 1 is simply a statement of the fact that nonaromatic hydrocarbons are not adsorbed with any of the solvents listed in Figure 5.5. Fluoroalkanes and fluoroethers have solvent strengths less than 0, resulting in some adsorption of olefins and saturated hydrocarbons (6). However, these solvents have found little use in adsorption chromatography, presumably because of their high cost.

Rules 2 and 3 arise from the very small group adsorption energies of alkyl groups generally and of halogens substituted on aromatics.

The graphical and rule-of-thumb approach presented here is an approximate but convenient method for quickly finding the mobile phase appropriate for the separation of a given mixture. Those who wish more accurate prediction should refer to the detailed calculations presented in Snyder's momograph (1).

The mobile phase strength data given in Tables 5.3 and 5.4 are based on results obtained with alumina. These data will be helpful in selection of a mobile phase. Table 5.3 gives solvent strength of binary mixtures of pentane with any of eight different solvents. Table 5.4 gives the data for benzene mixed with other solvents. Using these tables one can find different mobile phases with the same strength but different selectivities by simply following a line through the point of required mobile phase strength (ε°).

To summarize solvent selection: If the R_f is too high, choose a mobile phase of lower strength (ε°). If it is too low, go to a higher ε°. Once the desired mobility is attained, work on selectivity. If a pair does not separate with the mobile phase (misture of solvents) being used, select a mobile phase with the same strength but different components and thus different interactions. The tables included in this chapter provide a number of mobile phases of the same strength but of widely varying composi-

TABLE 5.3   SOLVENT STRENGTHS FOR MOBILE PHASES OF POLAR SOLVENTS IN PENTANE

Solvent A:	Pentane							
Solvent B:	CS_2	i-PrCl	Benzene	Ethyl ether	$CHCl_3$	CH_3Cl_2	Acetone	Methyl acetate
0.00	0	0	0	0	0	0	0	0
0.05	18	8	3.5	4	2	1.5	1.5	
0.10	48	19	8	9	5	4	3.5	2
0.15	100	34	16	15	9	8	6	3.5
0.20		52	28	25	15	13	9	5
0.25		77	49	38	25	22	13	8
0.30			83	55	40	34	19	13
0.35				81	65	54	28	19
0.40					100	84	42	29
0.45							61	44
0.50							92	65
0.55								100
0.80								

NOTE: Values refer to volume percent of solvent B in solvent A (pentane) for the given $\epsilon°$ value. These data are for alumina as sorbent. For silica gel the order will be the same, but the actual $\epsilon°$ value will be slightly different.

TABLE 5.4   SOLVENT STRENGTHS FOR MOBILE PHASES OF POLAR SOLVENTS IN BENZENE

Solvent A:	Pentane	Benzene					
Solvent B:	Diethyl-amine	Acetone	Methyl acetate	Diethyl-amine	Aceto nitrile	i-PrOH*	MeOH†
0.30	2.5						
0.35	5	6	4				
0.40	8	18	12	2			
0.45	13	36	24	7	1.5		
0.50	22	60	42	14	6		
0.55	38	93	66	26	14	4	
0.60	73		100	45	36	7	
0.65				77	100	12	
0.70						21	
0.75						∠0	4
0.80						75	8
0.85							18
0.90							44
0.95							100

NOTE: Values refer to volume percent of solvent B in solvent A (pentane or benzene) for the given ε° value. These data are for alumina as sorbent. For silica gel the order will be similar but ε° will be different

* Isopropanol
† Methanol

tions, which will provide different solute-mobile phase interactions.

Snyder, more recently, advanced the concept of polarity index (12). The data used was derived from the Rohrschneider constants (13). With this index it is possible to set up a more systematic approach to finding the constituents of a mobile phase that provides appropriate R_f values for the solute in analysis. Once an index (P') is obtained to give the median R_f it should be kept at that level and changes in selectivity made. This is done by use of the table published in Snyder's report. Selections of the more common solvents are given in Table 5.5. The solvent strength parameter, $\varepsilon°$, previously described, provides no means of determining the volumes of various solvents that can be added together to provide a specified R_f. The formula for "total P'" is P' = $F_aP'_a + F_bP'_b + F_cP'_c$, ..., where F is the volume fraction of the pure solvents a, b, or c while the P'_a, etc., is their polarity index. Using the table and this formula one can eliminate much guess work in deriving mobile phases. As discussed before, solvents may have the same P' or $\varepsilon°$ value but because of different constituents in their makeup will have different reactivity or interaction with the sorbent or solute. This difference is the basis of selectivity in solvent selection.

TABLE 5.5   POLARITY INDEX DERIVED FROM ROHRSCHNEIDER DATA

Solvent	P'	Solvent	P'
N-hexane	0.00	ethanol	5.2
carbon disulfide	1.00	pyridine	5.3
triethylamine	1.8	acetone	5.4
toluene	2.3	methoxyethanol	5.7
diphenyl ether	2.8	aniline	6.2
diethyl ether	2.9	methylformamide	6.2
benzene	3.0	acetic acid	6.2
methylene chloride	3.4	acetonitrile	6.2
t-butanol	3.9	dimethylformamide	6.4
tetratydrofuran	4.2	dimethylsulfoxide	6.5
ethyl acetate	4.3	methanol	6.6
chloroform	4.4	formamide	7.3
dioxane	4.8	water	9.0

5.8   ION EXCHANGE

Ion-exchange chromatography is widely used in the separation of
amino acids, proteins, protein hydrolysates, and other ionized
compounds. The theory for selection of solvents is not as well
advanced as in adsorption chromatography. However, advances in
resin technology in the past few years have resulted in the introduction of some very useful ion-exchange sorbents. This technique is more often used in column chromatography. A number of
ion-exchangers are available for thin layer separations (see Chapter 2 on sorbents).

Aqueous mobile phases containing various ionic components are
used for ion-exchange chromatography, with resolution and solute
mobility varied by changes in pH and ionic strength. Small
amounts of organic solvent are added to increase mobility of
strongly retained compounds as well as to increase solubility. In
ion-exchange TLC, the regeneration or activation of the layers and
development of the layer must be distinguished. Conversion to the
necessary ion form is usually carried out before the sample is applied to the plates. It can also be performed on the plate with
the solvent or buffer used to develop the chromatograms. In many
cases it is sufficient to develop the layer with water or 0.1N
sodium chloride solution. This type of pretreatment of the plate
often leads to better separation.

There are a few rather basic precepts learned from column
chromatography in terms of the mobile phase. The mobile phases
used can be divided into four classes:

1.   Buffer solutions of various ionic strengths within the pH
range 3.0-8.0; for example, 0.005M phosphate buffer of pH 8.0 or
2M sodium formate buffer of pH 3.4.
2.   Salt solutions; for example, 0.1-2.0M LiCl or NaCl.
3.   Distilled water.
4.   Any one of the above with organic solvent added.

Table 5.6 shows some buffers and pH ranges typically used to
separate mixtures of amino acids on cation exchangers. Several
different buffers are usually used for each analysis.

A large number of other buffers or salt solutions can be used,
depending on the nature of the compound to be separated: is it
acidic or basic? Is it a cation or an anion? In a general sense,
pH will be controlled by the use of buffer, and the ionic strength
can be adjusted by the addition of some other salt. An example of
this is the borate buffer of pH 9.2 with ionic strength adjusted
by adding 0.002N sodium nitrate. The selectivity is kept in narrow limits by controlling the ionic strength.

A basic understanding of the process taking place in the ion-
exchange phenomenon will facilitate selection of the mobile phase

TABLE 5.6   BUFFERS FOR USE WITH CATION EXCHANGE

	Protein hydrolysate	Physiologic fluid
Composition	Sodium citrate	Sodium or lithium citrate
Concentration	0.2-1.5M sodium	0.2-1.5M sodium or 0.3-0.8M lithium
	0.03-0.1M citrate	0.05-0.20M citrate
pH	2.8-6.0	2.5-6.0

for use with sorbents for ion exchange that have charge-bearing functional groups. The most common mechanism is simple ion-exchange as expressed in the following formulas:

$$S^- + M^+Y^- \longleftrightarrow Y^- + R^+X^- \quad \text{(anion exchange)}$$
$$S^+ + M^-Y^+ \longleftrightarrow Y^+ + R^-X^+ \quad \text{(cation exchange)}$$

Where: S = sample ion
         M = mobile phase ion
         R = ionic site on exchanger
         X = sample anion
         Y = sample cation

In anion exchange, the sample anion S^- is in competition with the mobile phase ion M^- for ionic sites on the exchanger, R^+. Conversely, in cation exchange, sample cation S^+ is in competition with the mobile phase cation M^+ for sites on the exchanger, R^-. Simple ion-exchange separations are based on the different strengths of the solute ion/resin interactions. Solutes that interact weakly with the exchanger in the presence of the mobile phase are poorly retained on the sorbent and show higher mobility. Conversely, solutes that interact strongly are strongly retained and move more slowly in a layer or column.

Both acids and bases can be separated by ion-exchange. A weak acid, HA, dissociates into its ionic form, H^+ and A^-:

$$HA \longleftrightarrow H^+ + A^-$$

The degree of this dissociation can be controlled by the pH of the system. By increasing the concentration of hydrogen ions (decreasing the pH), the dissociation equilibrium can be forced to the left, thus decreasing the concentration of A^- ions available for interaction with anion-exchange resin and decreasing the k'

value for solute HA. Change in pH can also control the availability of ionic forms of bases for ion-exchange separations:

$$B + H^+ \longleftrightarrow BH^+$$

In this case, increasing the concentration of hydrogen ions (decreasing the pH) shifts equilibrium to the right, thereby increasing the concentration of the charged (BH^+) ions available for ion-
exchange interaction and increasing the separation of basic solute
B.

Ion-exchangers in the acid or base form can also be used to
carry out separations via *acid-base reactions*:

$$1. \quad HA + R^+OH^- \longleftrightarrow R^+A^- + H_2O$$

$$2. \quad B + R^-H^+ \longleftrightarrow R^- + BH^+$$

where R = ion exchange resin.

This technique is useful for separating weak acids and bases
in nonaqueous systems. A weak acid, HA, undergoes interaction
with the anion exchanger, R^+OH^-, to form the resin anion salt,
R^+A^-, plus water. This is a reversible acid-base reaction. The
similar reaction of a weak base, B, reacting with the cationic
resin, R^-H^+, to form a salt, R^-BH^+ is also seen.

5.9 EFFECT OF IONS

Most ion-exchange chromatographic separations are carried out in
aqueous media because of the desirable solvent and ionizing properties of water. Mixed solvents, such as water-alcohol, and even
totally organic solvents have also been used. In ion-exchange
chromatography with aqueous mobile phases, mobility is controlled
by the total salt concentration (or ionic strength) and by the pH
of the mobile phase. An increase in the concentration of the salt
in the mobile phase increases the mobility of solutes. This is
due to the decreased ability of the solute ions to compete with
the mobile phase counterions for the ion-exchange groups in the
resin.

The type of ions in the mobile phase can significantly affect
the mobility of solute molecules, due to the varying interaction
of these different mobile phase ions with the ion-exchange resin.
The retention sequence of various anions for conventional cross-
linked polystyrene anion-exchange resin is as follows:

$$\text{citrate} > SO_4^{2-} > \text{oxalate} > I^- > NO_3^- > CrO_4^{2-} >$$

$$Br^- > SCN^- > Cl^- > \text{formate} > \text{acetate} > OH^- > F^-$$

This retention sequence varies somewhat for different resins, but it is a qualitative indication of the ability of various anions to interact with strong anion exchangers. In this sequence, citrate is very strongly bonded to the resin, while fluoride ions are very weakly bound. Thus, solute molecules will have higher R_f rates with citrate ions than with fluoride ions.

For many years, when sodium citrate buffers were used almost exclusively as the mobile phase, a complete analysis of physiologic fluids by ion-exchange chromatography was hampered by an inability to separate asparagine and glutamine from the other amino acids (14, 15). Either two separate analyses under different conditions were required to separate all of the acids, or two analyses were carried out under the same conditions, with one or two of the amino acids selectively destroyed between runs (16) and a difference of peak areas used as a means of quantitation. The use of lithium citrate buffers instead of sodium citrate has now made the complete separation of these two important amino acids possible (17). In this case, the lithium ion is not bound as strongly to the resin, yielding generally larger k' values. This, in addition to specific characteristics of the amino acids, gives some separation advantages with asparagine, glutamine, and the other amino acids in that region of the chromatogram.

There are some general guidelines for predicting the ability of an exchanger to complex with solute or mobile phase ions. The ion exchanger tends to prefer:

1. The ion of higher valence (charge).
2. The ion with a smaller (solvated) equivalent volume.
3. The ion with the greater polarizability.
4. The ion that interacts more strongly with the fixed ionic groups or with the matrix.
5. The ion that interacts weakest with other materials in the mobile phase.

Effect of pH separation in ion-exchange chromatography can also be controlled by changes in mobile phase pH. Thus, an increase in pH decreases mobility in cation exchange chromatography and increases it in anion exchange. The influence of pH on retention can be explained by the effect of hydrogen ion concentration on available sample ions:

1. $RCOOH \longleftrightarrow COO^- + H^+$ (anion exchange)
2. $RNH + H^+ \longleftrightarrow NH_2^+$ (cation exchange)

Here it is seen that a free carboxylic acid is in equilibrium with the carboxylic ion plus a proton. At highest pH (lower concentration of H^+) the concentration of carboxylic ions is higher; thus,

the retention of this sample ion is greater because of its increased ability to compete for the anionic sites of the resin.
The formation of the charged cationic form of a weak base is enhanced at higher hydrogen ion concentrations. Thus, a decrease in
pH results in lower R_f due to the increased competition of these
charged sample ions for the cation-exchange sites. Table 5.7 is
from the classical study by Hamilton (14) and shows the effect of
changes in pH on the elution order of amino acids. These data illustrate why changes in buffer pH are often used to control selectivity. However, changes in pH can also cause changes in the
order of separation of some amino acids.

Variation in pH can also cause changes in the *relative* retention of sample components. Therefore, changes in pH can not
only be used to control mobility but can also be useful in changing selectivity. Changes in selectivity as a function of pH are
difficult to predict and usually have to be found by a trial-and-
error process. When using anion-exchange resins, pH is normally
controlled by using cationic buffers (e.g., buffers containing
ammonia, pyridine, etc.); with cation-exchange resins, buffers
containing ions such as acetate, formate, and citrate are often
used to fix the pH in the system. However, other factors can
cause variation in pH. It is well known that temperature changes
can cause changes in the pH of Tris buffer systems.

With the basic amino acids, an increase in cation concentration improves resolution (band width) and decreases k' value.
However, changes in k' are nearly parallel, so these variables are
not of much help in selectively moving one acid with respect to
the others.

TABLE 5.7   SEPARATION OF ACIDIC AMINO ACIDS (12)

Order	At pH 4.15	At pH 4.80	At pH 5.25
1	Tyrosine	Tyrosine	β-Alanine
2	Phenylalanine	β-Alanine	β-Aminoisobutyric acid
3	β-Alanine	β-Aminoisobutyric acid	Tyrosine
4	β-Aminoisobutyric acid	Phenylalanine	Phenylalanine

5.10   EFFECT OF ADDITION OF ORGANIC SOLVENT

The differential solubility of threonine and serine in various organic solvents (methanol, thiodiglycol, methyl cellosolve) is an aid in separating them. The addition of up to 10% of one of these organics in the buffer makes a dramatic improvement in the separation of these amino acids. It has been observed (17, 18), however, that the addition of these solvents has some effect on the elution pattern of other sample components. All other conditions remaining the same, several amino acids (methionine sulfone, glutamic acid, proline, and citrulline) are eluted sooner in the presence of an organic, and others (glycine, alanine, and α-amino-n-butyric acid) are delayed.

The solvents used in the mobile phase can also provide changes in selectivity. As previously indicated, an aqueous carrier is normally used in ion-exchange chromatography. However, small amounts of organic solvents, such as ethanol, tetrahydrofuran, and acetonitrile, are sometimes added to these aqueous systems to increase the solubility of sample components, and these solvents also can provide changes in selectivity. The addition of organic solvents to aqueous buffers sometimes decreases the tendency of certain sample components to form tailing peaks. Weak acids and bases can sometimes be separated with conventional ion-exchange resins, using nonaqueous carrier systems. In this form of chromatography, decreasing the polarity of the organic decreases the mobility of sample components so that resolution is affected. Changing from methanol to benzene, then to pentane, greatly decreases the mobility of sample components with conventional macroreticular ion-exchange resins (19).

The foregoing points out the problems and some solutions to ion-exchange chromatography. However, it should be remembered that numerous ionizable compounds have also been separated by adsorption chromatography as previously discussed. The solvent systems used must include water, and in many cases acid or basic solvents are also included. Most of the theories of solvent selection in chromatography have been formulated using column chromatography. Therefore, the rules that have been presented may vary slightly when applied to TLC. Many times the investigator has done preliminary evaluations on TLC before going to high-pressure liquid chromatography and can benefit from this knowledge.

The separation scientist is often confronted with a large number of possible systems from which to make a choice for a particular separation. In TLC, qualitative analysis is often performed by using several different solvents in parallel chromatograms. The question arises as to which combination will provide the most information.

The problem in TLC is the choice of a number of mobile phases or sorbents or combination of them so that the separation factors

will be as different as possible. These problems are common, but
rarely are they stated explicitly, and there are no methods for
mathematical classification of chromatographic systems. Often the
mobile phases referred to in the literature must be improved to
fit the situation at hand.

More recently other attempts to place solvent selection on a
firmer basis have been reported. Numerical analysis was used by
Turina and Trbojevic (20). Using equations based on R_s values,
where R_s is the ratio between the distance of two spot centers according to Giddings (21), they were able to calculate the optimal
ratio of components in a TLC system. Numerical analysis was used,
and three experiments were required for a two-component system. A
three-component system required five to nine experiments. Turina
and Trbojevic concluded that the method may be more useful for
correcting commonly applied solvent systems than in developing new
systems.

Massart and DeClerq (22) classified systems according to mu-
tual resemblances by numerical taxonomy techniques. They illustrated the use of the technique by selecting a combination of
three solvent/stationary phases from a set of ten for the separation of 26 different dyes used in the food industry. Those who
are interested in improving the selectivity of solvents as well as
in more scientific solvent selection will find this article helpful.

5.11 LIQUID-LIQUID (PARTITION) TLC

The selection of solvents for liquid-liquid chromatography or partition chromatography is still largely empirical (23). There are
no universal partitioning systems for classes of solutes, but
there are infinite capabilities for separation simply by selecting
the appropriate pair of partitionary liquids. Since this type of
thin layer chromatography has seen relatively little use it will
not be discussed at length here, but a few basic generalizations
will be presented. The interested reader should refer to the discussions in the section on liquid-liquid chromatography in the
book *Modern Liquid Chromatography* (23). Some of the generalizations in paper chromatography are applicable. Partition chromatography is often used in the paper chromatography of steroids
using the system as described by Zaffaroni et al. (24), with propylene glycol as the stationary phase and toluene or benzene as
the mobile phase. The stationary phase can be sprayed on the
plate before using or mixed with the slurry before preparing the
layer. Many partition systems are described for a wide variety of
compounds in the book *Paper Chromatography* (25). Some of these
systems have been adapted to thin layer as well as column chromatography.

In liquid-liquid partition the mobile phase must be immiscible with the stationary phase. Both partitioning phases, but especially the mobile phase, should have as low a viscosity as possible for higher sorbent permeability. Most important is the selectivity of a liquid-liquid system for a given solute.

In partition TLC the resolution of solutes is generally controlled by changing the mobile phase. While it is possible to vary separation somewhat by selection of the stationary phase, this is often inconvenient in actual operation. The interaction of the two liquid phases (mobile and stationary) controls the overall separation for solutes that can be separated in the particular system. Thus, polar compounds may be separated using polar stationary phases and nonpolar mobile phases. Conversely, nonpolar solutes are best separated with a system consisting of a nonpolar stationary phase and a polar mobile phase. However, once the selection of the basic system is made, adjustment of the separation is normally carried out by varying the mobile phase composition.

In partition TLC, the sorbent acts merely as a support for the stationary phase. Kieselguhr is often used because of its low activity. Cellulose is often used. The sorbent can be impregnated in a number of ways depending on the type of partition system to be used. The plate can be equilibrated with the vapor of a solvent mixture consisting of the stationary phase saturated with the mobile phase, much as in paper chromatography. Or the plate can be sprayed with the stationary phase alone or in a volatile solvent. The plate can also be dipped in the solution, or the solution can be allowed to migrate up the plate. In each method the plate must be dried of solvent before application of the sample. The mobile phase must be saturated with the stationary phase (if they are immiscible) before use. In most cases the stationary phase is highly polar (water, formamide, or propylene glycol) and the mobile phase is relatively nonpolar.

In reversed-phase partition the stationary phase is nonpolar (paraffins) and the mobile phase is polar (water or alcohol or mixtures of these). In this case it is the nonpolar compounds that will remain near the origin.

5.12 THE MOBILE PHASE IN GEL THIN LAYER CHROMATOGRAPHY

Gel chromatography can be used for separating higher molecular species, particularly nonionic molecules such as proteins and nucleic acids. Mixtures of components of different molecular weights are routinely separated with this method. Hruska and Franek (26) used Tris buffer (0.2M, pH 8.0) to separate individual albumins on layers of Sephadex G-200 or Sephadex G-75. The separations were carried out in a moist chamber. Tween 80 (60 ml per 100 ml buffer) was added to reduce the adsorption of the protein

to the gel. The method gave chromatograms that could be scanned
densitometrically as well as by autoradiography.

Blessing and Gebele (27) used Sephadex-100, 150, and 200 in
thin layers for the separation of myoglobin and hemoglobin using
phosphate buffers as the mobile phase.

In gel chromatography, the mobile phase is *not* varied to control resolution. Rather, the mobile phase is chosen for low viscosity at the temperature of separation and for its ability to
dissolve the sample. The effect of the mobile phase on various
nonrigid sorbents for gel chromatography must be determined. Thus,
in gel filtration, a salt should be added to the solvent to maintain constant ionic strength. Otherwise the passage of ionic samples through the bed can lead to shrinkage of the gel and a decrease in permeability.

5.13   SUMMARY

This chapter shows that TLC can utilize a large number of separation parameters depending on the sorbent of choice. As further
knowledge is gained specific to sorbent-mobile phase, sorbent-
solute, and mobile phase-solute interactions, it may soon be possible to finally systematize mobile phase selection and eventually
to remove the trial-and-error methodology that has plagued the
separation scientist using chromatography as a tool. In TLC, sorbent characteristics are rather constant as opposed to the mobile
phase. This is the reason so much emphasis is placed on these
characteristics. Perhaps the use of liquid-liquid chromatography
or partition in TLC will increase and further the utility of this
technique in separations. However, as pointed out earlier, many
investigators first assess the applicability of a system by means
of TLC.

If the sorbent is to be deactivated with water, dry solvents
cannot be used. However, most TLC is done with plates that are
activated by heating just before use, and it is necessary to use
dry solvents to prevent deactivating the sorbent. Thus, the condition of the mobile phase itself can determine the effectiveness
of the separation. If activated plates are used, attention must
be given to the amount of water in the mobile phase. Different
solvents will have different water content. Therefore, it is necessary to bear these factors in mind when reporting the mobile
phase and the activity of the sorbent used in a specific separation. Some chromatographers use solvents containing a known
amount of water. The sorbent will dry the solvent as it moves up
the plate. The sorbent then is somewhat deactivated, and an equilibrium condition is set up that sometimes improves the separations.

REFERENCES

1. L. R. Snyder, *Principles of Adsorption Chromatography*, Marcel Dekker, New York, 1968.
2. W. Trappe, Biochem. Z., *305*, 150 (1940).
3. M. Steiger and T. Reichstein, Helv. Chim. Acta, *21*, 546 (1938).
4. J. Jacques and J. P. Mathieu, Bull. Soc. Chem. Frence, *94*, (1946).
5. E. Hecker, *Vereilungsnerfrahren in Laboratorium*, Supplements to Angermudle Chemig, Verlag Chemie, Weinheim, 1954.
6. R. Neher, in *Thin Layer Chromatography*, G. B. Marini-Bettolo, Ed., Elsevier, Amsterdam, 1964.
7. L. R. Snyder, J. Chromatogr., *63*, 15 (1971)
8. J. J. Kirkland, Ed., *Modern Practice of Liquid Chromatography*, Wiley-Interscience, New York, 1971.
9. P. R. Brown, *High Pressure Liquid Chromatography*, Academic Press, New York, 1973.
10. S. G. Perry, R. Amos, and P. I. Brewer, *Practical Liquid Chromatography*, Plenum Press, New York, 1972.
11. D. L. Saunders, Anal. Chem., *46*, 470 (1974).
12. L. R. Snyder, J. Chromatogr. Sci., *92*, 223 (1974).
13. L. Rohrschneider, Anal. Chem., *45*, 1241 (1973).
14. P. B. Hamilton, Anal. Chem., *35*, 2055 (1963).
15. J. C. Dickinson, H. Rosenbloom, and P. B. Hamilton, Pediatrics, *36*, 2 (1965).
16. S. Mone and W. H. Stein, J. Biol. Chem., *211*, 893 (1954).
17. J. V. Benson, M. J. Gordon, and J. A. Patterson, Anal. Biochem., *18*, 228 (1967).
18. G. Ertingshausen and H. A. Adler, Amer. J. Clin. Pathol., *53*, 680 (1970).
19. L. R. Snyder and B. E. Buell, Anal. Chem., *40*, 1295 (1968).
20. S. Turina and M. Trbojevic, Anal. Chem., *46*, 1988 (1974).
21. J. C. Giddings, Anal. Chem., *37*, 60 (1965).
22. D. L. Massart and H. DeClerq, Anal. Chem., *46*, 2988 (1974).
23. L. R. Snyder and J. J. Kirland, *Modern Liquid Chromatography*,
24. A. Zaffaroni, R. B. Burton, and E. H. Kentman, J. Biol. Chem., *177*, 109 (1949).
25. I. M. Hais and L. Macek, *Paper Chromatography*, Academic Press, New York, 1963.
26. K. J. Hruska and M. Franek, J. Chromatogr., *93*, 475 (1974).
27. M. H. Blessing and B. Gebele, Res. Exp. Med., *162*, 143 (1974).

CHAPTER 6
Developing Techniques

Development in thin layer chromatography is the process in which
the mobile phase moves across the sorbent layer to effect separation of the sample substances. It may be accomplished a number of
ways with several forms of apparatus, ranging from simple to complex. The most widely used types of apparatus and development
will be discussed in this chapter.

6.1 APPARATUS

The majority of thin layer development is done by allowing the
solvent (mobile phase) to migrate up (ascend) the plate, which is
standing in a suitable sized chamber. Developing chamgers may be
glass, plastic, or metal; glass chambers are the most common. Depending on plate size, budget, etc., the chamber may be a wide-
mouthed screw-cap jar, a battery or museum jar, or a specially
constructed chamber or tank for TLC purchased from a general laboratory supply firm or a TLC apparatus company. It can accommodate up to two 20×20 cm plates, or their equivalent, by standing
them face to face, and it generally requires a convenient volume
of 100 mL or less of mobile phase.

A tank with a raised glass hump running lengthwise on the inside bottom that acts as a tank divider has been introduced by

Camag, Inc. It requires only 20 mL of solution per side and allows two large plates to be developed simultaneously facing each
other; this design conserves solvent. The tank has the additional
feature of an unbreakable stainless steel cover, which is available separately for use on other tanks. The divided bottom allows
the tank to be saturated with two systems placed on either side of
the divider, and the plate may then be inserted on either side for
development. This tank is illustrated in Figure 6.1.

Brinkmann Instruments, Inc. offers four different sizes of
tanks specially designed for multiplate development. One tank
will hold five 20×20 cm plates, another holds up to ten 5×20 cm
plates, and a third size will accommodate sixteen 5×10 cm plates.
The largest tank is constructed of stainless steel and will hold
up to 17 analytical layer (250 μm) plates or eight preparative
(2000 μm) plates.

Figure 6.1 Glass developing tank for plates up to and including
20×20 cm sizes. Courtesy of A. H. Thomas Co.

Another commonly used developing apparatus, called a sandwich chamber, is shown in Figure 6.2. This particular apparatus can be used with 20×20 cm or 10×20 cm plates with several advantages, including small required solvent volume (25 mL), rapid chamber saturation, and fast development.

Figure 6.2. Sandwich chamber apparatus for development of 20×20 and 10×20 cm plates. Courtesy of A. H. Thomas Co.

   The chamber is formed from the 20×20 cm coated thin layer
plate to which the sample has been applied, a U-shaped cardboard
spacer, and a clean, uncoated 20×20 cm plate. The coated plate
must have an 8-10 mm wide strip of sorbent scraped off on the top
and sides before the sample is applied. The cardboard spacer is
placed on this cleared area, and the uncoated, clean plate is
placed on top of the spacer, thereby forming the chamber. The
sides are clamped together with metal clips, and the open end of
the chamber is placed into a glass solvent trough containing the
developing solvent. Other commercially available sandwich chambers will accommodate smaller (5×20 and 10×20 cm) plates. These
chambers are easy to make, or they can be purchased commercially
from a number of suppliers for about the same price as a normal
glass TLC tank.
   Analtech, Inc. supplies a "hanging TLC chamber" that allows
up to five plates to hang in the chamber containing the mobile
phase so that they may equilibrate before being lowered into the
solvent system. This may be accomplished without removing the top.
Analtech, Inc. also offers glass and stainless steel chambers
tailored for all common size plates.
   Applied Science Laboratories offers an anodized aluminum rack
for supporting and developing up to twelve plates at one time in a
normal sized tank. They also offer a plastic rack containing five
cylindrical tanks for developing 5×20 cm plates.
   Kontes Glass Co. manufactures two sizes of cylindrical tanks
with plastic cap lids for the development of 5×10 cm or 5×20 cm
plates. These smaller tanks are more convenient to use than the
large tanks when working with the 5 cm plates because they equilibrate quickly and require less mobile phase.
   Brinkmann Instruments, Inc. offers the "gas flushing tank"
made by Desaga of West Germany shown in Figure 6.3. The intake
and effluent stopcocks allow the tank environment to be controlled
according to the particular needs of the development process required.
   When working with heat-sensitive or unstable compounds, such
as many vitamins, it is desirable to work at low temperatures.
Thin layer chromatography may be performed at low temperatures by
working with the plates and development chamber in a cold room.
Brinkmann Instruments, Inc. has a "kryobox" low-temperature developing tank in their apparatus line, if cold-room facilities are not
available. The developing tank itself is surrounded by a plastic
jacket, which may be insulated, through which a constant-temperature liquid is passed. It is fitted with a dial-reading thermome-
ter. The developing tank contains a dual plate holder that may be
used to equilibrate the plate to the mobile-phase temperature and
atmosphere before it is lowered by external control into the mobile
phase.

Figure 6.3.   Gas flushing tank for flushing of the atmosphere in
the tank.   Courtesy of Brinkmann Instruments, Inc.

The Gelman Sciences, Inc. has a self-contained disposable TLC
system as part of the Instant Thin Layer Chromatography (ITLC)
line. This system is made up of a plastic "Seprachrom" microchro-
matography chamber, which contains one of the ITLC glass-fiber-
impregnated silica gel plates (see Chapter 3). In use, the top of
the Seprachrom chamber is pulled from the base and the sample is
applied through a hole in the chamber wall onto the plate layer.
The spots are allowed to dry. Four milliliters of mobile phase
is added to the base, and the chamber is attached, allowed to
equilibrate, and then carefully positioned for development. From
this point, the chromatogram may be treated in the usual manner.
A special developing chamber has been produced by Pharmacia
Fine Chemicals, Inc. to be used for the development of Sephadex-

coated thin layer plates. This apparatus is pictured in Figure
6.4. Special procedures have to be employed for the proper use of
Sephadex-coated plates, and a few of these will be discussed later
in the chapter. The special booklet published by Pharmacia Fine
Chemicals, Inc. should be consulted for further details (1).

6.2 DEVELOPMENT MODE

6.2.1 Ascending Development

The majority of thin layer chromatography is done by developing the
plate with the mobile phase, which is moving up (ascending) the
plate. This mode of development is popular because it is the simplest of the three modes available. The apparatus is readily
available commercially and is ready to use properly. Figures 6.1,
6.2, and 6.3 show apparatus that are used for ascending development. The special chamber pictured in Figure 6.4, which is generally used for the development of Sephadex gel plates, can be used
in the ascending mode as well as in the other two modes, descending
development and horizontal development. Any closed container that
will hold the plate upright is usable.
There are a number of factors to keep aware of when perform-
ing the development. It has been observed that the angle of development, i.e., the angle at which the plate is supported, affects

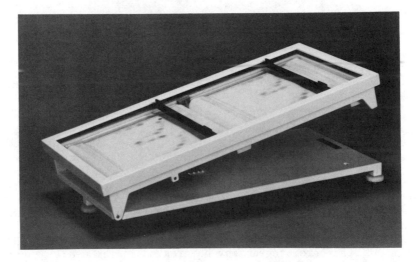

Figure 6.4. Special chamber for the development of Sephadex-
coated plates. Courtesy of Pharmacia Fine Chemicals, Inc.

the rate of development as well as the shape of the spots (2). As
the angle decreases toward the horizontal, the flow of the mobile
phase increases, but spot distortion also increases. An angle of
75° is recommended as optimum for development.

If it has been determined that the 75° development angle is
not critical for a particular separation, it is possible to develop a number of plates simultaneously if desired. A tank such as
the one shown in Figure 6.3 may be used. The glass channels on
the inside ends of this tank will support up to seven plates vertically for development at the same time. If a divided tank such
as this is not available, plate separator supports may be made by
bending a glass rod in the shape of a series of inverted connected
U's:⋂⋂⋂⋂ Each plate can then be supported by placing its top
into a U-section (3). The "hanging TLC chamber" as supplied by
Analtech, Inc., may be used to support up to five 20×20 cm plates
vertically for simultaneous development.

If a plate is developed in a chamber that is not saturated
with mobile-phase vapor, unsymmetrical flow up the plate may occur
because the mobile phase is evaporating off the plate into the
chamber atmosphere to establish an equilibrium condition. This
entire process depends on the volatility of the mobile phase, the
temperature, and the rate of development. It is for this reason
that the development chamber should be as small as conveniently
possible for the plate size and number of plates being developed
at one time. Niederwieser and Honegger (4) note that the separation process in TLC occurs in a three-phase system of stationary,
mobile, and gas phases, all of which are interacting with each
other until an equilibrium is reached. The following qualitative
equations sum up what is occurring:

(1) Adsorbent + Solvent $\xrightarrow{\text{ADSORPTION}}$ Stationary phase +
                                           mobile phase + heat

(2) Mobile phase + air + $\xrightarrow{\text{EVAPORATION}}$ Gas mixture
     heat

(3) Adsorbent + gas $\xrightarrow{\text{IMPREGNATION}}$ Impregnated adsorbent
     mixture   + air + heat

Equation 1 describes a very fundamental process for TLC.
While wetting the layer, one portion of the solvent becomes more
highly ordered and together with the layer becomes the stationary
phase. Equations 2 and 3 describe interactions of the mobile phase
that are insignificant with solvents of very low vapor pressure
but are important with volatile solvents. Evaporation of the mobile phase will increase with an increase in the amount of air or
heat in the system, and this will cause increased sample migration
because of greater solvent penetration into the layer.

For consistent, reproducible results it is important to maintain a constant-temperature environment about the development chamber. Benches that have the sun on them during part of the day should be avoided because of temperature variation. Development chambers are of such size that they may be conveniently placed in a controlled environment space, room or bath in order to maintain consistency.

It is advisable to weigh down the lid of the chamber so that the vapor pressure and heat produced by the mobile phase do not dislodge it and disturb the constant environment and equilibrium formed within the system.

6.2.2 Chamber Saturation

To aid saturation of the developing chamber atmosphere, the inner walls are normally lined with blotting paper or a thick chromatography paper such as Whatman 3MM. This is conveniently done by taking a single large sheet of paper (approximately 33 cm long × 21 cm high for a normal tank) and placing it into the dry tank so that the back and ends are lined. The front is left open so the plate may be observed. If cut large enough to line the three sides, the paper will stand by itself, and it is best not to use any glue or tape to support it. The paper should come to within 1 cm of the top of the tank. Be careful that the paper is not so high that it prevents proper seating of the lid on the tank. Some separations are best done in an unlined tank.

After carefully measuring the individual components of the mobile phase to be used, mix them thoroughly in a mixing cylinder, Erlenmeyer flask, or other container *before* adding them to the developing chamber. Most solvent proportions are measured by volume, unless stated otherwise in the literature. It is convenient to pour the mixed mobile phase over the lining paper as it is poured into the developing chamber. This will speed the saturation of the paper and the chamber. This is usually done about an hour before the plates are to be developed.

Plate development may be measured in terms of either time or distance. It may be allowed to proceed for a given period of time, after which the plate is carefully lifted out of the chamber and the solvent front marked with a pencil or a sharp object such as a syringe needle. If the mobile phase is to travel a fixed distance, for example, 10 cm, this distance is marked on the plate during the time of sample application. Then when the plate is developing, it is easy to note when the mobile phase has gone the desired distance and remove the plate from the chamber at this point. A fixed development distance is used to standardize R_f value, and because R_f values are between 0.0 and 1.0, 10 cm is a convenient distance.

   Sometimes the solvents present in a mobile phase are not completely miscible with each other. This condition may not be no-
ticed until a second solvent front becomes visible on the plate
during plate development. This occurs because of a process called
solvent demixing, which is more prevalent in adsorption chromatography than in partition chromatography. It results from the preferential retention by the stationary phase of the polar components of the mobile phase (4). A two-component (binary) mobile
phase will produce two zones of fairly constant polarity (polar
near the origin, nonpolar near the solvent front) separated by a
transition zone. If this transition zone is abrupt, it becomes
noticeable as a secondary solvent front. Great differences in polarity of the components in a mobile phase are responsible for
such secondary fronts. This occurrence is not necessarily bad,
and the chromatographic process, including visualization, should
be completed before any judgment is made. Often a mobile phase of
this sort is the only system that will provide a given separation,
with the substance in question migrating with the secondary solvent
front.
   Pittoni and Sussi (5) designed a saturation chamber for TLC
that is smaller in volume than a normal development tank and has
provisions for putting in the mobile phase without removing the
lid. The major drawback to the design, however, is that this
chamber will not accommodate 20×20 cm plates.
   DeZeeuw (6,7) and others have studied the influence of chamber saturation on separation. Some of these factors are mentioned
in Chapter 9.
   Dhont (8) has described an apparatus that allows TLC plates
to reach equilibrium with a well-defined vapor phase without
coming into contact with atmospheric moisture. The apparatus
is a nice, easy-to-use design that will accept only 5×20 cm plates.
It may be possible to scale up the apparatus to accept larger
plates.

6.2.3 Descending Development

Descending development is widely employed in paper chromatography
but is very seldom used in TLC. Reasons for this include the complexity of the apparatus, lack of commercial apparatus, and no apparent advantages other than speed of separation for most separations. The same can be stated for horizontal development in TLC,
which is used mostly in Sephadex development. Discussion of these
two development modes is therefore confined to Section 6.6, Sephadex development.

6.3 DEVELOPMENT TECHNIQUES

The simplest development technique in thin layer chromatography is that of developing a plate once in a pure solvent used as the mobile phase. An example of this would be the separation of a simple mixture of colored dyes on an alumina plate using pure benzene as the mobile phase.

If the nature of the sample substance is completely unknown, it is best to use a simple, single-component mobile phase in the primary experiments to determine the separation characteristics of the unknown. If a single-component mobile phase cannot be found to provide the separation desired, a multicomponent system may be necessary, or possibly a different type of development using the single-component system will work. The following sections will deal with the development options available to achieve the separation desired.

6.3.1 Multiple Development

The technique of multiple development was first applied to thin layer chromatography in 1955 (9). If a single development of a plate indicates a partial separation, the plate may be allowed to dry and then may be reinserted into the developing chamber for redevelopment. This may be repeated any number of times until a satisfactory separation is achieved. In effect, it increases the effective length of the layer, so as to increase the likelihood that a separation will occur because the molecules have been able to interact over a greater distance and will resolve themselves in the process. As the mobile phase encounters a substance with a low R_f value, it will begin to move it along the layer until it reaches another substance with a greater R_f value and begins to move it. As the mobile phase continues to migrate along the layer, this process is repeated a number of times, and the travel length of the substance with the lower R_f keeps increasing until the mobile phase reaches the substance with the higher R_f value. The distance the mobile phase travels after it reaches this substance is continually decreasing.

A number of researchers have discussed the theoretical aspects of multiple development (10-13). Thoma (12) has exhaustively determined the number of developments necessary to separate two substances of known R_f values by various degrees. These are given in Tables 6.1-6.4. The degree of separation desired should first be decided. Using the table for the degree of separation desired, the R_f value of the faster moving substance is located in the left-hand column. To the right of this value, in the same row, is then located the R_f value of the slower moving substance. The number of

developments necessary to separate these two substances is found at
the head of the column for the slower moving substance.

Longer than normal (20×40 cm) plates with precoated layers are
available commercially from a number of suppliers. Often these can
be used to provide a greater development distance for a difficult
separation. Both analytical (250 μ) and preparative (500, 1000,
1500, 2000 μ) layers are available, as listed in Chapter 3. One
problem is encountered when it is desired to use the 20×40 cm
plates; tanks large enough to hold the plates during development
are not widely available from commercial sources. Brinkmann Instruments, Inc. appears to be the only supplier of tanks specially
made for the 20×40 cm plates, and here the development is done over
the short distance (20 cm) and not the long one. If it is desired
to develop over the 40 cm distance, a container of suitable size,
such as a large paper chromatography tank, which is available from
the general laboratory supply houses, may suffice. Line the tank
with a large sheet of heavy filter or chromatography paper, and allow the tank to equilibrate with the mobile phase for an hour before development.

6.3.2   Stepwise Development

When a mixture contains compounds that differ considerably in polarity, a single development with one mobile phase may not provide
the desired separation. However, if the plate is developed successively with different mobile phases that have different selectivities or strengths, a separation will probably be forthcoming.
This procedure was first used by Stahl (14,15).

The mobile phases may be used in a number of ways. The less
polar phase may be used first, followed by a more polar phase, or
vice versa. More than two developments may be carried out, and
the development distance on the plate may be changed with each development.

For example, the first mobile phase used to separate a given
sample mixture may be very polar, such as methanol, in order to
separate the polar substances. It is allowed to develop part way
up the layer before the plate is removed from the chromatography
tank and dried. The second mobile phase, a nonpolar solvent such
as cyclohexane, is then used to develop the entire length of the
plate in an attempt to resolve in the upper portion of the plate
those nonpolar substances that were not resolved with the first,
polar-phase, development. Depending on the separation, these two
developments could be carried out in reverse sequence.

6.3.3   Continuous Development

Continuous development is exactly what the name implies: a continuous flow of mobile phase along the length of the plate, with

TABLE 6.1 NUMBER OF SOLVENT PASSES REQUIRED TO SEPARATE TWO SOLUTES 0.1 × LENGTH OF SUPPORT

R_f of Faster Moving Solute × 100	R_f of Slower Moving Solute × 100								
	Number of Solvent Passes for Required Separation								
	2	3	4	5	6	7	8	9-14	Impossible
30	23-21								24-29
29	22-20								23-28
28	21-19	22							23-27
27	20-18	21							22-26
26	19-17	20							21-25
25	18-16	19							20-24
24	17-15	18							19-23
23	16-14	17	18						19-22
22	15-13	16	17						18-21
21	14-12	15	16						17-20
20	13-11	15-14							16-19
19	13-10	14							15-18
18	12-9	13			14				15-17
17	11-8	12		13					14-16
16	10-7	11	12						13-15
15	9-6	10	11						12-14
14	8-5	9	10						11-13
13	7-4	8	9			10			11-12
12	6-3	7	8		9				10-11
11	5-2	6	7	8					9-10
10	4-1	6-5		7					8-9
9	3-1	5-4		6					7-8
8	2-1	4-3		5				6	7
7	1	3-2	4				5		6
6		2-1	3			4			5
5		1	2			3			4
4			1		2			3	
3					1			2	
2								1	

NOTES: To determine the number of passes for separation, locate the R_f of the faster moving component in the left-hand column. In the same row to the right of this figure locate the R_f of the slower moving component. The number of passes to separate these two components will then be found at the head of the column of the slower moving component.

From J. A. Thoma, J. Chromatogr., *12*, 441 (1963). Reproduced with permission of the author and Elsevier Publishing Company.

TABLE 6.2 NUMBER OF SOLVENT PASSES REQUIRED TO SEPARATE TWO SOLUTES 0.08 × LENGTH OF SUPPORT

R_f of Faster Moving Solute × 100	R_f of Slower Moving Solute × 100								
	Number of Solvent Passes for Required Separation								
	2	3	4	5	6	7	8	9-14	Impossible
30	24-23								25-29
29	23-22	24							25-28
28	22-21	23							24-27
27	21-20	22							23-26
26	20-19	21							22-25
25	19-18	20							21-24
24	18-17	19							20-23
23	17-16	18							19-22
22	17-15		18						19-21
21	16-14		17						18-20
20	15-13	16							17-19
29	14-12	15							16-18
18	13-11	14							15-17
17	12-10	13							14-16
16	11-9	12		13					14-15
15	10-8	11		12					13-14
14	9-7	10	11						12-13
13	8-6	9	10						11-12
12	7-5	8	9						10-11
11	6-4	7	8						9-10
10	5-3	6	7				8		9
9	4-2	5	6			7			8
8	3-1	4	5		6				7
7	2-1	4-3		5					6
6	1	3-2		4					5
5		2-1		3				4	
4		1		2				3	
3				1				2	

NOTE: See footnote to Table 6.1 for directions.

From J. A. Thoma, J. Chromatogr., *12*, 441 (1963). Reproduced with permission of the author and Elsevier Publishing Company.

TABLE 6.3   NUMBER OF SOLVENT PASSES REQUIRED TO SEPARATE TWO SOL-
UTES 0.06 × LENGTH OF SUPPORT

R_f of Faster Moving Solute × 100	R_f of Slower Moving Solute × 100								
	Number of Solvent Passes for Required Separation								
	2	3	4	5	6	7	8	9–14	Impossible
30	25	26							27–29
29	24	25							26–28
28	23	24							25–27
27	23–22								24–26
26	22–21								23–25
25	21–20								22–24
24	20–19								21–23
23	19–18								20–22
22	18–17		19						20–21
21	17–16		18						19–20
20	16–15		17						18–19
19	15–14	16							17–18
18	14–13	15							16–17
17	13–12	14							15–16
16	12–11	13							14–15
15	11–10	12							14–13
14	10–9	11				12			13
13	9–8	10			11				12
12	8–7	9		10					11
11	7–6	8		9					10
10	6–5	7	8						9
9	5–4	6	7						8
8	4–3	5	6						7
7	3–2	4	5						6
6	2–1	3	4					5	
5	1	2	3					4	
4		1	2				3		
3			1			2			
2						1			

NOTE: See footnote to Table 6.1 for directions.

From J. A. Thoma, J. Chromatogr., *12*, 441 (1963).   Reproduced with permission of the author and Elsevier Publishing Company.

150

TABLE 6.4 NUMBER OF SOLVENT PASSES REQUIRED TO SEPARATE TWO SOLUTES 0.04 × LENGTH OF SUPPORT

R_f of Faster Moving Solute × 100	R_f of Slower Moving Solute × 100								
	Number of Solvent Passes for Required Separation								
	2	3	4	5	6	7	8	9-14	Impossible
30	27								28-29
29	26								27-28
28	25								26-27
27	24								25-26
26	23								24-25
25	22								23-24
24	21								22-23
23	20								21-22
22	19								20-21
21	18		19						20
20	17		18						19
19	16	17							18
18	15	16							17
17	14	15							16
16	13	14							15
15	12	13							14
14	11	12							13
13	10	11							12
12	9	10							11
11	8	9							10
10	7	8						9	
9	6	7					8		
8	5	6			7				
7	4	5			6				
6	3	4			5				
5	2	3		4					
4	1	2		3					
3		1		2					
2				1					

NOTE: See footnote to Table 6.1 for directions.

From J. A. Thoma, J. Chromatogr., *12*, 441 (1963). Reproduced with permission of the author and Elsevier Publishing Company.

the mobile phase normally allowed to evaporate off as it reaches
the end of the plate. This continuous flow development with evaporation was first used by Mottier and Potterat (9). The evaporation was encouraged by raising the lid of the chromatography tank
on one edge. Van den Eijnden (16) has used a chromatography tank
that is not as tall as a normal tank (18.5 cm vs. 23 cm) to allow
a normal sized TLC plate to stick out through the halves of a
split lid. As the mobile phase evaporates off the end of the
plate, a continuous flow of phase is maintained up the layer. Although Van den Eijnden, in the title to his article (16), calls
this "a new technique," an identical split-lid procedure was used
by Truter some ten years earlier (17). Other workers (18) used a
piece of Saran film for their slotted lid.

Regis Chemical Co. markets a simple apparatus for continuous
development, the SB/CD chamber. SB/CD is the abbreviation for
short bed continuous development. The apparatus is shown in Figure 6.5. It consists of a shallow (2 cm) glass dish about 24 cm
long and 10.5 cm wide with a ground glass lip, a lid, and two
Teflon cover wings. On the bottom of the tank (dish) are four
raised glass ridges running nearly the entire width. Add the mobile phase (about 75 mL) to the chamber, and in the chamber against
one of the ridges, place the plate to which the samples have been
applied. Five different positions are available. A portion of
the plate will be projecting out of the top, and the lid and Teflon
wings are placed in position sealing the chamber and giving a saturated system. When the solvent front migrates up the layer and

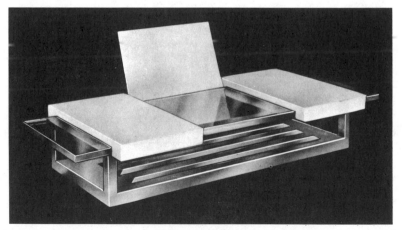

Figure 6.5. Regis apparatus for continuous development. Courtesy
of Regis Chemical Co.

emerges from the chamber, the mobile phase evaporates. This provides a continuous rapid flow of the mobile phase up the plate.

The SB-CD chamber is supposed to offer the following advantages: (a) increased spot separation through continuous development; (b) simplified solvent selection because of the utilization
of the high separation ability of low polarity solvents without
the associated problem of reduced R_f; (c) rapid development; (d)
increased sensitivity because spot diffusion is reduced through
smaller development distances; (e) increased reproducibility.

Brenner and Niederwieser (19) devised a chamber for the continuous development of a thin layer plate in the horizontal position. It is often called the BN chamber. The sample to be separated is first applied to the plate and the apparatus is assembled.
The mobile phase is passed to the origin end of the plate by means
of a paper wick from a reservoir. The plate is in contact with a
cooling block that circulates cold water to prevent vapor condensation in the chamber. The open end of the chamber, through which
the evaporation occurs, may be heated to further the evaporation
of high boiling solvents. The apparatus can be used horizontally
for normal or multiple development if the open end is sealed off
to allow an equilibrium to become established. The BN chamber is
supplied commercially by Brinkmann Instruments, Inc.

6.3.4 Two-Dimensional Development

This developing technique is the application of normal, multiple,
or stepwise development in two dimensions. A very versatile method, it has been widely used in paper chromatography and with certain compound areas in TLC. The best-known applications are those
for the separation of clinically important carbohydrates (20-23)
and amino acids (24). Some of the two-dimensional techniques for
amino acids involve an electrophoretic separation in one dimension
and a TLC separation in the second dimension. Kirchner and co-
workers (25) were the first to apply this method to TLC.

The single sample to be separated is spotted in one corner of
a plate and developed in the normal way for a fixed distance. The
plate is removed from the developing chamber and dried. The plate
is turned 90° and placed into a chamber containing a different mobile phase so that it develops in a direction perpendicular to that
of the first development. The line of the partially resolved sample components from the first development forms the origin for the
second development. This procedure is illustrated in Figure 6.6.
The initial sample application should be done carefully, so that
it is not too near to the edge. A distance of 2.5-3 cm is satisfactory; otherwise the line on which the developed spots lie may
be below the solvent level when the plate is turned to be developed
the second time. Versatility of the technique lies in the second

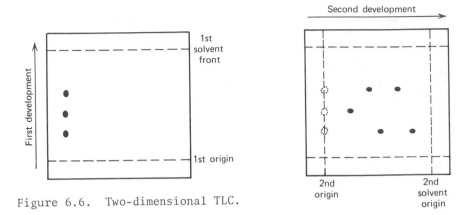

Figure 6.6.   Two-dimensional TLC.

development with a different mobile phase and in the opportunity to treat or modify the layer and samples before the second development. Stahl first demonstrated such versatility with his separation-reaction-separation (SRS) procedure, in which he carried out a chemical reaction on the plate before the second development (26). Known standards can be applied to the second origin before the second development to use for comparison and calculation of R_f values after development and visualization.

6.4   ADDITIONAL DEVELOPMENT TECHNIQUES

A few other development techniques have been reported in the literature once or twice but are seldom used, primarily because most of them are impractical in operation.   It is easier to substitute a more practical technique.

6.4.1   Centrifugal Chromatography

First applied to paper chromatography, centrifugal chromatography has also been used in TLC (27,28).   Here the TLC plates are made from circular glass or aluminum with a hole in the center to bolt onto the centrifuge.   The mobile phase is delivered from a reservoir to the plate at a preset rate while the plate spins at 500-700 rpm.   In one instance (27), the development time was decreased from 35 min to 10 min.

6.4.2   Gradient Elution Development

Gradient elution development is commonly used in column chromatography and occasionally in TLC.   In this form of development, the

solvent composition of the mobile phase is changed either in steps
or continuously. Wieland and Determann designed a development
chamber for gradient elution TLC (29). It does not appear to be
commercially available. The thin layer plate is supported on a
perforated tray near the bottom of the chamber, which contains a
magnetic stirrer bar. The mobile phase is delivered beneath the
perforated tray by means of a connecting tube, and an overflow out-
let about a centimeter above the perforated tray keeps the volume
of mobile phase constant. Wieland and Determann used this apparatus to continuously increase the strength of a buffer solution (29),
as did Hofmann (30). It could be used to change the solvent composition or pH of the mobile phase or to effect any other change
necessary.

Niederwieser and Honegger have written a review on gradient
techniques in thin layer chromatography (4). It is possible to use
a gradient in the mobile phase, as discussed above, or a gradient
in the stationary phase, as Stahl and Dumont have done with defined
pH layers (31,32). They point out that a gradient layer offers
three different directions for mobile phase flow. Chromatograms
can be developed with or in the gradient, against the gradient, or
across the gradient. Stahl and Dumont have developed a plate-
coating apparatus for the preparation of the gradient layer. This
gradient-mixer (GM) applicator is manufactured by Desaga and is
distributed by Brinkmann Instruments, Inc. The researcher interested in gradient layers should refer to these papers.

Gradient TLC has been used to separate lipid mixtures having
a wide range of polarity (33,34). Reference 34 describes a simplified gradient procedure requiring only the addition of a glass
solvent trough to a normal-sized TLC developing tank. The paper-
lined developing tank is first saturated with a volatile nonpolar
solvent, referred to as the atmospheric solvent. The trough is
placed in the tank and filled with a relatively polar solvent,
called the trough solvent. The plate to be developed is placed in
the trough. As the trough solvent migrates up the plate and evaporates, it is slowly replaced by atmospheric solvent. To produce
gradients that are steep enough for lipid analysis, both evaporation and influx are necessary. The more polar sample components
will be retained on the lower portion of the plate as the polarity
of the developing solvent decreases.

Neutral and total lipids have been separated by this procedure. Many combinations of polar trough solvent and nonpolar atmospheric solvent will produce a suitable gradient for the separation
of neutral lipids. Cyclohexane, toluene and alkanes with five to
eight carbon atoms are good atmospheric solvents. Suitable trough
solvents may be made by mixing two or more polar solvents or a
nonpolar with a polar solvent. The author states that this method
has been successfully combined with reversed-phase TLC and with

argentation TLC. In reversed-phase TLC the trough contains the
less polar solvent. A stationary phase of 50% paraffin oil on
kieselguhr with 2:1 butanone-acetonitrile as the trough solvent and
methanol as the atmospheric solvent has proved successful.
The following practical points may be made:

1. The trough solvent should be volatile enough to evaporate
off the plate but not so volatile that it escapes completely from
the trough.

2. Temperature should be closely controlled.

3. The greater the polarity difference between the trough
system and the atmospheric system, the greater will be the gradient.

4. The steepness of the gradient is affected by the rate of
development. This depends on the mesh size of the sorbent. Once
a suitable sorbent has been found for a given separation, it should
be retained.

5. Plate layer thickness should be carefully controlled because of its affect on influx and efflux of the mobile-phase solvents.

6.4.3 Partition Thin Layer Chromatography

Partition TLC is of form of liquid-liquid chromatography in which
the "solid" coated on the plate is impregnated with the liquid
phase before chromatography occurs, and the separation is caused
by a partitioning effect between the liquid phase on the sorbent
and the mobile phase. The impregnated liquid phase is called the
stationary phase. To prevent this phase from changing during development, the mobile phase is usually saturated with the stationary phase.

This is the primary separation process in paper chromatography, where, unless it is intentionally displaced by another liquid,
water from the atmosphere is the stationary phase. This process
occurs on cellulose thin-layer plates, and many of the development
systems and visualization procedures employed in paper chromatography may be directly transposed to cellulose TLC. Cellulose TLC
has some advantages compared to paper chromatography, which include ease of handling, smaller space requirements for apparatus,
smaller sample size, generally increased detection sensitivity,
and faster development. These reasons account for the greater emphasis on TLC compared to paper chromatography.

There are a number of ways of impregnating the sorbent with
the stationary phase. Using a hanging TLC chamber the plate can
be equilibrated with the vapor of a mixture of the stationary
phase. The stationary phase may also be sprayed on the plate, either by itself or in a volatile solvent as a vehicle. The plate
may be dipped in such a solution or allowed to develop in it be-

fore the samples are applied. Any volatile solvent is allowed to
evaporate before sample application. In normal partition TLC, the
stationary phase is a polar substance such as water or formamide
and the mobile phase is relatively nonpolar, such as cyclohexane
or heptane. Scott has reviewed the theoretical and practical aspects of the stationary phase in TLC (35).

Table 5.3 (in Chapter 5) presents a mixotropic solvent series
that will serve as a guide for the selection of stationary and mobile phases for partition TLC. (See also the appropriate section
in Chapter 5.) The polar solvents at the top of the table may be
used as stationary phases and the solvents further down, which are
less polar, as mobile phases. Remember that the closer two solvents are in the series, the more similar are their solubilities
with a given stationary phase and the more similar will be the R_f
values of the separated substances.

6.4.4 Reversed-Phase Partition Development

Reversed-phase development employs a relatively nonpolar stationary phase, which makes the layer hydrophobic, rather than hydrophilic as in normal partition TLC. The phase would be chosen from
the lower portion of Table 5.3. The mobile phase is relatively
polar and may be water or an alcohol. Most commonly used TLC sorbents such as cellulose, silica gel, and kieselguhr have been used
for reversed-phase development. Starch has also been used (36).
This technique is generally useful for the separation of polar
compounds, which often remain at the origin during conventional
development. With reversed-phase development, it is usually the
nonpolar compounds that remain near the origin.

Silica gel treated with dichlorodimethylsilane has been used
for the separation of fatty acids as the free compounds, the
esters, and the hydroxamates (37). Many suppliers offer
plates that are precoated with silanized silica gel 60, both with
and without fluorescent indicator. These plates should prove useful in the separation of long-chain molecules such as triglycer-
ides, cholesterol esters, fatty acids and esters, and carotinoids.
For development of the silanized silica gel plates, it is best to
use solvents of medium to high polarity such as acetonitrile, acetone, ethanol, methanol, acetic acid, formic acid, and water in
various combinations. Higher R_f values will result as the amount
of lipophilic solvent in the eluent is increased. The layer behaves as a macroporous silica gel or as a deactivated medium porosity silica gel if it is used for adsorption chromatography.

The following points should be observed when using silanized
silica gel plates (38):

1. Eluents (stationary or mobile) containing more than 80%
water should not be used.
2. When the plate is to be impregnated with a stationary
phase containing more than 60% water, it is suggested that the
plate be dried at 120° before the chromatography. This makes the
layer more resistant to water. Prolonged heating should be avoided, to prevent driving the silane groups off the layer.
3. The use of polar eluents results in longer development
times than those normally found with regular silica gel and a less
polar eluent.
4. Eluents containing alkaline or ammoniacal components are
usable.
5. Eluents containing mineral acid may release minute
amounts of polymeric binding agent from a precoated plate. This
may interfere with separation or visualization. It is advisable
to develop the plate with the acid-containing eluent and dry the
plate at 120° for 15 min prior to the chromatography of samples.
6. The fluorescent indicator (F-254) is sensitive to acid,
which will reduce the fluorescence. Avoid using acid-containing
eluent whenever indicating plates are necessary for visualization.

An apparatus for small-scale reversed-phase TLC that may
prove useful has been developed (39).

6.5 DEVELOPMENT OF SEPHADEX PLATES

After the plate is spread with the gel, it is immediately put into
the thin layer gel (TLG) chamber for equilibration, *before* the
sample is applied. The layer should be connected to the mobile
phase reservoirs at both ends by means of filter paper bridges,
which should be thoroughly saturated with the eluent before they
are connected. The paper bridges should be thick and porous to
allow a high flow of solvent. Paper equivalent to Whatman 3MM is
most suitable.
Equilibration is carried out with a flow maintained by at
least a 10° angle, for a minimum of 12 hr. The major function of
equilibration in this process is to normalize the ratio between
the volumes of the stationary and mobile phases. If the plate has
been prepared with a slurry that is too thin, the layer weight
will decrease during equilibration, indicating a loss of mobile
phase. If the slurry is too thick, the weight of the layer increases.
After equilibration the mobile-phase flow is temporarily
stopped to allow the sample to be applied. The plate must be horizontal during this application, otherwise elongated sample spots

will result. If the gel layer is disturbed during application, the flow of the mobile phase will be distorted.

The concentration of the applied sample is very important in TLG, as it is in TLC. A high concentration (viscosity) of applied sample will cause zone distortion and influence the migration rate. If different samples have different viscosities, the migration of the substances relative to each other will vary. Distortion of each sample zone leads to elongated spots, which make it difficult to precisely compare and calculate R_f values and migration distances. Since the flow at the periphery of a concentrated spot area is faster than at the center, this outer material will move ahead of the major portion of the spot zone, producing a leading edge and an elongated spot. It is recommended that protein concentration not exceed 20 mg/mL.

After sample application, flow can be resumed for development of the plate. The flow is maintained by allowing a difference in the levels of the mobile phase between the top and bottom reservoirs. This is done by tilting the plate or, in the case of horizontal operation, by adjusting the levels in the two reservoirs. The TLG apparatus is designed so that the plate may be developed horizontally or at angles between 10° and 45° in 5° intervals. The flow rate is dependent on the angle, the layer thickness, and the type of gel in the layer.

A flow rate of about 3 cm/hr, obtained at an angle of 10-20° is generally most suitable. Higher flow rates can cause broadening of the separated zones in the direction of the development, with a resulting decrease in resolution. The superfine grades of Sephadex G-100, G-150, and G-200 will flow between 2.5 and 5 cm/hr when coated in 0.6 mm layers and set at an angle between 10 and 30°. Superfine grades of G-50 and G-75 will flow 3 cm/hr at even a 10° angle.

To follow the development process, it is convenient to apply a marker substance, which may be colored, in a number of positions across the plate. This substance will show any variation in solvent flow and can also be used for calculation of relative migration distance. Suitable marker substances include myoglobin, cytochrome c, hemoglobin, Bromphenol blue, labeled albumin, and Amido Black labeled serum.

6.6   DIAGNOSING A DEVELOPED CHROMATOGRAM

Any of five or six major problems may be encountered if development conditions have not been optimum for a given separation. Each problem is usually dependent upon three or four major factors. These problems and the factors controlling them will be discussed here.

Beginners often find that no separation has taken place. This may be due to the use of an incorrect sorbent, to sample zones that were not dry before development, to the sample being lost during preparation, or to the polar component of the mobile phase not being sufficiently fresh.

A distorted solvent front may be due to the movement of the chamber after development has started, or the polar mobile phase component may not be highly enough concentrated or may not be fresh.

A distorted separation, with individual known standards not matching corresponding substances in mixtures, for example, may be due to inadequate mixing of the mobile-phase components before their addition to the development chamber. Other causes include nonalignment of sample zones at the origin when application is being carried out, nonequilibrium in the system, or uneven placement of the plate in the mobile phase.

Undermigration of samples, where R_f values are all low and incomplete or inadequate separation has occurred, may be due to the mobile phase not being polar enough. An inadequate volume of mobile phase, or one that is not reasonably fresh, can also cause this problem. Nonequilibrium may also be a causative factor.

Overmigration of sample zones with resultant high R_f values and inadequate separation often results when the operator does not remove the plate from the chamber after a given development distance. The second most common cause is a mobile phase that is too polar.

REFERENCES

1. "Thin-Layer Gel Filtration with the Pharmacia TLC Apparatus," Pharmacia Fine Chemicals, Inc., Uppsala, Sweden, 1971.
2. D. C. Abbott, H. Egan, E. W. Hammond, and J. Thomson, Analyst, 89, 480 (1964).
3. M. Brenner and A. Niederwieser, Experientia (Basle), 16, 378 (1960).
4. A. Niederwieser and C. C. Honegger, in Advances in Chromatography, Vol. 2, J. C. Giddings and R. A. Keller, Eds., Marcel Dekker, New York, 1966, p. 125.
5. A. Pittoni and P. L. Sussi, J. Chromatogr., 32, 422 (1968).
6. R. A. deZeeuw, J. Chromatogr., 32, 43 (1968).
7. R. A. deZeeuw, J. Chromatogr., 33, 222 (1968).
8. J. H. Dhont, J. Chromatogr., 90, 203 (1974).
9. M. Mottier and M. Potterat, Anal. Chim. Acta, 13, 46 (1955).
10. A. Jeans, C. S. Wise, and R. J. Dimler, Anal. Chem., 23, 415 (1951).
11. J. A. Thoma, Anal. Chem., 35, 214 (1963).

12. J. A. Thoma, J. Chromatogr., *12*, 441 (1963).
13. L. Starka and R. Hampl, J. Chromatogr., *12*, 347 (1963).
14. E. Stahl, Arch. Pharm., *292/64*, 411 (1959).
15. E. Stahl and V. Kaltenbach, J. Chromatogr., 5, 458 (1961).
16. De. H. Van den Eijnden, Anal. Biochem., *57*, 321 (1974).
17. E. V. Truter, J. Chromatogr., *14*, 57 (1964).
18. L. M. Libbey and E. A. Day, J. Chromatogr., *14*, 273 (1964).
19. M. Brenner and A. Niederwieser, Experientia, *17*, 237 (1961).
20. R. L. Jolley and M. L. Freeman, Clin. Chem., *14*, 538 (1968).
21. R. L. Jolley and C. D. Scott, Clin, Chem., *16*, 687 (1970).
22. A. S. Saini, J. Chromatogr., *61*, 378 (1971).
23. M. Ghebregzabher, S. Rufini, G. Ciuffini, and M. Lato, J. Chromatogr., *95*, 51 (1974).
24. R. M. Scott, *Clinical Analysis by Thin-Layer Chromatography Techniques,* Ann Arbor-Humphrey, Ann Arbor, 1969, p. 84.
25. J. G. Kirchner, J. M. Miller, and G. J. Keller, Anal. Chem., *23*, 420 (1951).
26. E. Stahl. Arch. Pharmacol., *293/65*, 531 (1960).
27. B. P. Korzun and S. Brody, J. Pharm. Sci., *53*, 454 (1964).
28. J. Rosmus, M. Pavlicek, and Z. Deyl, in *Thin-Layer Chromatography,* G. B. Marini-Bettolo, Ed., Elsevier, Amsterdam, 1964, p. 119.
29. T. Wieland and X. Determann, Experientia, *18*, 431 (1962).
30. A. F. Hofmann, Biochem. Biophys. Acta, *60*, 458 (1962).
31. E. Stahl and E. Dumont, Talanta, *16*, 657 (1969).
32. E. Stahl and E. Dumont, J. Chromatogr. Sci., 7, 517 (1969).
33. L. R. Snyder and D. L. Saunders, J. Chromatogr., *44*, 1 (1969).
34. G. E. Tarr, J. Chromatogr., *52*, 357 (1970).
35. R. M. Scott, J. Chromatogr. Sci., *11*, 129 (1973).
36. J. Davidek, in *Thin-Layer Chromatography,* G. B. Marini-Bettolo, Ed., Elsevier, Amsterdam, 1964, p. 117.
37. D. Heusser, J. Chromatogr., *33*, 62 (1968).
38. M. Gurkin, personal communication.
39. G. De Vries and V. A. Th. Brinkman, J. Chromatogr., *64*, 374 (1972).

CHAPTER 7
Visualization Procedures

7.1 INTRODUCTION

An ideal visualization or location procedure for thin layer chromatograms should be able to do the following:

1. Visualize microgram quantities of separated substances.
2. Give a visualized area that is firm in its appearance.
3. Give a satisfactory contrast between the visualized area and the background.
4. Give a visualized area that is stable enough and suitable enough for quantitative measurement, if desired.

Very few visualizing agents will do all of these, but it is generally possible to find one that will be suitable for the substances being studied.

7.2 TYPES OF METHODS

Methods for the location (visualization) of substances are of two major types: physical methods, such as the use of ultraviolet (UV) light; and chemical methods, such as reaction with sulfuric acid to produce a brown or black charred area. Two further sub-

classifications are made within each of these two major types:
destructive methods are those that permanently change the chemical
identity of the substance being visualized, for example, sulfuric
acid charring; and nondestructive methods, which do not produce
any permanent change in the chemical identity of the substance being visualized. Visualization with UV light is nondestructive
with most substances; ultraviolet light can rearrange some steroid
and vitamin molecules. Long-wave UV light (366 nm) is used to
identify substances with either inherent or reagent induced fluorescence. Short-wave UV light (254 nm) renders visible substances
which absorb at 254 nm. Provided the plate sorbent contains a
fluorescent indicator, these substances appear as dark spots on a
bright background.

The use of iodine vapor is an example of a nondestructive
chemical method (1) and the colored complex formed is generally
nonpermanent and reversible, which leaves the substance unchanged
chemically after the process has occurred. This type of visualization reaction can be beneficial when it is desired to follow
multi- or two-dimensional development procedures, for instance, or
when it is necessary to isolate a substance unchanged from the
plate. The substance can be visualized, its location marked on
the layer physically with a scribe such as a syringe needle or
spatula, the reaction can be allowed to subside, and the substance may then be scraped from the plate and eluted from the layer.

7.3 APPARATUS

The majority of visualization procedures may be carried out with
only a few basic pieces of apparatus that are easily afforded by
most laboratories.

For visualization by ultraviolet (UV) light, a number of
long-wave and short-wave lamps are available. Both wavelengths
may be obtained from portable lamps, or from lamps fixed in a
light-tight cabinet, which becomes a convenient, small-scale dark-
room for those laboratories not having such a room to work in.
Figure 7.1 shows such a cabinet in use. For best results, particularly in the submicrogram range, it is necessary to have such
a dark area or dark-room available for ultraviolet visualization.
This cabinet contains both long- and short-wave lamps as well as
white light. The area set aside for such a purpose should be kept
free of dust at all times, and as chemically clean as possible, so
that these forms of contamination may be kept to a minimum.

Visualization reagents are usually placed in contact with the
substances being detected by spraying. Dipping procedures and
impregnation of the plate layer with a visualization reagent be-

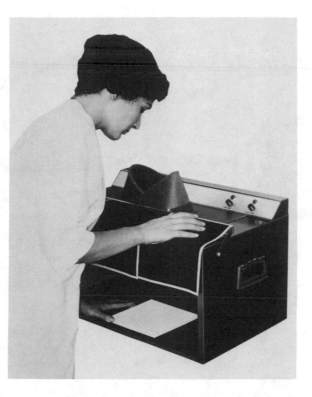

Figure 7.1. Ultraviolet light viewing cabinet for the visualization of fluorescent substances on chromatograms. Courtesy of
Brinkmann Instruments, Inc.

fore development (2-4) are also used. More will be said about
these later. A number of different spray apparatuses are commercially available. Glass atomizers connected to compressed air
are used, when the air is available. Figure 7.2 shows one such
glass atomizer with a rubber bulb attachment, which can be used
when compressed air is not available.

Spray guns using cans of propellant are commercially available. Usually a 4-oz wide-mouthed glass jar is used to contain
the reagent and is screwed onto the spray head. This makes it a
convenient sprayer. The jar used is a standard jar that may be
purchased separately with caps. Such jars may be used to store
the different reagents until they are needed and can then be
attached directly to the spray head without transfer and risk of
contamination. The spray head is thoroughly cleaned after use
with distilled water or other suitable solvent.

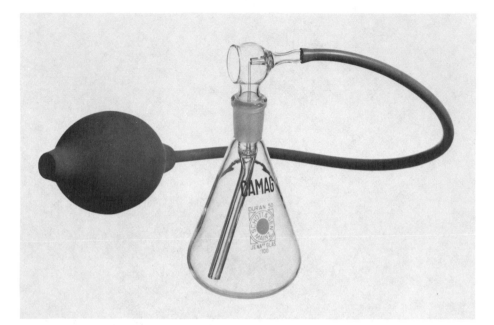

Figure 7.2. Atomizer spray bottle for applying detection reagents
to chromatograms. Courtesy of Camag, Inc.

Some of the thin layer chromatography suppliers offer a collection of prepared, commonly used reagents in aerosol spray cans.
These are convenient, but often the spray nozzles do not produce
a mist fine enough for uniform coverage.

It is convenient to set aside one section of a fume hood to
be used for the plates. In order to avoid spraying everything in
the hood, a good-sized cardboard box with its top and one side removed may be used to contain the plate while it is being sprayed.
It is best to line this box with heavy filter or chromatography
paper to absorb excess spray and prolong the life of the box. The
plate to be sprayed is simply stood against the back wall of the
box; the open top allows fumes to go up the hood, and the sides
contain the spray within the box. The most inexpensive way to obtain a box is to locate one being discarded from a store. A number of TLC suppliers have cardboard or formed plastic spray boxes
available.

When it is preferable for the plate to be dipped into the
visualization reagent, as in the recommended procedure in quanti-

tative thin layer chromatography (5), then it is necessary to have
a container large enough to completely immerse the plate. A regular TLC tank may be used, but a large volume of reagent is necessary. More convenient containers that do not have such a large
volume are glass baking dishes and photographic developing trays
of a size suitable to hold the particular size plate being used.
Two different size trays, one for 10×20 cm plates, the other for
20×20 cm plates, may be kept on hand. It will be necessary to
prepare 200-500 mL of reagent to adequately cover the plate. The
preferred technique for dipping is to fill the container with the
reagent before the plate is added. Placing a plate into the container first and then adding the reagent usually produces bad results. The reagent will unevenly penetrate the layer, producing
a nonuniform distribution that often results in poor visualization
and/or quantitation. If the plate layer is very soft, or the reagent contains too much water, the layer may bubble or flake off
when the reagent is added. Kontes Glass Co. supplies a small-
volume dipping tank for TLC plates. It is shown in Figure 7.3.

If the plate is impregnated with the visualization reagent
before sample application, the choice of the mobile phase for development must be made very carefully. An incorrect mobile phase
will wash out the reagent during development by carrying it up the
plate with the solvent front. Because of this, the plate is usually dipped after development.

Heating is often necessary to speed the visualization reaction. Most often this can be done at 100-115° for a short period of time such as 10-15 min. Occasionally, higher temperatures
and longer periods of time are needed to bring about reaction,
such as in some charring techniques. When this is the case, plate
layers containing an inorganic binder such as gypsum, rather than
those made with organic binder, should be used. Organic binders

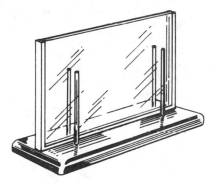

Figure 7.3. Small-volume tank used to dip TLC plates in visualization procedures. Courtesy of Kontes Glass Co.

will char. Since gypsum loses its binding capabilities at 130°,
care must be exercised when heating layers containing gypsum (such
as silica gel G) at temperatures this high.

The plate can be conveniently heated in a small oven placed
near the spraying/dipping location. This oven should be used only
for chemical work and not for delicate operations such as drying
glassware because of the fumes and residues trapped within the
oven. Camag, Inc., offers a hot plate heating apparatus for heating plates uniformly. It is shown in Figure 7.4.

7.4 REVERSIBLE COLOR REACTIONS

7.4.1 Iodine

It is often advantageous to visualize a substance by a method that
will not permanently alter the substance being visualized. Such
methods would be nondestructive, and many are reversible chemical
reactions in which a colored product is formed to allow location
of the substance. Often, because of instability, the addition of
heat, or other conditions, the visualization reaction may reverse
to the condition prior to treatment and thus allow subsequent
operations to be carried out with the plate. Such necessary operations may include chromatographic development in a second, or
the same, dimension, elution of the substance from the layer, or
in situ visualization reaction or quantitation.

Elemental Iodine is a simple, **rapid**, inexpensive, sensitive
and generally nondestructive and reversible visualization reagent
that will stain substances on a chromatogram more readily than the
background. The easiest way to perform Iodine visualization is to
place a few crystals of Iodine in a covered chromatography tank.
The volatile nature of Iodine is such that the tank is readily
filled with purple vapor. The chromatogram to be visualized is
generally dried of any developing solvents and placed into the
iodine-containing tank for a given period of time. Located substances normally turn light brown. The plate is lifted out of the
tank and, if desired, the areas may be physically marked with a
scribe for future reference; otherwise the areas will normally
fade, losing their color completely in a few minutes.

Mangold and Malins (6-8) have demonstrated the use of Iodine
for the visualization of lipids. The reversible reaction has also
been demonstrated for amino acid derivatives (9), psychoactive
drugs (10), indoles (11), and steroids (12). Many additional compound types have been located with this technique; Barrett reviewed these (13).

Iodine may react with some compounds. Sensitive ^{131}I techniques (14) and GLC of material extracted from plates (15) have

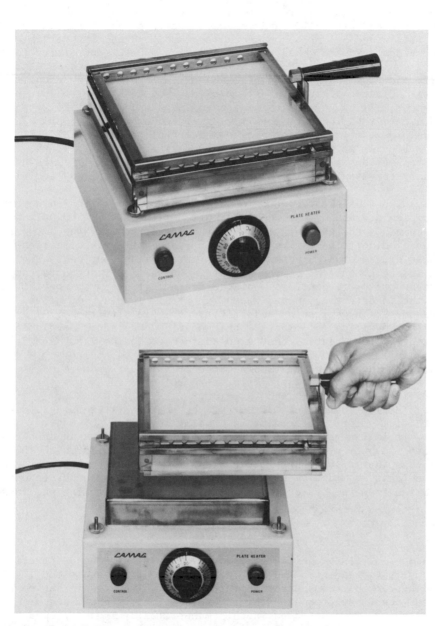

Figure 7.4 Hot plate heating apparatus for promoting reactions on TLC plates. Courtesy of Camag, Inc.

been used to show the iodination of olefinic double bonds in poly-
unsaturated fatty acids and esters. Lecithin has also been shown
to retain Iodine for a considerable time (16), and irreversible
reactions may occur when compounds having a high sensitivity toward Iodine are being detected. Phenolic steroids such as estrone
(17) and drugs such as morphine and oxymorphone (13) apparently
form iodinated derivatives after exposure. In both of these instances, silica gel plates were being used and it is likely that
this catalyzed iodination (17). Prolonged exposure (18 hr) of
alkaloids to Iodine vapor produced numerous products (18).

The Iodine reagent may be sprayed on the plates using a di-
        Because of the possibility of undesirable reactions, the
plate should be exposed to the Iodine vapor as briefly as is necessary to visualize the substances sought; often only 30 sec will
suffice.

        High sensitivity detection is often possible with Iodine.
Examples are 0.5 μg estrone (17), 0.1 μg lipid (19), and 0.1 μg
of codeine and morphone (10). In the latter report, the authors
found Iodine to be equivalent or superior in the visualization of
these drugs compared with the other three visualization reagents
including ammoniacal silver nitrate, potassium permanganate, and
Dragendorff's reagent. Ammoniacal silver nitrate was superior to
Iodine in detecting chlordiazepoxide at all levels. Other researchers, investigating sensitivity of lipid detection (19), impregnated silica gel plates with Rhodamine 6G. After developing
and drying the plates, they exposed them to Iodine vapor for 2-5
min, followed by UV illumination to reveal intense blue spots.
This combined Iodine-UV fluorescence techniques was more sensitive
than either method alone. Milborrow (20) also found greater sensitivity with the combined technique.

        The Iodine reagent may be sprayed on the plates using a di-
lute (1%) solution of Iodine in a volatile, inert solvent.

        It is interesting to note that Bromine vapor was an unsatis-
factory substitute for Iodine in the detection of lipids on TLC
(16). Other researchers (21) working in paper chromatography
noted that bromination occurred, with the formation of hydrogen
bromide from the hydrocarbons being visualized. Bromine was thus
a destructive reagent in this instance.

        7.4.2 Water as a Visualization Agent

A number of researchers have used water spray as a nondestructive
visualization agent. The sprayed plate is held against the light,
and hydrophobic compounds are revealed as white areas areas on a
translucent background. Steroids such as cholestanone and α-
cholestanol were detected in this manner (22). A "heavy spray"
technique consists of saturating the plate with spray and then allowing it to dry slowly. Sometimes, clearer zones are obtained in

in this fashion. This method has been used for cyclohexanols (23) and hydrocarbons (24). Water visualization has also been used for bile acids (25, 26).

Often, while pulling a wet, developed plate out of a chromatography tank, it will be noted that some separated areas are visible. As the plate dries, however, these areas are not as visible. If they are rewetted with solvent, they reappear. Lower limits of detection are often obtained by this method, even after visualizing a substance by spraying it with a reactive, destructive reagent.

7.4.3 pH Indicators

Commonly used pH indicators such as Bromocresol Green and Bromphenol Blue are useful for the detection of acidic and basic compounds on TLC plates, through a reversible acid-base equilibrium. The polar nature of the indicator should be such that it can be removed from the sample area after visualization, if this is desired.

A dilute (0.01-1.0%) aqueous or aqueous alcohol solution of the indicator is made, and its pH is adjusted to be close to the endpoint before use. Bromphenol Blue (0.5% in 0.2% aqueous citric acid) produces yellow areas on a blue background with aliphatic carboxylic acids (27). Methyl red may also be used (28). Bromocresol Green (0.3% in 80% aqueous methanol containing 8 drops 30% NaOH per 100 mL solution) produces yellow spots on a green background with aliphatic carboxylic acids (29, 30).

7.5 REAGENTS FOR VISUALIZATION

The preceding section dealt with reagents for visualization that are essentially nondestructive to the substances being visualized. The present section will deal with the majority of visualization reagents used, those that are "destructive" in their action, chemically changing the substances being visualized.

To review, the commonly used procedure for visualization is to spray the chromatogram after it is developed with the detection reagent. Other methods include dipping the plate in the reagent before the sample is applied or after the plate is developed. The dipping procedures are generally recommended when sensitive *in situ* quantitation is to be performed (5).

There are a few reagents used generally for the visualization of organic compounds. The most widely used is sulfuric acid (31), which sometimes reacts immediately or upon heating to form brown or black charred areas. Often a 50% aqueous solution is used, rather than the concentrated acid. For very unreactive substances,

5% nitric acid can be added to the concentrated acid to increase its oxidizing capabilities.

Solutions of sulfuric acid-acetic anhydride in ratios of 1:4 (32) and 5:95 (33) have been used. This is the Liebermann-Burchard reagent. Touchstone et al. (4) have impregnated silica gel thin layer plates with sulfuric acid (4% v/v) in MeOH, by dipping prior to sample application and development. These plates were used for the *in situ* quantitation of steroids and fatty acids. Charring is done by heating at 140° for 20 min after development. Very similar results are obtained by impregnating the gel plates with ammonium bisulfate (3) or ammonium sulfate.

Phosphomolybdic acid, a reagent suitable for many organics, has also been impregnated into silica gel before sample application. These plates were used for the *in situ* quantitation of steroids (2).

In the following pages, reagents and test procedures are listed according to the compound class being detected, with universal reagents and organic nondestructive reagents listed first. The remainder are grouped alphabetically by compound class. These were compiled from many sources (34-38).

Universal Reagents

1. *Sulfuric Acid*

    Reagent: Concentrated or 50% aqueous.
    Procedure: Spray, heat at 110-120° for a few minutes
    Results: Brown to black spots for most organics; often the
                charred products fluoresce.

2. *Sulfuric Acid-Acetic Anhydride*

    Reagent: 1 part acid; 3 parts anhydride.
    Procedure: Spray, heat at 110-120°.
    Results: Brown to black charred areas.

3. *Sulfuric Acid-Sodium Dichromate*

    Reagent: Dissolve 3 g sodium dichromate in 20 mL H_2O,
                dilute with 10 mL conc. H_2SO_4.
    Procedure: Spray, heat at 110°.
    Results: Brown to black charred areas.

4. *Sulfuric Acid-Nitric Acid*

    Reagent: 1:1 solution.

Procedure: Spray, heat at 110°.
Results: Brown to black charred areas.

Organic Acid, Nondestructive Reagents

5. *Bromcresol Green*

Reagent: 0.3% bromcresol green in 1:4 H_2O:methanol, con-
              taining 8 drops 30% NaOH/100 mL.
Procedure: Spray.
Results: Yellow areas for aliphatic carboxylic acids on a
              green background.

6. *Bromcresol Purple*

Reagent: 0.04 g bromcresol purple in 100 mL 50% C_2H_5OH.
              Adjust to pH 10 with alkali.
Procedure: Spray.
Results: Yellow areas on blue background with dicarboxylic
              acids.

7. *Bromphenol Blue*

Reagent: 0.5% bromphenol blue in 0.2% aqueous citric acid
Procedure: Spray.
Results: Yellow areas for aliphatic carboxylic acids on a
              blue background.

Alcohols

8. *Cericammonium Nitrate*

Reagent: 6% cericammonium nitrate in 2N HNO_3.
Procedure: Dry plate 5 min at 105°, cool before spraying.
Results: Polyalcohols produce brown areas on yellow back
              ground.

9. *2,2-Diphenylpicrylhydrazyl*

Reagent: 15 mg reagent in 25 mL $CHCl_3$.
Procedure: Spray. Heat at 110° for 5-10 min.
Results: Alcohols produce yellow spots on purple back-
              ground. Also used for aldehydes and ketones.

10. *4,4'-Methylenebis(N,N-dimethylaniline)*

    Reagent: Solution (a): 0.25% reagent in acetone.
                    Solution (b): 1% ammonium ceric nitrate in 0.2N
                    HNO_3.
                    Mix equal portions of each solution before spraying.
    Procedure: Spray with 1:1 solution, heat at 105° for 5 min.
    Results: Light blue spots on a blue background.

11. *Vanadium Oxinate*

    Reagent: Dissolve 0.4 g β-hydroxyquinoline in 50 mL of 1:1
                    xylene-glacial acetic acid. Heat on a steam bath
                    to 55°. Add 0.2 g ammonium vanadate while stir-
                    ring. Cool and filter. Solution is satble 3
                    days.
    Procedure: Spray.
    Results: Alcohols produce light red spots on a blue-black
                    background.

12. *Vanillin-Sulfuric Acid*

    Reagent: Dissolve 3 g vanillin in 100 mL abs. C_2H_5OH.
                    Add 0.5 mL $H_2S)_4$, stirring well.
    Procedure: Spray and heat at 120°.
    Results: Higher alcohols and ketones produce blue-green
                    spots. Also detects bile acids and steroids.

Aldehydes and Ketones

13. *2,4-Dinitrophenylhydrazine*

    Reagent: Dissolve 0.4 g reagent in 100 mL 2N HCl.
    Procedure: Spray
    Results: Yellow-red spots.

14. *2,2'-Diphenylpicrylhydrazyl*

    Reagent: 15 mg reagent in 25 mL $CHCl_3$.
    Procedure: Spray. Heat at 110° for 5-10 min.
    Results: Yellow spots on a purple background.

15. *Hydrazine Sulfate*

    Reagent: 1% solution of hydrazine sulfate in 1N HCl.

Procedure: Spray. Observe in daylight and under UV. Heat
            at 100° and observe under UV.
Results: Detects aldehydes.

Alkaloids

16. *Bromcresol Green*

Reagent: 0.05% bromcresol green in C_2H_5OH.
Procedure: Spray. Expose to NH_3 vapor if no reaction, after
            which the NH_3 is removed.
Results: Green or blue spots as the blue background fades.
            May develop at once or within 30 min.

17. *Ceric Sulfate-Trichloroacetic Acid*

Reagent: Boil 0.1 g ceric sulfate in 4 mL H_2O containing
            1 g trichloroacetic acid, adding H_2SO_4 dropwise
            until the solution clarifies.
Procedure: Spray and heat at 110°.
Results: Defects spomorphine, brucine, colchicine, papav-
            erine, and physostigmine. Also detects organic
            iodine compounds.

18. *Cinnamaldehyde-Acid*

Reagent: Prepare a fresh solution of 1 g cinnamaldehyde in
            100 mL CH_3OH.
Procedure: Spray plate and place in a tank containing a
            beaker with a 1:1 solution of conc. H_2SO_4:conc.
            HCl.
Results: Detects curane alkaloids.

19. *Cobolt(II) Thiocyanate*

Reagent: Dissolve 3 g ammonium thiocyanate and 1 g cobalt
            (II) chloride in 20 mL H_2O.
Procedure: Spray.
Results: Alkaloids and amines appear as blue spots on a
            white to pink background. The colors will fade
            after 2 hr, but can be revived by spraying with
            water.

20. *Dimethylaminobenzaldehyde*

Reagent: Dissolve 1 g reagent in 30 mL C_2H_5OH, 3 mL conc.
            HCl, and 180 mL n-butanol.

Procedure: Spray
Results: Ergot alkaloids produce blue spots.

21. *Dragendorff Reagent-Munier Modification*

Reagent: Solution (a):   Dissolve 1.7 g bismuth sub-
                             nitrate and 20 g tartaric acid
                             in 80 mL H_2O.
            Solution (b):   Dissolve 16 g K1 in 40 mL H_2O.
            Stock solution: Mix equal parts of solution (a)
                             and solution (b). Stable 1
                             month or more.
Procedure: The spray reagent is prepared by mixing 5 mL of
            stock solution with a solution of 10 g tartaric
            acid in 50 mL H_2O. Stable a week.
Results: For detection of alkaloids and other nitrogen-
            containing compounds, including antihistamines,
            cyclohexylamines, lactams, lipids.

22. *Dragendorff Reagent-Munier and Macheboeuf Modification*

Reagent: Solution (a):   Dissolve 0.85 g bismuth sub-
                             nitrate in a solution of 10 mL
                             acetic acid and 40 mL H_2O.
            Solution (b):   Dissolve 8 g K1 in 20 mL H_2O.
            Stock solution: Mix equal parts of solution (a)
                             with solution (b).
Procedure: The spray reagent is prepared by mixing 1 mL of
            stock solution with 2 mL acetic acid and 10 mL
            H_2O.
Results: For detection of alkaloids and other nitrogen-
            containing compounds.

23. *Formic Acid Vapor*

Reagent: Formic acid.
Procedure: Expose plate to formic acid vapor for 1 min, ob-
            serve under UV.
Results: Detects quinine and quinidine as fluorescent blue
            spots.

24. *Iodine-Potassium Iodide*

Reagent: Prepare a 5% solution of I_2 in 10% K1 solution.
            A spray solution is made from 2 parts of this
            solution, 3 parts H_2O, and 5 parts 2N acetic acid.
Procedure: Spray.
Results: Detects alkaloids.

25. *Iodoplatinate Reagent*

Reagent: Solution (a):   5% platinic chloride in water.
                Solution (b):   10% aqueous K1.
                Spray solution: Mix 5 mL of solution (a) with 45 mL of solution (b), dilute to 100 mL with H_2O. The addition of conc. HCl, 1 part to 10 parts of spray solution, will often increase the sensitivity of the reagent toward certain drugs such as caffeine.

Procedure: Spray.
Results: Most basic drugs give blue or blue-violet spots, which turn brownish-yellow. Compound containing only primary or secondary amine groups produce bluish white spots.

Amines, Amides, and Related Compounds (see reagent 45)

26. *Acetoacetylphenol*

Reagent: 1% solution of acetoacetylphenol in n-butanol.
Procedure: Spray and observe under UV light.
Results: Fluorescent areas for adrenaline and similar compounds.

27. *Alizarin*

Reagent: 0.1% alizarin in ethanol.
Procedure: Spray.
Results: Aliphatic amines and amino alcohols produce violet areas on a faint yellow background.

28. *Calcium Nitrate*

Reagent: 5% calcium nitrate in 95% ethanol.
Procedure: Spray and observe under UV.
Results: Diphenylamine produces a yellow-green spot.

29. *p-Dimethylaminobenzaldehyde*

Reagent: Spray solution (a):   1 g p-Dimethylaminobenzaldehyde dissolved in a mixture of 25 mL conc. HCl and 75 mL CH_3OH.

Spray solution (b):   1 g *p*-Dimethylamino-
benzaldehyde dissolved in
100 mL of 96% ethanol.
Procedure: Spray plate with solution (a). Warm plate.
Spray with solution (b), then place in a tank
saturated with HCl vapor for 3-5 min or spray
with 25% HCl.
Results: Detects amines such as tryptamine and related
citrulline, urea, and tryptophan.

30. *Diphenylamine-Palladium Chloride*

Reagent: Solution (a): 1.5% diphenylamine in ethanol.
Solution (b): 0.1% palladium chloride in 0.2%
N Cl solution.
Procedure: Spray with a solution of 5 parts solution (a) to
1 part solution (b). Expose moist plate to 240
mμ UV.
Results: Nitrosamines produce blue-violet spots.

31. *Ferric Chloride-Potassium Ferricyanide*

Reagent: Solution (a):   0.1 M ferric chloride.
Solution (b):   0.1 M potassium ferricyanide.
Spray solution: Prepare a 1:1 mixture of solution (a) to solution (b) im-
mediately before use.
Procedure: Spray. May have to heat plate at 110°.
Results: Aromatic amines produce blue spots, which turn
darker, against a light blue background. Also
detects phenols and phpnolic steroids.

32. *Glucose-Phosphoric Acid*

Reagent: Dissolve 2 g glucose in a mixture of 10 mL of
85% H_3PO_4 and 40 mL H_2O. Add 30 mL ethanol and
30 mL n-butanol to this solution and mix well.
Procedure: Spray. Heat plate 10 min at 115°.
Results: Aromatic amines are detected.

33. *Malonic Acid*

Reagent: Dissolve 0.2 g malonic acid and 0.1 g sali-
cylaldehyde in 100 mL absolute ethanol.
Procedure: Spray plate and heat for 15 min at 120°. Observe
under UV.
Results: Amines produce yellow spots.

34. *Picric Acid (Jaffe Reagent)*

Reagent: Spray solution (a): 1% picric acid in ethanol.
             Spray solution (b): 5% KOH in ethanol.
Procedure: Spray the plastic with solution (a) and allow it
             to dry. Then spray with solution (b).
Results: Substances such as creatinine, glycocyamidine and
             lactams of other α-guanidino acids produce orange
             spots on a yellow background.

35. *Potassium Ferrocyanide-Cobalt Chloride*

Reagent: Spray solution (a): 1% aqueous potassium ferro-
                                  cyanide.
             Spray solution (b): 0.5% aqueous cobaltous chlo-
                                  ride.
Procedure: Spray with solution (a), dry briefly, then spray
             with solution (b).
Results: Choline yields a green spot.

36. *Potassium Iodate*

Reagent: 1% aqueous potassium iodate.
Procedure: Spray, heat at 110° for 2 min.
Results: Detects phenylethylamines (sympathomimetic
             amines).

37. *Potassium Persulfate-Silver Nitrate*

Reagent: 1% aqueous potassium persulfate containing
             1×10^{-3}M silver nitrate.
Procedure: Spray and heat to 45° if necessary.
Results: Many compounds will react without heating. Var-
             ious aromatic amines may be detected; o-toluidine
             produces blue color changing to yellow; benzidine,
             brown; p-anisidine, green; p-toluidine, pink; and
             p-aminophenol, violet.

38. *p-Quinone*

Reagent: 0.5 g p-Benzoquinone is dissolved in a mixture of
             10 mL pyridine and 40 mL n-butanol.
Procedure: Spray.
Results: Ethanolamine produces red spots immediately after
             spraying; choline does not react.

39. *Sodium Nitroferricyanide (Sodium Nitroprusside)*

  Reagent: Solution (a):   5% sodium nitroferricyanide in
                                 10% acetaldehyde.
                Solution (b):   1% aqueous sodium carbonate.
                Spray solution: Mix equal parts of each solu-
                                 tion.
  Procedure: Spray.
  Results: Secondary aliphatic amines, morpholine, and
              diethanolamine produce blue to violet spots.

40. *Sulfanilic Acid (Pauly's Reagent)*

  Reagent: Solution (a):   1 g sulfanilic acid in 10 mL
                                 conc. HCl plus 90 mL H_2O.
                Solution (b):   5% aqueous sodium nitrite.
                Solution (c):   10% aqueous sodium carbonate.
                Spray solution: Mix one part of (a) with one
                                 part (b). Allow to stand for 5
                                 min, then add 2 parts solution
                                 (c). Allow to cool.
  Procedure: Spray. If plate has been developed in an acidic
              mobile phase, it is sprayed first with solution
              (c) and then dried before spraying with the diazo
              spray.
  Results: Catecholamines, other bases, and hydroxycyclic
              acids produce red-orange or yellow positive re-
              actions. Also detects iodophenols and imidazoles.

Amino Acids, Peptides, Proteins, Enzymes

41. *Ceric-Arsenite Reagent*

  Reagent: Solution (a):   10% ceric sulfate tetrahydrate
                                 in cold H_2SO_4. The cloudy so-
                                 lution should be refrigerated
                                 and then filtered. Store in the
                                 cold.
                Solution (b):   Dissolve 5 g sodium arsenite in
                                 100 mL cold (0°) 1N H_2SO_4, with
                                 stirring. Do not let tempera-
                                 ture rise during this step, or
                                 arsenious oxide will precipi-
                                 tate.
                Spray solution: Mix equal parts of each solution
                                 immediately before use.

Procedure: Place a sheet of filter paper equal in size to
the TLC plate being visualized onto a clean glass
plate. Use a small amount of the freshly mixed
reagent to evenly wet the filter paper. Place
the layer of the plate upon the wetted paper and
place a second clean plate on top, pressing down
firmly. Leave in place for 30 min and visualize
under UV. Spraying reduces the sensitivity of
the reagent and exposes the user to the toxic
reagent.

Results: Amino acids containing iodine produce white spots
on a yellow background and also fluoresce. Other
organic iodine compounds and iodide react. The
sensitivity is 0.1 µg for triiodothyronine and
thyroxine and 0.01 µg for iodine.

42. Dehydroascorbic Acid

Reagent: Dissolve 0.1 g dehydroascorbic acid in 5 mL of
H_2O at 60°. Dilute to 100 mL with butanol.

Procedure: Spray and heat at 100° for 5 min.

Results: Hydroxyproline, violet-blue, 25 µg; proline, pale
yellow, 25 µg; phenylalanine, tyrosine, tryptophan, pale pink, 10 µg; other amino acids, pink
to red, 1-3 µg.

43. N-Ethylmaleimide

Reagent: Solution (a): 0.05M N-ethylamaleimide in absolute isopropanol.
Solution (b): 1.4% aqueous KOH.

Procedure: Spray plate well with solution (a), allow to dry
15 min, then spray with solution (b).

Results: Detects amino acids containing -SH groups, as
pink-red spots, often in amounts as low as 0.1
µg. Also detects S-acetyl derivatives and thio-
lactones, but not disulfides.

44. Fluorescamine (Fluram, Roche)

Reagent: Dissolve 25 mg Fluram in 100 mL of dimethylsulfoxide or dimethylformamide.

Procedure: Spray, observe under UV while wet.

Results: Amino acids fluoresce white.

45. *1,2,3-Indantrione (Ninhydrin)*

Reagent: Solution (a):   0.2% ninhydrin in n-butanol.
              Solution (b):   10% aqueous acetic acid.
              Spray solution: Mix 95 parts (a) with 5 parts
                               (b).
Procedure: Spray and heat at 110°.
Results: Amino acids detected as pink-red spots; proline
              is yellow. Detects amines.

46. *1,2,3-Indantrione (Ninhydrin) Cadmium Acetate*

Reagent: Dissolve 0.1 g ninhydrin, 0.25 g cadmium acetate,
              and 1.0 mL glacial acetic acid in 49 mL methanol.
Procedure: Spray and heat at 120°.
Results: Detects amino acids.

47.   *Isatin-Zinc Acetate*

Reagent: Dissolve 1 g isatin and 1.5 g zinc acetate in
              100 mL 95% isopropanol by warming to 80°. Cool,
              then add 1 mL acetic acid and store the reagent
              in the refrigerator.
Procedure: Spray and heat 30 min at 90°. Good color de-
              velopment is also obtained by not heating, al-
              lowing the plate to stand at ambient temperature
              for 20 hr.
Results: Detects amino acids.

48. *Morin (2',3,4',5,7-Pentahydroxyflavone)*

Reagent: 0.005-0.05% morin in methanol.
Procedure: Spray, dry plate at 100° for 2 min, and examine
              immediately under UV.
Results: Amino acids produce yellow-green fluorescent or
              dark areas on a fluorescent background.

49. *Naphthalene Black (stain for Sephadex thin layers)*

Reagent: Dissolve 1 g naphthalene black in 100 mL of a
              mixture of 10 mL glacial acetic acid, 40 mL H_2O,
              and 50 mL methanol.
Procedure: Cover developed Sephadex plate with a sheet of
              Whatman 3MM filter paper, being careful to ex-
              clude air bubbles. Dry for 30 min at 80-90°.
              Immerse in naphthalene black solution for 30 min,
              then rinse in the acetic acid-methanol-H_2O solu-

tion without the dye, to remove excess naphthalene black.

Results: Detects proteins.

50. *1,3-Naphthoquinone-4-Sulfonic Acid Sodium Salt*

Reagent: Dissolve 0.2 g reagent in 100 mL 5% sodium carbonate.

Procedure: Spray after fresh preparation of reagent.

Results: Amino acids produce pink-red spots, which may be
intensified by spraying with a solution of 2 mL
5N NaOH in 98 mL ethanol.

51. *Vanillin-Potassium Hydroxide*

Reagent: Solution (a):   2% vanillin in n-propanol.
Solution (b):   1% KOH in ethanol.

Procedure: Spray with (a) and heat at 110° for 10 min, observe under long-wave UV. Spray with (b) and
heat again.

Results: Detects amino acids and amines. After spraying
with (a), ornithine will exhibit a bright green-
yellow and lysine a weak green-yellow fluores-
cense. After spraying with (b), ornithin will
possess a salmon color that will fade; proline,
hydroxyproline, pipecolinic (pipecolic) acid, and
sarcosine will turn red after a few hours.
Glycine will produce a green-brown spot, and
other amino acids will turn faint brown.

Antibiotics

52. *Sodium Azide*

Reagent: Solution (a):   0.5% soluble starch.
Solution (b):   3.5% sodium azide in 0.1N iodine
solution.

Procedure: Spray with (a), dry, spray with (b).

Results: Detects penicillin, sensitivity 0.2 µg. Penicillin and penicilloic acids are also detected by
starch-iodine-iodine reagents, as are cephalo-
sporins.

Barbiturates

53. *Diethylamine*

Reagent: Dissolve 0.5 g copper sulfate in 100 mL methanol.
             Add 3 mL diethylamine, stirring well.
Procedure: Spray.
Results: Thiobarbituric acids produce green spots.

54. *s-Diphenylcarbazone*

Reagent: 0.1% s-diphenylcarbazone in 95% ethanol.
Procedure: Spray.
Results: Barbiturates produce purple spots.

55. *Silver Nitrate*

Reagent: Solution (a): 1% aqueous silver nitrate.
             Solution (b): Reagent 54.
Procedure: Spray with (a), then with (b).
Results: Barbiturates produce purple spots.

Carbohydrates

56. *c-Aminodiphenyl*

Reagent: Dissolve 0.3 g o-aminodiphenyl and 5 mL o-phos-
             phoric acid in 95 mL ethanol.
Procedure: Spray and heat at 110° for 20 min.
Results: Brown spots with carbohydrates.

57. *Aminohippuric Acid*

Reagent: 0.3% aminohippuric acid in ethanol. Add 3%
             phthalic acid if reducing disaccharides are sus-
             pected of being present.
Procedure: Spray and heat 8 min at 140°. Observe under UV.
Results: Pentoses and hexoses produce orange-red spots
             that are more readily seen under UV.

58. *Aminophenol*

Reagent: 0.15 g o- or p-aminophenol in 10 mL 50% H_3PO_4,
             prepared before use.
Procedure: Spray and heat at 105°.
Results: Aldoses and ketoses detected 1-5 µg sensitivity.

59. *p-Anisaldehyde*

Reagent: Dissolve 1 mL *p*-anisaldehyde and 1 mL H_2SO_4 in
             18 mL ethanol.
Procedure: Spray and heat at 110°.
Results: Sugar phenylhydrazones produce green-yellow spots
             in 3 min. Sugars will produce blue, green,
             violet spots in 10 min. Also detects digitalis
             glycosides.

60. *p-Anisidine Hydrochloride*

Reagent: 3% *p*-anisidine hydrochloride in n-butanol.
Procedure: Spray and heat at 100° for 2-10 min.
Results: Aldohexoses, green-brown spots; ketohexoses, yel-
             low spots; aldopentoses, green spots; uronic
             acids, red spots.

61. *p-Anisidine-Phthalic Acid*

Reagent: Dissolve 1.23 g *p*-anisidine and 1.66 g phthalic
             acid in 100 mL methanol.
Procedure: Spray.
Results: Hexoses, green; pentoses, red-violet, sensitivity
             0.5 µg; methylpentoses, yellow-green; uronic
             acids, brown, sensitivity 0.1-0.2 µg.

62. *Anthrone*

Reagent: Dissolve 0.3 g anthrone in 10 mL boiling glacial
             acetic acid. Add 20 mL ethanol, 3 mL H_3PO_4, and
             1 mL H_2O.
Procedure: Spray and heat 5 min at 110°.
Results: Ketoses and oligosaccharides produce yellow spots.

63. *Benzidine*

Reagent: 0.2% benzidine in acetic acid.
Procedure: Spray and heat for 10 min at 100°.
Results: Monoaldoses give brown spots.

64. *Bromphenol Blue-Boric Acid*

Reagent: Dissolve 40 mg Bromphenol blue (or Bromcresol
             green) in 10 mL ethanol. Add 0.1 g boric acid
             and 7.5 mL of 1% sodium tetraborate, and dilute
             to 100 mL with ethanol.

Procedure: Spray
Results: Sugar alcohols detected as yellow spots on a
              blue background.

65. 3.5-Diaminobenzoic Acid

Reagent: Dissolve 1 g 3,5-diaminobenzoic acid dihydro-
              chloride in a mixture of 60 mL H_2O and 25 mL of
              80% H_3PO_4.
Procedure: Spray, heat at 110°, and observe under UV.
Results: 2-deoxy sugars produce a green-yellow fluores-
              cence.

66. 3,5-Dinitrosalicylic Acid

Reagent: Prepare a 0.5% solution of 3,5-dinitrosalicylic
              acid in 4% NaOH.
Procedure: Spray and allow plate to air dry. Heat in oven
              at 100° for 5 min.
Results: Reducing sugars will produce brown spots on a
              yellow background; sensitivity is 1 μg.

67. Diphenylamine

Reagent: 2.3% solution of diphenylamine in H_2O saturated
              with n-butanol.
Procedure: Spray and allow plate to air dry. Heat in oven
              at 130° for 20 min.
Results: Aldoses and ketoses produce blue spots.

68. Hydroxylamine Hydrochloride

Reagent: Solution (a): 1 g hydroxylamine hydrochloride
                             in 9 mL H_2O.
              Solution (b): 2 g HaOH in 8 mL H_2O.
              Solution (c): 4 g ferric nitrate in 60 mL H_2O
                             and 40 mL acetic acid.
Procedure: Spray with 1:1 mixture of (a) and (b). Dry for
              10 min at 110°. Spray with mixture of 45 mL of
              (c) and 6 mL conc. HCl.
Results: Detects sugar acetates.

69. Malonic Acid-Aniline

Reagent: Dissolve 1 g malonic acid and 1 mL aniline in 100
              mL methanol.
Procedure: Spray, allow plate to air dry, and then heat at
              110°. Observe under UV.

Results: Sugars produce yellow, gray, or brown spots fluorescent in UV.

70. *Naphthoresorcinol*

Reagent: Dissolve 0.2 g naphthoresorcinol in 100 mL ethanol plus 10 mL H_3PO_4.
Procedure: Spray and heat at 110° for 10 min.
Results: Detects glucuronides as blue spots.

71. *Periodic Acid-Benzidine*

Reagent: Solution (a):   2.28% aqueous periodic acid (stable in cold). Mix 1:19 with acetone (Stable a few hours).
                Solution (b):   Dissolve 184 mg benzidine in 95 mL acetone. Add 4.4 mL H_2O and 0.6 mL acetic acid. This will turn yellow but is stable indefinitely.
Procedure: Spray with (a), allow to stand 4 min, then spray with (b).
Results: Sugars will produce white spots on a blue background. Nearly all sugars and sugar alcohols react. Sucrose, trehalose, and glycosamines do not.

72. *Phenol-Sulfuric Acid*

Reagent: Dissolve 3 g phenol and 5 mL H_2SO_4 in 95 mL ethanol.
Procedure: Spray and heat 10-15 min at 110°. Additional heating may intensify spots.
Results: Carbohydrates produce brown spots.

73. *m-Phenylenediamine Hydrochloride*

Reagent: 3.6% *m*-phenylenediamine hydrochloride in 70% ethanol.
Procedure: Spray and heat briefly at 105°. Observe under UV.
Results: Detects reducing sugars as fluorescent spots.

74. *Starch-Iodine*

Reagent: Solution (a):   2% aqueous starch.
                Solution (b):   Dissolve 50 mg I_2 in 100 mL 1% KI (aqueous).

Procedure: Spray with (a), and place plate into a moist
             chamber at 45-50° for 1 hr. Dry at room tem-
             perature and spray with (b).
Results: Amylases detected as white spots on a violet or
             brown background.

75. *Vanillin-Perchloric Acid*

Reagent: Solution (a):   1% vanillin in ethanol.
             Solution (b):   3% aqueous perchloric acid.
             Spray solution: Mix (a) and (b) (1:1).
Procedure: Spray and heat 10 min at 85°.
Results: Deoxy sugars produce various colors.

Carboxylic Acids

See reagents 1-7, 21, 40.

76. *Benzidine-Metaperiodate*

Reagent: Solution (a):   0.1% aqueous sodium metaperio-
                              date.
             Solution (b):   Dissolve 2.8 g benzidine in 80
                              mL of 96% ethanol. To this add
                              a solution of 70 mL H_2O, 30 mL
                              acetone, and 1.5 mL 1N HCl.
Procedure: Spray with (a); dry partially, then spray moist
             plate with (b).
Results: Detects organic acids, sugars, and sugar alco-
             hols.

77. *2,6-Dichlorophenol-Indophenol*

Reagent: 0.1% solution of 2,6-dichlorophenol-indophenol in
             95% ethanol.
Procedure: Spray. Brief heating may aid color development.
Results: Organic acids produce pink spots on a sky blue
             background.

78. *Diphenylcarbazone*

Reagent: Solution (a):   0.1-0.2% s-diphenylcarbazone in
                              95% ethanol.
             Solution (b):   0.05N HNO_3 in ethanol.
Procedure: Spray with (a). Allow to dry. Spray with (b)
             to intensify colors.

Results: Purple spots are produced with additional com-
                pounds of unsaturated acids (e.g., with mercury).
                Produces purple spots with acetoxymercuric-
                methoxy derivatives of unsaturated esters and
                barbiturates.

79. *Glucose-Aniline (Schweppe Reagent)*

Reagent: Solution (a):   Dissolve 2 g glucose in 2 mL
                                 H_2O.
                Solution (b):   Dissolve 2 mL aniline in 2 mL
                                 ethanol.
                Spray solution: Mix total solution (a) with
                                 total solution (b) in a 100 mL
                                 volumetric flask. Dilute to
                                 volume with n-butanol.
Procedure: Spray and heat 5-10 min at 125°.
Results: Organic acids produce dark brown spots on a white
                background.

80. *Methyl Red-Bromthymol Blue*

Reagent: Dissolve 0.2 g bromthymol blue and 0.2 g methyl
                red in a mixture of 400 mL of 95% ethanol and
                100 mL of formaldehyde. Adjust pH to 5.2 with
                0.1N NaOH.
Procedure: Spray and place plate in NH_3 vapor.
Results: Organic acids detected as yellow spots on a pink
                background changing to red-orange spots on a
                green background after exposure to NH_3.

81. *o-Phenylenediamine-Trichloroacetic Acid*

Reagent: Dissolve 50 mg *o*-phenylenediamine in 100 mL 10%
                aqueous trichloroacetic acid.
Procedure: Spray and heat at 100° for not more than 2 min.
                Observe under UV.
Results: Detects α-keto acids.

82. *Silver Nitrate-Pyrogallol*

Reagent: Solution (a):   Dissolve 0.17 g silver nitrate in
                                 1 mL H_2O. Add 5 mL NH_4OH and di-
                                 lute to 200 mL with ethanol.
                Solution (b):   Dissolve 6.5 mg pyrogallol in 100
                                 mL of ethanol.
Procedure: Spray with (a), then with (b).
Results: Detects organic acids.

Drugs

See also barbiturates section.

83. p-Dimethylaminobenzaldehyde

Reagent: 1% p-dimethylaminobenzaldehyde in 5% HC1.
Procedure: Spray
Results: Detects sulfonamides.

84. FPN Reagent

Reagent: Solution (a): 5% ferric chloride.
            Solution (b): 20% perchloric acid.
            Solution (c): 50% nitric acid
            Spray reagent: Mix (a), (b), and (c) in pro-
                           portions 5:45:50.
Procedure: Spray.
Results: Phenothiazines located as orange, red, or blue
            spots.

85. Furfural-Hydrochloric Acid

Reagent: Solution (a): 10% furfural in ethanol.
            Solution (b): Conc. HC1.
Procedure: Spray lightly with (a), then (b). Exposure to
            HCL vapor is an alternative to spraying with HCL.
Results: Purple to black areas are produced by carbamates
            with free NH_2 groups. n-Substituted carbamates
            do not react. Ethinamate, meprobamate, and
            styramate react.

86. Mandelin's Reagent

Reagent: 1% ammonium vanadate in conc. H_2SO_4. Shake be-
            fore use.
Procedure: Spray and observe. Heating at 85° for 5 min will
            intensify or change the color of reacting drugs.
Results: Detects antihistamines as various colored spots.

87. Sodium 1,2-Naphthaquinone-4-Sulfonate (NZS Reagent)

Reagent: Solution (a): 0.1N NaOH
            Solution (b): Saturated solution of sodium
                           1,2-naphthaquinone-4-sulfonate in
                           1:1 ethanol-H_2O.
Procedure: Spray with (a), then (b).

Results: Thiazide drugs appear as orange spots within 15
             min. Basic drugs with primary amine groups also
             react. Barbiturates do not.

General Reagents

See also reagents 1-4.

88. *Antimony Trichloride*

Reagent: Saturated solution of antimony trichloride in
             chloroform.
Procedure: Spray and heat 10 min at 100°.
Results: Varied colors produced with various compounds.

89. *Sodium Fluorescein*

Reagent: Dissolve 50 mg sodium fluorescein in 100 mL 50%
             methanol.
Procedure: Spray plate and observe under UV.
Results: Many aromatic and heterocyclic compounds will
             fluoresce.

90. *Fluorescein-Bromine*

Reagent: Solution (a): Reagent 89.
             Solution (b): Bromine
Procedure: After spraying with (a), expose to bromine vapor.
Results: Ethylenic unsaturated or other types of compounds
             that react with bromine produce yellow spots on a
             pink background.

91. *Iodine*

Reagent: Iodine crystals (I_2) or saturated solution of I_2
             in hexane.
Procedure: Place plate in covered tank containing a few
             crystals of iodine that have vaporized, or spray
             plate with I_2 solution.
Results: Many compounds absorb iodine reversibly to pro-
             duce yellow to brown spots on a faint yellow
             background. Particularly useful for detection
             of unsaturated fatty acids.

Glycosides

92. *Diphenylamine*

Reagent: The spray solution consists of 20 mL 10% diphe-
             ylamine in ethanol, 100 mL HCl, and 80 mL acetic
             acid.
Procedure: Spray lightly, cover sprayed plate with another
             clean glass plate, and heat 30-40 min at 110°
             until positive areas appear.
Results: Glycolipids produce blue spots.

93. *Orcinol (Bials Reagent)*

Reagent: Dissolve 0.1 g orcinol in 40.7 mL conc. HCl, add
             1 mL 1% $FeCl_3$ solution, and dilute to 50 mL.
Procedure: Spray and heat at 80° for 90 min.
Results: Glycolipids produce violet spots.

94. *Trichloroacetic Acid*

Reagent: 25% trichloroacetic acid in $CHCl_3$.
Procedure: Spray and heat 2 min at 100°; observe under UV.
Results: *Stropanthus* glycosides produce a yellow fluores-
             cence with a sensitivity of 0.4 µg.

Heterocyclic Oxygen Compounds.

95. *Aluminum Chloride*

Reagent: 1% aluminum chloride in ethanol.
Procedure: Spray and view under long-wave UV.
Results: Flavonoids produce yellow fluorescent spots.

96. *p-Toluenesulfonic Acid*

Reagent: 20% solution of *p*-toluenesulfonic acid in meth-
             anol or chloroform.
Procedure: Spray and heat plate at 80° for 10 min. Observe
             under long-wave UV.
Results: Flavonoids, indoles, and steroids produce fluo-
             rescent spots.

97. *Trichloroacetic Acid*

Reagent: 4% trichloroacetic acid in $CHCl_3$.

Procedure: Spray and let plate stand 10-30 min.
Results: Menthofuran produces a pink spot.

Hydrocarbons

See reagents 89, 90.

98. *Chlorosulfonic Acid*

Reagent: Prepare a solution of 1:2 chlorosulfonic acid-
             glacial acetic acid.
Procedure: Spray and heat plate 1 min at 130°.
Results: Detects olefins and sapogenins.

99. *Osmium Tetroxide*

Reagent: Osmium tetroxide crystals.
Procedure: Expose plate to osmium tetroxide vapors in a
             clased tank. For isolated double bond compounds,
             expose 5-10 min; for conjugated double bonds, ex-
             pose 60 min or more.
Results: Brown to black spots with double-bond hydrocar-
             bons. May be used for lipids and steroids.

100. *Tetracyanoethylene*

Reagent: 10% tetracyanoethylene in benzene.
Procedure: Spray plates as soon as they are removed from
             development chamber. Heat at 100°.
Results: Aromatic hydrocarbon show different colors.

Inorganic Compounds

See reagents 41, 78.

101. *Alizarin*

Reagent: Solution (a): Saturated solution of alizarin in
                            ethanol.
             Solution (b): 25% NH_4OH.
             Solution (c): Glacial acetic acid.
Procedure: Spray with (a), then with (b), and then with (c)
             to eliminate background color.
Results: Many ions produce violet to red spots, including
             Ba, Ca, Mg, Al, Ti, Fe, Zn, Li, Th, Zr, Se, NH_4.

102. *Ammonium Molybdate-Stannous Chloride*

      Reagent: Solution (a): 1% aqueous ammonium molybdate.
                   Solution (b): 1% stannous chlorideiin 10% HCl.
      Procedure: Spray with (a), dry, and spray with (b).
      Results: Phosphate and phosphite ions produce blue spots.

103. *Ammonium Sulfide*

      Reagent: Saturated aqueous H_2S made alkaline with NH_4OH.
      Procedure: Spray.
      Results: Cations detected as black: Ag, Hg(1), Hg(III),
                   Co, Ni; as brown: Au, Pd, Pt, Pb, Bi, Cu, V,
                   Ti; as yellow: Cd, As, Sn; as yellow-orange:
                   Sb.

104. *Aurintricarboxylic Acid-Ammonium Salt*

      Reagent: 1° solution of aurintricarboxylic acid-ammonium
                   salt in 1° aqueous ammonium acetate.
      Procedure: Spray and expose plate to NH_3 vapors.
      Results: Detects Al, Cr, and Li.

105. *Benzidine*

      Reagent: Dissolve 50 mg benzidine base or hydrochloride
                   in 10 mL acetic acid, and dilute to 100 mL with
                   H_2O filter.
      Procedure: Spray and heat as 85° for 30 min.
      Results: Detects Au(III), Ti(III), chromate, Mn(IV).

106. *Bromocresol Purple*

      Reagent: Dissolve 40 mg bromocresol purple in 100 mL of
                   50% ethanol. Adjust to pH 10 with alkali.
      Procedure: Spray.
      Results: Detects halogen anions except F^-. Also detects
                   carboxylic acids (test 6).

107. *Brucine*

      Reagent: Solution (a): 0.02% brucine in 2N H_2SO_4.
                   Solution (b): 2N NaOH.
      Procedure: Spray with (a), warm and note colors, spray with
                   (b).

Results: Detects BrO_3, deep red (a), blood red (b); NO_3
                red then yellow (a), orange-red (b); ClO_3,
                red-brown (a), blood red (b).

108. *Cinchonine-Potassium Iodide*

Reagent: Prepare a fresh solution of 1 g cinchonine in
                100 mL hot H_2O that contains a few drops of
                HNO_3. Cool and add 2 g KI.
Procedure: Spray.

Results: Detects Bi, orange; Ag, Hg(II), Pb, Sb, V, Tl,
                yellow; Cu, brown; Pt, pink.

109. *Copper Sulfate-Mercury Ammonium Sulfate*

Reagent: Solution (a): 0.1% cupric sulfate-2N H_2SO_4
                               (9:1).
                Solution (b): 2.7% mercuric chloride and 3%
                               ammonium thiocyanate in H_2O.
Procedure: Spray with (a), then with (b).
Results: Detects: Zn, red to violet; Cu, yellow; Fe(III),
                          red; Au, orange-pink; and Co, blue.

110. *Dimethylglyoxime*

Reagent: 10% dimethylglyoxime in ethanol containing a
                small amount of NH_4OH.
Procedure: Spray.
Results: Ni ion produces a pink-red spot.

111. *Diphenylcarbazide*

Reagent: Solution (a): 0.1% *s*-diphenylcarbazide in 96%
                               ethanol.
                Solution (b): 25% NH_4OH.
Procedure: Spray with (a), then with (b).
Results: Heavy metals detected including Ni, blue; Co,
                orange-brown; Ag, Pb, Cu, Sn, Mn, brown; Zn,
                purple.

112. *Potassium Thiocyanate*

Reagent: 1% aqueous potassium thiocyanate.
Procedure: Spray
Results: Metal cations detected: Au, orange; Pt, Mo,
                orange-red; Hg(I), black; Bi, U, yellow; V, Co,

blue; Ni, green; Fe, deep red; Cr, purple; Cu,
green-black.

113. *Pyrocatechol Violet*

Reagent: 0.1% pyrocatechol violet in ethanol.
Procedure: Spray with reagent after exposing to UV light
              for 20 min.
Results: Organotin compounds produce dark blue spots on a
              gray-black background.

114. *Silver Nitrate*

Reagent: 0.05N silver nitrate.
Procedure: Spray, expose to UV light for 10 min. Observe
              immediately, as zones darken with time.
Results: Halogens detected as brown-black areas.

115. *Violuric Acid (Acid Violet)*

Reagent: 1.5% aqueous violuric acid.
Procedure: Spray and heat at 100° for 20 min.
Results: Alkali and alkaline earth metals detected: Li,
              red-violet; Na, violet-red; K, violet; Be,
              yellow-green; Mg, yellow-pink; Ca, orange; Sr,
              red-violet; Ba, light red; Co, green-yellow; Cu,
              yellow-brown.

Lipids, Phospholipids

See reagents 1-3, 88, 91, 92.

116. *Ammonium Molybdate-Perchloric Acid*

Reagent: Solution (a):   3 g ammonium molybdate in 25 mL
                               H_2O.
              Solution (b):   1N HCl.
              Solution (c):   60% $HClO_4$.
              Spray solution: Mix total solution (a) with 30
                               mL (b) and 15 mL (c).
Procedure: Spray and heat at 105°
Results: Lipids produce blue-black spots.

117. *Bromthymol Blue*

Reagent: 0.1% bromthymol blue in 10% aqueous ethanol made
              just alkaline with NH_4OH.

Procedure: Spray.
Results: Lipids and phospholopids produce blue-green
             areas. Sensitivity 0.1-1 µg.

118. α-Cyclodextrin

Reagent: 1% α-cyclodextrin in 30% ethanol.
Procedure: Spray plate and allow it to dry. Place it in a
             humidity chamber for 1 hr at room temperature,
             then expose it to iodine vapor.
Results: Fatty acids and esters, monoglycerides, and sat-
             urated alcohols remain white. Corresponding un-
             saturated compounds become yellow or brown.

119. Dragendorff's Reagent

Reagent: Solution (a):   17% basic bismuth nitrate in
                              20% aqueous acetic acid.
             Solution (b):   40% aqueous KI.
             Solution (c): H_2O.
             Spray solution: 4:1:14 (a):(b):(c). Store (a)
                              and (b) at 4°. Mix before use.
Procedure: Spray.
Results: Phospholipids containing choline produce orange
             or red-orange spots.

120. Ferric Chloride-Sulfosalicylic Acid

Reagent: Dissolve 0.1 g $FeCl_3 \cdot 6H_2O$ and 7 g sulfosalicylic
             acid in 25 mL H_2O, and dilute to 100 mL with 95%
             ethanol.
Procedure: Spray.
Results: Phosphate groups in lipids and other compounds
             are detected as white fluorescent spots on a
             purple background.

121. Gentian Violet-Bromine

Reagent: 0.1% gentian violet (crystal violet) in methanol.
Procedure: Spray and place in tank containing bromine vapor.
Results: Lipids produce blue spots on a yellow background.

122. Phosphotungstic Acid

Reagent: 10% phosphotungstic acid in ethanol.
Procedure: Spray and heat at 100° for 5-15 min.
Results: Cholesterol and its esters produce red spots.

123.   *Rhodamine B*

        Reagent: Solution (a):   0.05-0.1% rhodamine B in ethanol
                                     or 0.2% in H_2O.
                     Solution (b):   3% hydrogen peroxide.
                     Solution (c):   10N KOH.
        Procedure: Spray with (a) and observe. Also observe under
                     UV. Spraying with (b) may enhance the color.
        Results: Higher fatty acids detectdd. A bright red fluo-
                     rescence is produced by many lipids. For gly-
                     cerides, spray with (a) and then (c); sensitiv-
                     ity may be increased by repeat spraying with (c)
                     after a few minutes. Lipids produce purple
                     spots on a pink background; glycerides yield
                     bright white spots on a pink-red to blood-red
                     background. Food preservatives produce purple
                     spots intensified by spraying with (b).

Nitro and Nitroso Compounds

124.   *Dimethylaminobenzaldehyde-Tin Chloride-Hydrochloric Acid*

        Reagent: Solution (a):   Prepare fresh before use. Mix 3
                                     mL 15% stannous chloride solution
                                     with 15 mL HCl. Add 180 mL H_2O.
                     Solution (b):   Dissolve 1 g 4-dimethylamino-
                                     benzaldehyde in a mixture of 30
                                     mL ethanol, 3 mL HCl and 180 mL
                                     n-butanol.
        Procedure: Spray with (a), air dry, spray with (b).
        Results: Nitro compounds produce yellow spots. 3,5-Di-
                     nitrobenzoyl derivatives of aliphatic amines are
                     also detected.

125.   *Diphenylamine*

        Reagent: 1% diphenylamine in 95% ethanol.
        Procedure: Spray and expose to 254 nm UV.
        Results: Nitrate esters produce yellow-green spots. Ex-
                     plosives will produce various colors when a 5%
                     solution is used.

126.   *Malonic Acid Diethylester*

        Reagent: Solution (a):   10% malonic acid diethylester in
                                     ethanol.

Solution (b):   10% aqueous NaOH.
Procedure: Spray with (a), then with (b). Heat at 95° for
5 min.
Results: Compounds produce red-violet spots. 3,5-Dinitro-
benzoic acid esters are visible under 254 nm UV
as dark violet spots. Also visible in daylight.

127. α-Naphthylamine

Reagent: 1% α-naphthylamine in ethanol.
Procedure: Spray.
Results: 3,5-Dinitrobenzoates detected as yellow ɔ or-
ange spots. If the plate is sprayed with 10%
KOH in methanol following the naphthylamine,
these compounds will produce red-brown spots.

Nitrogen Heterocyclic Compounds

[Pyrroles, pyrazole derivatives, imidazoles, indoles, pyridine
derivatives, phenoxazines, quinoline derivatives]

128. Anisidine

Reagent: Solution (a): 1% p-anisidine in ethanol containing 1% HCl.
Solution (b): 2% amyl nitrite in ethanol.
Solution (c): 0.4 or 2% aqueous NaOH.
Procedure: Spray with (a), then (b), then (c).
Results: Polyphenols and imidazoles produce red or brown
spots. Hydroxy indoles also react. The 2% NaOH
is used when applied after Ehrlich reagent in
multiple reagent sequences.

129. Boric Acid-Citric Acid

Reagent: Dissolve 0.5 g each of boric acid and citric
acid in 20 mL methanol.
Procedure: Spray and heat 10 min at 100°. Observe under UV.
Results: Quinolines detected; 8-hydroxy quinoline gives
a yellow-green fluorescence.

130. Dimethyldihydroresorcinol

Reagent: Solution (a): 10% dimethyldihydroresorcinol in
ethanol.
Solution (b): 5% aqueous ferric chloride.

Procedure: Spray with (a), dry, and spray with (b).
Results: Detects aldehydes and ketones of pyridine bases as
           violet or red-brown spots.

131.   2,4-Dinitrophenylhydrazine

Reagent: 0.4% 2,4-dinitrophenylhydrazine in 2N HCl.
Procedure: Spray and heat at 105° for 10 min and observe in
              visible and UV light.
Results: Tests for α,β- unsaturated ether lipids (plas-
              malogens). Aldehyde hydrazones are formed when
              the plasmalogens are heated in the presence of
              HCl. These react to produce brown, red, or
              orange-red spots against a yellow background.

132.   Ehrlich Reagent

Reagent: 10% -dimethylaminobenzaldehyde in conc. HCl.
              Spray solution: Mix the 10% solution 1:4 with
                               acetone before use.
Procedure: Spray. Color development occurs within 20 min.
Results: Detects indoles, purple; hydroxyindoles, blue;
              aromatic amines and ureides, yellow; tyrosine,
              purple-red. Colors may change and fade with
              time. These may serve as additional identifica-
              tion aids.

133.   Ferric Chloride-Perchloric Acid

Reagent: Solution (a):   0.05M ferric chloride.
              Solution (b):   5% $HClO_4$.
              Spray solution: Mix 1:50, (a):(b).
Procedure: Spray.
Results: Indoles produce red spots that turn blue-yellow
              after 24 hr.

134.   Formaldehyde-Schiff's Reagent

Reagent: Solution (a): 1% formaldehyde.
              Solution (b): Schiff's reagent: 1 g fuchsin,
                             110 mL 1N HCl, and 5 g sodium
                             bisulfite are mixed are diluted
                             to 1 liter with H_2O.
Procedure: Spray with (a), evaporate excess at 110°, then
              spray with (b).
Results: Triazines produce red spots.

135. *Sodium Nitroferricyanide*

Reagent: Solution (a): 2% sodium nitroferricyanide
                            (nitroprusside).
             Solution (b): 5% sodium carbonate.
             Solution (c): 50% acetic acid.
Procedure: Spray with (a), then with (b). Dry partially
             and spray with (c).
Results: Indoles unsubstituted in positions 1, 2, or 3
             produce blue spots.

Peroxides

136. *Ammonium Thiocyanate-Ferrous Sulfate*

Reagent: Solution (a):   0.2 g ammonium thiocyanate in
                                15 mL acetone.
             Solution (b):   4% aqueous ferrous sulfate.
             Spray solution: Add 10 mL of (b) to (a) before
                                use.
Procedure: Spray.
Results: Brownish-red spots from peroxides.

137. *N,N-Dimethyl-p-phenylenediamine Dihydrochloride*

Reagent: Dissolve 1.5 g N,N-dimethyl-p-phenylenediamine
             dihydrochloride in 128 mL methanol, 25 mL H_2O,
             and 1 mL acetic acid.
Procedure: Spray.
Results: Organic peroxides produce purple-red spots.

138. *Potassium Iodine-Starch*

Reagent: Solution (a): Mix 10 mL of a 4% KI solution with
                            40 mL glacial acetic acid. Add a
                            small pinch of zinc dust.
             Solution (b): Prepare fresh 1% aqueous starch.
Procedure: Filter solution (a) free of zinc dust immediate-
             ly before use, then spray the plate. After 5
             min, spray the plate heavily with (b) until the
             layer is transparent.
Results: Blue spots for peroxides due to free iodine re-
             acting with the starch.

Pesticides

139. *Brilliant Green*

Reagent: 0.5% Brilliant green in acetone.
Procedure: Spray and expose plate immediately to bromine
             vapor.
Results: Organophosphorus and triazine herbicide com-
             pounds produce dark green spots.

140. *Bromine-Congo Red*

Reagent: Solution (a): 0.4% Congo Red dye in 50% etha-
                            nol.
             Solution (b): 10% bromine in carbin tetrachlo-
                            ride.
Procedure: Expose plate to vapors from solution (b) for 20
             sec. Aerate to get rid of the bromine and spray
             with (a).
Results: Thiophosphate pesticides produce red spots on a
             blue background. Sensitivity 0.5 µg.

141. *Cupric Chloride*

Reagent: Dissolve 2 g cupric chloride in 11 mL ethanol
             containing 2.5 mL conc. HCl.
Procedure: Spray.
Results: Detects Systox and Meta-Systox.

142. *Diphenylamine-Zinc Chloride*

Reagent: Dissolve 0.5 g each of diphenylamine and zinc
             chloride in 100 mL acetone.
Procedure: Spray and heat at 200° for 5 min.
Results: Detects chlorinated pesticides.

143. *Ferric Chloride-Sulfosalicylic Acid*

Reagent: Solution (a): 1% 5-sulfosalicylic acid in 80%
                            ethanol.
             Solution (b): 0.1% ferric chloride in 80% eth-
                            anol.
Procedure: Expose plate to bromine vapor for 10 min. Spray
             with (b), dry 15 min, then spray with (a).
Results: Thiophosphate pesticides produce white spots on
             a mauve background. Sensitivity 5 mg.

144. *Methylumbelliferone*

Reagent: Solution (a): 0.5% iodine in ethanol.
             Solution (b): Dissolve 0.075 g 4-methyl-
                            umbelliferone in 100 mL 50:50
                            ethanol-H_2O. 10 mL 0.1N NH_4OH.
Procedure: Spray with (a), observe positive areas, then
             spray with (b) and observe under UV.
Results: Detects organophosphorus pesticide.

145. *Potassium Iodine-Phosphoric Acid*

Reagent: Solution (a): 5N KI.
             Solution (b): 85% H_3PO_4 (conc.).
             Spray solution: Mix 1 part (a) with 15 parts
                            (b)before use.
Procedure: Spray. Color development may take between ½ and
             6 hours.
Results: Rotenone produces a blue spot.

146. *Silver Nitrate-Bromphenol Blue*

Reagent: Solution (a): 0.5% silver nitrate in ethanol.
             Solution (b): 0.2% bromphenol blue plus 0.15%
                            silver nitrate in 1:1 ethanol-
                            ethyl acetate.
Procedure: Spray with (a) and heat at 100° for 5 min.
             Spray with (b) and heat at 100° for 10 min.
Results: Chlorinated pesticides produce yellow spots on a
             blue background.

147. *Silver Nitrate-Phenoxyethanol*

Reagent: Dissolve 0.1 g silver nitrate in 1 mL H_2O and
             add 10 mL of 2-phenoxyethanol. Dilute to 200 mL
             with acetone.
Procedure: Spray and dry in hood for 5 min. Heat at 75°
             for 15 min. Expose to UV for a short period of
             time and observe.
Results: Chlorinated pesticides produce black spots.
             Sensitivity 0.01-0.1 µg.

148. *o-Toluidine*

Reagent: 0.5% *o*-toluidine in ethanol.
Procedure: Spray and allow to dry. Observe under 254 nm
             UV.

Results: Detects chlorinated pesticides as green spots.
                Sensitivity 0.5-1 μg.

149. *1,3,5-Trinitrobenzene*

Reagent: Solution (a): Dissolve 1 g KOH in 10 mL H_2O and
                               dilute to 100 mL with 95% ethanol.
                Solution (b): Saturate solution of 1,3,5-trini-
                               trobenzene in acetone.
Procedure: Spray with (a) and heat at 150° for 5 min. Wash
                cooled plate with acetone to remove organic re-
                sidues and spray with (b).
Results: Organic sulfite pesticides detected as pink to
                red spots.

Phenols, including Plant

See reagents 31, 40, 128, 163.

150. *Ammonium Vanadate-Anisidine*

Reagent: Solution (a): Saturated, aqueous ammonium vana-
                               date.
                Solution (b): Dissolve 0.5 g p-anisidine in 2
                               mL H_3PO_4, dilute to 100 mL with
                               ethanol and filter.
Procedure: Spray with (a); while plate is still wet, spray
                with (b). Heat at 80°.
Results: Phenols produce various colored spots on a pink
                background.

151. *Benzidine, Tetrazotized*

Reagent: Solution (a):   0.5% benzidine in dilute HCl.
                                  Triturate 5 g benzidine with 15
                                  mL of conc. (12N) HCl. Dissolve
                                  in 980 mL H_2O. Stable one week.
                Solution (b):   10% sodium nitrite.
                Spray solution: Mix equal volumes of (a) and
                                  (b) before use. The solution
                                  should be clear and yellow.
Procedure: Spray.
Results: Phloroglucinol-resorcinol type of phenols pro-
                duce spots. Sensitivity 2-3 μg.

152. *p-Dimethylaminobenzaldehyde-Acetic Anhydride*

    Reagent: Solution (a):   5% p-dimethylaminobenzaldehyde
                                        in acetic anhydride.
                Solution (b):   Acetone.
                Spray solution: Mix 1 part (a) with 4 parts (b).
    Procedure: Spray and heat briefly at 130°.
    Results: Aroyl-glycines produce orange and orange-red
                spots that fluoresce yellow under UV. Useful for
                location of citric acid cycle acids.

153. *Mercury-Nitric Acid (Millon's Reagent)*

    Reagent: Digest 1 part mercury with 2 parts fuming HNO_3,
                and dilute solution with 2 parts H_2O.
    Procedure: Lightly spray and heat at 35°. Repeat if re-
                quired.
    Results: Detects phenols, phenol ether glycosides, and
                aromatic methoxy compounds.

154. *p-Nitrosodimethylaniline*

    Reagent: 1% p-nitrosodimethylaniline in 50% ethanol.
    Procedure: Spray.
    Results: Phenols produce spots of various colors.

155. *Silver nitrate, Alkaline*

    Reagent: Solution (a):   Saturated aqueous silver nitrate-
                                          acetone, 1:20.
                Solution (b): 0.5% NaOH in 80% ethanol.
    Procedure: Spray with (a), allow to dry, spray with (b).
    Results: Dihydroxy and many polyhydroxy compounds produce
                gray or gray-brown spots. Simple monohydroxy
                compounds usually react more slowly. Many
                dihydroxy phenols react before alkali is applied.

156. *Starch-Iodate*

    Reagent: Solution (a): 11% starch.
                Solution (b): 1% potassium iodate.
    Procedure: Spray with (a), then (b); while still wet expose
                plate to UV for 3-4 min.
    Results: Iodophenolic and iodoamino acid compounds produce
                blue spots that fade quickly.

157. *Ultraviolet Light*

    Procedure: Examine chromatogram under long-wave (360 nm)
                UV light.
    Results: Some phenolic acids fluoresce blue, green, pur-
                ple, or yellow. Identification should not be
                made solely on this basis; special chemical tests
                must also be made.

    Plasticizers

158. *p-Nitroaniline*

    Reagent: Solution (a);   0.5 N KOH in ethanol.
                Solution (b):   1 g *p*-nitroaniline in 200 mL
                                   2N HCl.
                Solution (c):   5% sodium nitrite.
                Spray solution: Add (c) to 10 mL of (b) until
                                   solution is colorless.
    Procedure: Spray plate with (a) and heat at 60° for 15 min.
                Then spray with solution of (b) and (c).
    Results: Detects plasticizers.

159. *Resorcinol-Sulfuric Acid*

    Reagent: Solution (a):   20% resorcinol in ethanol, with
                                   a pinch of zinc chloride added.
                Solution (b):   4N H_2SO_4.
                Spray solution: 1:1 mixture (a):(b).
    Procedure: Spray and heat at 120° for 10 min. Cool and ex-
                pose to NH_3.
    Results: Phthalate ester plasticizers detected. Sensi-
                tivity 20 µg.

160. *Thymol-Sulfuric Acid*

    Reagent: Solution (a): 1% thymol in ethanol.
                Solution (b): 4N H_2SO_4.
    Procedure: Spray with (a), heat at 90° for 10 min. Spray
                with (b), and heat at 120° for 10 min.
    Results: Plasticizers detected.

    Steroids, Sterols, Bile Acids

See reagents 1-4, 12, 21, 31, 70, 88, 91, 94, 99.

161. *Allen Test*

Reagent: Prepare a solution of 40 parts conc. H_2SO_4, 9
           parts ethanol, and 1 part H_2O.
Procedure: Spray carefully. May have to heat.
Results: 16-hydroxy steroids and their acetates produce
           mauve, rose, purple, or yellow spots.

162. *Anisaldehyde-Antimony Trichloride*

Reagent: Mix 1 mL p-anisaldehyde with 100 mL saturated
           antimony trichloride in chloroform. Add 2 mL
           conc. H_2SO_4. Keep solution at room temperature
           in the dark for 1.5 hr.
Procedure: Take off upper layer of reagent mixture and use
           it to spray plate. Dry 5 min in dark, and heat
           at 90° for 3 min. Observe in regular and UV
           light.
Results: Detects steroids.

163. *Anisaldehyde-Sulfuric Acid*

Reagent: 1 mL conc. H_2SO_4 is added to a solution of 0.5
           mL anisaldehyde in 50 mL acetic acid. Prepare
           fresh before use.
Procedure: Spray and heat at 110° until spots fully develop.
           The pink background may be bleached by exposure
           to steam.
Results: Detects steroids, phenols, terpenes, and sugars.

164. *Benzoyl Chloride*

Reagent: Solution (a): Dissolve 20 g zinc chloride in 30
                          mL glacial acetic acid.
           Solution (b): 50 g benzoyl chloride in enough
                          chloroform to make 100 mL solu-
                          tion.
Procedure: Spray with (a), heat for 5 min at 90°. Cool and
           spray with (b). Heat at 90° for 2-3 min. Ob-
           serve under visible and UV light.
Results: Detects steroids.

165. *Formaldehydogenic Reagent*

Reagent: Solution (a):   15 g ammonium acetate in 85 mL
                           80% methanol.
           Spray solution: Before use, add the following

to (a): 1 mL acetylacetone, 0.3
mL glacial acetic acid, and 0.1
mL 50% $HClO_4$.

Procedure: Spray and observe under UV. Fluorescent spots
will develop after 10 min and are most intense
after 60 min.

Results: 20-hydroxy- and 20-ketonic-21-ols produce green-
yellow fluorescent spots.

166. *Glycolic Acid*

Reagent: 50% aqueous glycolic acid.

Procedure: Spray and heat at 80° for 45 min. Observe under
254 nm UV.

Results: Detects ketals of α,β-unsaturated steroid
ketones.

167. *Hydroxylamine-Ferric Chloride*

Reagent: Solution (a):   14% hydroxylamine in methanol.
Solution (b):   3.5N KOH in methanol.
Solution (c):   2% ferric chloride in 10% HCl.
Spray solution: 5 parts (a), 4 parts (b),
shake, filter off KCl.

Procedure: Spray plate, allow to dry 10 min, then spray
with (c). If brown $Fe(OH)_3$ appears, let dry and
spray again with (c).

Results: Steroid alcohol esters (other than formates),
steroid acid methylesters, and steroid lactones
produce purple spots.

168. *Iodoplatinate Reagent*

Reagent: Solution (a):   5% platinum chloride in 1N HCl.
Solution (b):   10% KI.
Spray solution: Mix 5 mL (a), 45 mL (b), and
100 mL H_2O. Store in dark.

Procedure: Spray.

Results: Girard's t-hydrazones produce red to orange
spots. Ketosteroids also detected.

169. *Methylene Blue*

Reagent: Dissolve 25 mg methylene blue in 100 mL 0.5N
H_2SO_4. Before use, dilute 1 part with an equal
part of acetone.

Procedure: Spray.

Results: Steroid sulfates produce white zones on a blue background. Further spraying with 0.01% rhodamine 6G in $CHCl_3$ yields red spots on a blue background. Generally, sensitivity is increased.

170. *Perchloric Acid*

Reagent: 2% aqueous perchloric acid.
Procedure: Spray and heat 10 min at 150°.
Results: Steroids produce brown spots.

171. *Phosphomolybdic Acid*

Reagent: 5% phosphomolybdic acid in absolute ethanol. Filter. Refrigerate.
Procedure: Spray and heat 10-15 min at 120°.
Results: Detects steroids, lipids, antioxidants as blue spots on a yellow background. May also be used as a dipping solution.

172. *Phosphoric Acid*

Reagent: Dilute 1 part H_3PO_4 with 1 part H_2O.
Procedure: Spray until plate appears transparent and heat at 120° for 20 min. Observe in visible and UV light.
Results: Steroids and bile acids produce various colors under visible and UV light.

173. *Porter-Silber Reagent (Phenylhydrazine·HCl)*

Reagent: Dissolve 43 mg phenylhydrazine hydrochloride in a solution of 34 mL ethanol, 41 mL H_2SO_4, and 25 mL H_2O.
Procedure: Spray.
Results: Steroid-21-aldehydes react immediately with the reagent.

174. *Resorcylaldehyde (2,4-Dihydroxybenzaldehyde)*

Reagent: Solution (a):   0.5% resorcylaldehyde in glacial acetic acid.
             Solution (b):   5% H_2SO_4 in glacial acetic acid.
             Spray solution: Mix equal parts of (a) and (b).
Procedure: Spray and heat at 110° until maximum color intensity develops.

Results: 16-dehydrosteroids and their acetates produce
             blue, mauve, red, or orange spots.

175. *Sodium Hydroxide*

Reagent: 10% NaOH in 60% aqueous methanol.
Procedure: Spray and heat 10 min at 80°. Observe under 360
             nm UV.
Results: Δ^4-3-ketosteroids produce yellow fluorescence in
             long-wave UV.

176. *Tetrazolium*

Reagent: Solution (a):   0.5% blue tetrazolium in meth-
                              anol.
             Solution (b):   6N NaOH in H_2O or methanol.
             Spray solution: Equal parts of (a) and (b) are
                              mixed before use.
Procedure: Spray. Heat may be necessary.
Results: Corticosteroids, 16-hydroxy-17-keto steroids and
             2-hydroxy-3-keto steroids produce blue spots.

177. *p-Toluenesulfonic Acid*

Reagent: 20% p-toluenesulfonic acid in $CHCl_3$.
Procedure: Spray and heat at 100° for a few minutes. Ob-
             serve under 360 nm UV.
Results: Steroids, flavonoids, and catechins fluoresce in
             long-wave UV light.

178. *2,3,5-Triphenyltetrazolium Chloride*

Reagent: Solution (a):   4% triphenyltetrazolium chlo-
                              ride in methanol.
             Solution (b):   1N NaOH in methanol.
             Spray solution: Mix equal parts of (a) and (b).
Procedure: Spray and heat at 110° for 5-10 min.
Results: Steroids produce red spots. Also detects
             glycosides, reducing sugars, and thio acids.

179. *Xanthydrol*

Reagent: Dissolve 0.2 g zanthydrol in 90 mL ethanol and
             10 mL conc. HCl.
Procedure: Spray.
Results: Some steroids react. Also detects phenolic
             acids and pyrolles.

180. *Zimmermann Reagent (m-Dinitrobenzene)*

   Reagent: Dissolve 3 g m-dinitrobenzene in 190 mL methanol.
               Add 10 mL propylene glycol, 5 mL H_2O, and a sol-
               ution of 2.5 g KOH in 20 mL methanol.
   Procedure: Spray. May have to heat at 110°.
   Results: 3- and 17-oxosteroids are detected.

   Steroid Glycosides

See reagents 59, 88, 94, 184.

181. *1,3-Dinitrobenzene*

   Reagent: Solution (a): 10% 1,3-dinitrobenzene in
                              benzene.
               Solution (b): 6 g NaOH in 25 mL H_2O. Add 45
                              mL methanol.
   Procedure: Spray with (a), heat at 60°, then spray with
               (b).
   Results: Cardiac glycosides produce purple spots, turning
               to blue. They fade rapidly.

182. *Trichloroacetic Acid-Chloramine T*

   Reagent: Mix 10 mL of freshly prepared 3% aqueous
               chloramine T with 40 mL of 25% trichloroacetic
               acid in 95% ethanol.
   Procedure: Spray and heat 5-10 min at 110°. Observe under
               UV.
   Results: Digitalis glycosides detected as blue spots
               under UV.

   Sulfur Compounds

See reagents 25, 53.

183. *Benzidine*

   Reagent: 50 mg benzidine in 100 mL 1N acetic acid.
   Procedure: Spray and heat at 110°.
   Results: Persulfates produce blue spots.

184. *Copper Sulfate*

    Reagent: 10% copper sulfate.
    Procedure: Spray and heat at 120° for 20 min.
    Results: Sulfur-containing glycosides produce brown spots
                on a green background.

185. *p-Dimethylaminobenzaldehyde*

    Reagent: 1% p-dimethylaminobenzaldehyde in 5% HCl.
    Procedure: Spray.
    Results: Detects sulfonamides.

186. *Formaldehyde*

    Reagent: 40% formaldehyde-H_2O-H_2SO_4 (1:45:55).
    Procedure: Spray and heat at 120° for 10 min.
    Results: Detects phenothiazines.

187. *Iodine-Azide*

    Reagent: Solution (a):   1.27% I_2 in ethanol.
                Solution (b):   3.25% sodium azide in 1:3 H_2O-
                              ethanol.
                Spray solution: Mix equal volumes of (a) and
                              (b).
    Procedure: Spray.
    Results: Thiols and disulfides produce white spots on
                brown iodine background, which fades. Spots
                will remain visible in UV even with background
                color gone. Thio-ethers do not react. Detects
                thioureas and penicillins.

188. *Isatin*

    Reagent: 0.4% isatin in conc. H_2SO_4.
    Procedure: Spray, observe, and heat at 120°. Observe again.
    Results: Thiophene derivaties detected.

189. *2-Naphthol-Sodium Nitrite*

    Reagent: Solution (a): 1% sodium nitrite in 1N HCl. Pre-
                               pare fresh.
    Procedure: Spray with (a), allow to stand 1 min, then spray
                with (b). Heat at 60°.
    Results: Sulfonamides and aromatic amines detected.

190. *Periodic Acid-Perchloric Acid*

   Reagent: Dissolve 10 g periodic acid and a few mg vana-
                dium pentoxide in 100 mL 70% $HClO_4$. Mix well.
   Procedure: Spray.
   Results: Thiophosphoric acid esters detected.

191. *Potassium Ferricyanide-Ferric Chloride*

   Reagent: Solution (a): 1% potassium ferricyanide.
                Solution (b): 1% ferric chloride.
   Procedure: Spray with (a) and then (b). Observe under UV.
   Results: Detects thiosulfates.

192. *Resorcinol-Ammonia*

   Reagent: 10% resorcinol.
   Procedure: Spray plate and expose to NH_3 vapor.
   Results: Sulfonic acids detected.

193. *Sodium Periodate-Benzidine*

   Reagent: Solution (a): 1% sodium metaperiodate.
                Solution (b): 0.5% benzidine (CAUTION!) in
                               butanol-acetic acid (4:1).
   Procedure: Spray with (a). Wait 4 min. Then spray with
                (b).
   Results: Bivalent sulfur compounds produce white spots on
                a blue background.

Terpenes (Azulenes, Terpene Acids)

194. *p-Dimethylaminobenzaldehyde-Phosphoric Acid*

   Reagent: Dissolve 1 g p-dimethylaminobenzaldehyde in 5 g
                H PO and 50 mL acetic acid.
   Procedure: Spray and heat at 100°.
   Results: Azulenes produce blue spots.

195. *Phenol-Bromine*

   Reagent: 50% phenol in carbon tetrachloride.
   Procedure: Spray and expose to bromine vapor.
   Results: Terpenes detected.

196. *Phosphotungstic Acid*

  Reagent: 25% phosphotungstic acid in ethanol.
  Procedure: Spray and heat at 115° for 2 min.
  Results: Triterpenoids detected. Sensitivity 3-5 µg.

197. *Stannic Chloride*

  Reagent: 30% stannic chloride in chloroform.
  Procedure: Spray and heat at 90° for 5-20 min. Observe in
              visible and UV light.
  Results: Detects triterpenes, sterols, steroids, phenols,
              and polyphenols.

198. *Vanillin*

  Reagent: 1% vanillin in conc. H_2SO_4 or 1% vanillin in 50%
              H_3PO_4.
  Procedure: Spray and heat at 120° for 10-20 min. Observe
              in visible and UV light.
  Results: Terpenes detected.

  Vitamins

See reagent 88.

199. *Cacotheline*

  Reagent: 2% aqueous cacotheline.
  Procedure: Spray and heat at 100°.
  Results: Vitamin C detected as purple spot.

200. *4-Chloro-1,3-Dinitrobenzene*

  Reagent: Solution (a): 1% 4-chloro-1,3-dinitrobenzene in
                             methanol.
              Solution (b): 3N NaOH.
  Procedure: Spray with (a), then with (b).
  Results: Nicotinic acid, nicotinamide, and pyridoxol
              detected.

201. *o-Dianisidine, Diazotized*

  Reagent: Solution (a): 5% sodium carbonate.
              Solution (b): 0.5 g o-dianisidine·2HCl in 60
                             mL H_2O.

Solution (c):   5% sodium nitrite.
Solution (d):   5% urea.
Spray solution: Add 6 mL conc. HCl, 12 mL (c),
                 and 12 mL (d) (after 5 min) to
                 (b). Allow the solution to
                 stand 24 hr before use. It is
                 stable 10 days.

Procedure: Spray heavily with (a) and then lightly with the
            spray solution.

Results: T-Tocopherol and δ-tocopherol produce purple and
            red spots, respectively. Other tocopherols do
            not react.

202. *Dipicrylamine*

Reagent: Add 1 g dipicrylamine and 0.12 g magnesium
            carbonate to 15 mL H_2O. Heat the mixture 15
            min on a boiling H_2O bath and filter.
Spray solution: Using 0.2 mL of the above sol-
                 ution, add 50 mL methanol, 49
                 mL H_2O, and 1 mL NH_4OH.
Procedure: Spray.
Results: Detects vitamin B_1.

203. *α,α'-Dipyridyl-Ferric Chloride*

Reagent: Solution (a): 2% α,α'-dipyridyl in chloroform.
            Solution (b): 0.5% aqueous ferric chloride.
Procedure: Spray with (a), then spray with (b).
Results: Detects vitamin E, phenols, and other compounds
            with reducing properties.

204. *Iodine-Starch*

Reagent: 0.001-0.005% I_2, with KI, in 0.4% starch solu-
            tion.
Procedure: Spray.
Results: Ascorbic acid detected as white area on blue
            background.

205. *Methoxynitroaniline Sodium Nitrate*

Reagent: Solution (a): Dissolve 0.25 g 4-methoxy-2-
                           nitroaniline in 62 mL glacial
                           acetic acid. Dilute to 125 mL
                           with 10N H_2SO_4.
            Solution (b): 0.2% sodium nitrite.

Solution (c):   1:1 solution of (a) to (b).
Solution (d):   2N NaOH.
Procedure: Spray with (c) and then (d).
Results: Vitamin C produces a blue spot on an orange
             background.

206. *Potassium Ferricyanide*

Reagent: Solution (a):   1% potassium ferricyanide.
             Solution (b):   20% NaOH.
             Spray solution: Mix 1:15:5, (a):H_2O:(b).
Procedure: Spray, allow to dry. Observe under UV.
Results: Vitamin B_1 detected.

207. *N,2,6-Trichloro-p-benzoquinoneimine*

Reagent: 0.1% reagent in ethanol.
Procedure: Spray and expose to NH_3.
Results: Vitamin B_6 produces blue spots.

7.6 *IN SITU* CHEMICAL REACTIONS

The generally inert characteristics afforded by the layer sorbents
commonly used make them ideal media for *in situ* chemical reactions
other than those normally done for visualization by spraying and
dipping. Such reactions may include oxidation, reduction, acetylation, esterification, hydrolysis, and others. Miller and
Kirchner (39) were probably the first to describe *in situ* reaction
techniques on TLC plates. The sample to be reacted is applied to
the plate, and the reagent is placed on top of this spot. The reaction is allowed to proceed and then the plate is developed in the
normal manner. Applying known reference compounds to the plate
before development will allow comparison of reaction products and
unreacted compounds with the knowns. Reactions may also be run on
a microscale in a small tube or spot test plate and the mixture applied to the TLC plate along with known reference compounds. This
procedure is often used by the researcher doing synthesis to monitor the progress of the reaction. One nice feature is the microscale on which the reactions are carried out.

    There are two broadly classified types of reaction products.
In the first, the product is identifiable via a defined reaction.
One such example is the reduction of cinnamaldehyde by lithium
aluminum hydride to form cinnamyl alcohol (39). The second product type would include those that are complex and unidentifiable
and that may serve as merely "fingerprint methods."

Dallas (40) has summarized the many reactions that have been carried out on plates as shown in Table 7.1. The many different ways in which these reactions may be carried out are summarized in Table 7.2 (40). A few comments about some of these procedures follow.

TABLE 7.1 *IN SITU* REACTIONS ON TLC PLATES

Reaction	Reference
1. Acetylation	41, 42, 43, 44
2. Dehydration by H_2SO_4	39
3. Derivative formation	
(a) Acetate	41, 42, 43, 44
(b) Dinitrobenzoate	39
(c) Dinitrophenyl (of amino acid)	45, 46
(d) Dinitrophenyl hydrazone	42, 43, 47
(e) Methyl ester	42
(f) Phenyl isocyanate	39
(g) Semicarbazone	39
4. Diazotization	48
5. Esterification of acids	42
6. Halogenation	43, 49, 50, 51
7. Hydrogenation, catalytic	43
8. Hydrolysis with acid	52, 53, 54, 55, 56
9. Hydrolysis with alkali	39, 43
10. Isomerization	57
11. Methanolysis	54, 55, 58
12. Nitration	43
13. Oxidation with chromic acid	39, 43
14. Oxidation with O_2, O_3	51
15. Oxidation with peroxide, peracid	6, 41
16. Photochemical reaction	57
17. Pyrolysis	60
18. Reduction with $FeSO_4$	50
19. Reduction with $LiAlH_4$	39
20. Reduction with $NaBH_4$	43, 60
21. Reduction with Sn + HCl	48

7.6.1 Oxidations

The sample to be oxidized may be applied at the origin and covered with a solution of chromic anhydride (saturated) in glacial acetic acid. Application of 30% hydrogen peroxide and exposure to UV for 10 min may also work (39).

TABLE 7.2   TECHNIQUES FOR *IN SITU* REACTIONS (40)

Code	Method
Reaction Types	
0	Organized, intended
S	Spontaneous, unintended
D	Defined reaction
F	Fingerprint method
Methods of Reaction Application	
AR	Radiation, heat, light
AG	Gas, vapor
AL	Liquid spray
AM	In mobile phase
AS	In stationary phase
Thin Layer Chromatography	
X1	One-dimensional
X2	Two-dimensional
X2 RSS	Two-domensional with reaction before first development
X2 SRS	Two-dimensional with reaction between developments
SX RSRS	Two-dimensional with reaction before each development
M	Miscellaneous

Malins and Mangold (6) separated and oxidized unsaturated methyl esters in the presence of saturated compounds by developing the plate in peracetic acid-acetic acid-water (2:15:3). The oxidizing agent was thus present in the mobile phase (method type, Table 7.2 0/D/AM/X1).

7.6.2   Reductions

A 10% solution of lithium aluminum hydride in anhydrous ether may be used to effect reduction of a sample applied to the TLC plate. Scotney and Truter (50) have used ferrous salts for the *in situ* reduction of peroxides (method type 0/D/AL/X2 SRS) to identify the products of autooxidation of lanostenyl acetate.

Yasuda (61) incorporated a zinc reductor in the layer sorbent to reduce nitro groups.

Kaess and Mathis (60) used sodium borohydride for reduction *in situ*, and nitric acid or chromic acid for exidation in the

identification of alkaloids (method type o/D/AL/X1). Other reactions, carried out in capillaries before application to the
layer, are also discussed.

Elgamal and Fayez (43), working with pharmacological compounds, have described methods for many *in situ* reactions on both
alumina and silica gel, including reductions with sodium borohydride.

7.6.3 Esterification

Working with natural waxes, Holloway and Challen (42) used *in situ*
acetylation and esterification with methanol reactions to aid in
identification (method type O/D/AL/X1). Bennett and Heftmann (62)
esterified steroidal sapogenines *in situ* with trifluoroacetic
anhydride. The trifluoroacetic acid formed as a by-product was
removed in a hood. Riess (63) employed *in situ* methylation for
organophosphorus acids.

7.6.4 Methanolysis

In situ methanolysis of lipids and related compounds has proved of
use in this area. Kaufmann et al (64) have used 12% KOH in methanol for the methanolysis of phospholipids on silica gel for both
qualitative and quantitative analysis of mixtures (method types
o/D/AL/X1, X2 SRS). Oette and Doss (58) have compared *in situ* and
"off-plate" methanolysis of lipids (method type O/D/AL/X1).
Viswanathan et al. (54, 55) have developed an interesting procedure involving two-dimensional separation and the hydrolysis of
plasmalogens and the methanolysis of acyl phosphatide residues *in
situ* (method type O/D/AG + AL/X2 RSRS).

7.6.5 Miscellaneous Reactions

Cyclopropene acids were identified and determined by means of
their specific reaction with silver nitrate on the plate (method
type O/D/AS/X1) (65).

Dallas (44) described the partial acetylation of polyglycer-
ols *in situ* (method type O/D/AL/X2 SRS).

Stedman (47) formed the dinitrophenylhydrazine (DNPH) derivatives of the bromophenacyl esters of some aromatic acids (method
type O/D/AL/X1).

Aromatic nitro compounds have been reduced on silica gel by
hydrochloric acid and tin, and the amines formed diazotized (48).
Baggiolini and Dewald (56) have used HCl vapor for the partial
hydrolysis of p-aminobenzoic esters and sulfonamides on silica gel
(method type O/D/AG/X1).

7.7 DERIVATIVE FORMATION PRIOR TO TLC

It is often convenient to prepare derivatives of the substances to
be separated before they are applied to the plate. A suitable reaction procedure must be established to produce the desired derivative, and then a suitable mobile phase must be found to separate
the derivative.

Procedures for the formation of various derivatives of different compound classes may be found in textbooks of qualitative
organic analysis, the chromatography literature, and in the literature of the reagent specialty and chromatography suppliers.
These commercial firms offer the chemicals and apparatus necessary
to produce many derivatives and describe the procedures to go with
them.

A number of examples of derivative formation prior to TLC are
presented here to illustrate the applicability of these procedures.

The 2,4-dinitrophenylhydrazones of urinary amino acids (66),
urinary ketosteroids (67), and plasma ketosteroids (68) have been
prepared from the extracted compounds prior to their chromatographic separation. In the two ketosteroid procedures, thin layer densitometry was performed upon the separated derivatives for quantitative analysis.

For urinary ketosteroids (67) the following procedure is applicable: 5 mL of urine is acidified to pH 1 with 5N HCl, saturated with solid ammonium sulfate, and extracted with 2×20 mL of
ethyl acetate. The extract is made neutral with 0.5 mL NH_4OH and
evaporated to dryness. The residue is shaken in 40 mL of 1%
$HClO_4$ in diethyl ether and incubated for at least 15 hr at 39°.
A short boiling should precede the tight stoppering of the tube or
flask. After incubation, the ether is washed with 5 mL of 5N NaOH
and 3×5 mL H_2O, dried over anhydrous sodium sulfate, and evaporated to dryness.

The derivatization reaction is performed by adding to the
residue 0.1 mL of 0.2% (w/v) 2,4-dinitrophenylhydrazine in ethyl
acetate. The ethyl acetate is evaporated and 1 mL of 0.03% (w/v)
trichloroacetic acid in dry benzene is added and kept for 40 min at
40°. After this period, the solution is diluted with benzene to the
original volume of the urine so that aliquots can be easily taken.

The hydrazones may be chromatographed on silica gel G using
unidimensional double development in chloroform-carbon tetrachloride (2:1) followed by chloroform:dioxane (94:6). Quantities as
low as 1 μg are visible as yellow spots.

Stable yellow derivatives of estrogens may also be formed
from Fast Violet Salt B (6-benzoylamino-4-methoxy-t-toluidine,
diazonium salt) (69,70). These may be separated by TLC and quantitated if desired. The formation of the derivatives allows the
determination of as little as 0.05 μg of steroid.

7.8 ULTRAVIOLET LIGHT

Compounds absorbing ultraviolet light may often be located on thin
layer chromatograms if they show differential UV absorption relative to the background, or if they reemit the incident light.
Such compounds may be visualized by illuminating the plate with a
weak source of monochromatic UV light. A number of UV lamps and
lamp cabinets are available commercially; one such cabinet is
shown in Figure 7.1. These are essentially the "mineral lights"
that come in rock-collecting kits. Two wavelengths are commonly
used, 254 nm and 366 nm, as different substances may respond best
to only one wavelength.

Many compound types may be visualized with UV light including
aromatic amines, 366 nm (7:1); alkaloids, 254 nm (72); chlorinated
pesticides with silver nitrate, 254 nm (73); and polycyclic aromatic hydrocarbons, 366 nm (74-76). See the section on spray reagents for further applications of UV light in conjunction with
chemical reaction. The visualization of UV-absorbing compounds is
best achieved by a fluorescence-quenching technique, in which a
fluorescent substance is incorporated in the sorbent layer before
development or the chromatogram is sprayed with a fluorescent
agent between development and observation under UV light.

Mangold and Malins (6,8) first used 2',7-dichlorofluorescein
as a "nondestructive" spray reagent. The reagent is usually made
up as a 0.2% solution in ethanol, and most organic compounds yield
yellow-green fluorescent spots on a purple background when viewed
under 254 nm UV light. This reagent has been widely used because
of its general applicability. Kirchner et al. (31) first used inorganic phosphors, such as zinc silicate and zinc cadmium sulfide,
incorporated into the layer sorbents.

As mentioned in Chapters 2 and 3, commercially prepared sorbents and thin layer plates are available with incorporated fluorescent indicators: 254 nm, 366 nm, or both.

Commonly used fluorescent spray solutions include, in addition to the previously noted 2',7-dichlorofluorescein, 0.004-0.04%
sodium fluorescein (75,77); 0.05% aqueous Rhodamine B (78-81);
0.005-0.05% Morin in methanol; Rhodamine 6G (82); and fluorescamine (Fluram, Roche) (83-86). The use of some of these is discussed in the spray reagents sections.

Often the intensity of fluorescence on a TLC plate may be
increased by exposure to ammonia vapor or by spraying with ammoniacal fluorescein solution (87,88). This occurs as a result of
differential absorption or reemission of the incident light. Other
workers (84,86,89) have found that the intensity may be increased
by spraying with a less volatile organic amine such as triethanolamine (10% in methylene chloride) (84,86), or 50% aqueous morpholine (89).

REFERENCES

1. Z. Gregorowica and J. Sliwiok, Microchem. J., *16*, 480 (1971).
2. J. C. Touchstone, A. K. Balin, T. Murawec, and M. Kasparow, J. Chromatogr. Sci., *8*, 433 (1970).
3. J. C. Touchstone, T. Murawec, M. Kasparow, and W. Wortmann, J. Chromatogr., *66*, 171 (1972).
4. J. C. Touchstone, T. Murawec, M. Kasparow, and W. Wortmann, J. Chromatogr. Sci., *10*, 490 (1972).
5. J. C. Touchstone and T. Murawec, in *Quantitative Thin Layer Chromatography*, J. C. Touchstone, Ed., Wiley-Interscience, New York, 1973.
6. D. C. Malins and H. K. Mangold, J. Amer. Oil Chemists' Soc., *37*, 576 (1960).
7. H. K. Mangold, J. Amer. Oil Chemists' Soc., *38*, 708 (1961).
8. H. K. Mangold and D. C. Malins, J. Amer. Oil Chemists' Soc., *37*, 383 (1960).
9. G. C. Barrett, Nature, *194*, 1171 (1962).
10. J. H. Copenhauser, D. R. Cronk, and M. J. Carver, Microchem. J., *16*, 472 (1971).
11. O. Hutziger, J. Chromatogr., *40*, 117 (1969).
12. J. S. Matthews, V. A. L. Pereda, and P. A. Aquilera, J. Chromatogr., *9*, 331 (1962).
13. G. C. Barrett, in *Advances in Chromatography*, Vol. 11, J. C. Giddings and R. A. Keller, Eds., Marcel Dekker, New York, 1974, p. 151.
14. M. Z. Nichaman, C. C. Sweeley, N. M. Oldham, and R. E. Olsen, J. Lipid Res., *4*, 484 (1963).
15. E. Vioque and R. T. Holman, J. Amer. Oil Chemists' Soc., *39*, 63 (1962).
16. R. P. A. Sims and J. A. G. Larose, J. Amer. Oil Chemists' Soc., *39*, 232 (1962).
17. W. Brown and A. B. Turner, J. Chromatogr., *26*, 518 (1967).
18. M. Wilk and U. Brill, Arch. Parm., *301*, 282 (1968).
19. H. E. Vroman and G. L. Baker, J. Chromatogr., *18*, 190 (1965).
20. B. V. Milborrow, J. Chromatogr., *19*, 194 (1965).
21. L. Khin and G. Szasz, J. Chromatogr., *11*, 416 (1963).
22. R. J. Gritter and R. J. Albers, J. Chromatogr., *9*, 392 (1962).
23. R. J. Gritter and R. J. Albers, J. Org. Chem., *29*, 728 (1964).
24. M. E. Tate and C. T. Bishop, Can. J. Chem., *40*, 1043 (1962).
25. H. Ganshirt, Arch. Pharm., *296*, 73 (1963).
26. F. K. Grutte and H. Gartner, J. Chromatogr., *41*, 132 (1969).
27. H. J. Petrowitz and G. Pastuska, J. Chromatogr., *7*, 128 (1962).
28. A. Lynes, J. Chromatogr., *15*, 108 (1964).
29. D. Braun and H. Geenen, J. Chromatogr., *7*, 56 (1962).
30. S. Miyazaki, Y. Suhara, and T. Kobayashi, J. Chromatogr., *39*, 88 (1969).

31. J. G. Kirchner, J. M. Miller, and G. J. Keller, Anal. Chem.,
     23, 420 (1951).
32. L. L. Smith and T. Foell, J. Chromatogr., 9, 339 (1962).
33. W. L. Anthony and W. T. Beher, J. Chromatogr., 13, 367 (1964).
34. J. G. Kirchner, Thin Layer Chromatography, Wiley-Interscience,
     New York, 1967, pp. 147-191.
35. "Guide to TLC Visualization Reagents," J. T. Baker Chemical
     Co., Phillipsburg, N.J.
36. "TLC Visualization Reagents and Chromatographic Solvents,"
     Publication JJ-5, Eastman Kodak Co., Rochester, N.Y., 1973.
37. G. Zweig and J. Sherma, Ed., Handbook of Chromatography, Vol.
     2, CRC Press, Cleveland, 1972, pp. 111-172.
38. K. G. Krebs, D. Heusser, and H. Wimmer, in Thin Layer Chro-
     matography, E. Stahl, Ed.; Springer-Verlag, New York, 1969,
     pp. 855-905.
39. J. M. Miller and J. G. Kirchner, Anal. Chem., 25, 1107 (1953).
40. M. S. J. Dallas, J. Chromatogr., 48, 193 (1970).
41. G. Mathis and G. Ourisson, J. Chromatogr., 12, 94 (1963).
42. P. J. Holloway and S. B. Challen, J. Chromatogr., 25, 336
     (1966).
43. M. H. A. Elgamal and M. B. E. Fayez, Anal. Chem., 266, 408
     (1967).
44. M. S. J. Dallas, J. Chromatogr., 48, 225 (1970).
45. G. Pataki, J. Chromatogr., 16, 541 (1964).
46. G. Pataki, J. Boroko, H. Ch. Curtius, and F. Tancredi, Chro-
     matographia, 1, 406 (1968).
47. E. D. Stedman, Analyst, 94, 594 (1969).
48. A. R. Thawley, J. Chromatogr., 38, 399 (1968).
49. M. Wilk and U. Brill, Arch. Pharm., 301, 282 (1968).
50. J. Scotney and E. V. Truter, J. Chem. Soc., 199, (1968).
51. M. Wilk, U. Hoppe, W. Taupp, and J. Rochlitz, J. Chromatogr.,
     27, 311 (1967).
52. H. H. O. Schmid and H. K. Mangold, Biochim. Biophys. Acta,
     125, 182 (1966).
53. L. A. Horrocks, J. Lipid Res., 9, 469 (1968).
54. C. V. Viswanathan, F. Phillips, and W. O. Lundberg, J.
     Chromatogr., 38, 267 (1968).
55. C. V. Viswanathan, M. Basilio, S. P. Hoevet, and W. O.
     Lundberg, J. Chromatogr., 34, 241 (1968).
56. M. Baggiolini and B. Dewald, J. Chromatogr., 30, 256 (1967).
57. W. Schunack and H. Rochelmeyer, Arch. Pharm., 298, 572 (1965).
58. K. Oette and M. Doss, J. Chromatogr., 32, 439 (1968).
59. R. N. Rogers, Anal. Chem., 39, 730 (1967).
60. A. Kaess and C. Mathis, Ann. Pharm. France, 24, 753 (1966).
61. S. K. Yasuda, J. Chromatogr., 13, 78 (1964).
62. R. D. Bennett and E. Heftmann, J. Chromatogr., 9, 353 (1962).
63. J. Riess, J. Chromatogr., 19, 527 (1965).

64. H. P. Kaufmann, S. S. Radwan, and A. K. S. Ahmad, Fette
    Seifen Anstrichmittel, *68*, 261 (1966).
65. A. R. Johnson, K. E. Murray, A. C. Fogerty, B. H. Kennett,
    J. A. Pearson, and F. S. Shenstone, Lipids, *2*, 316 (1967).
66. K. Figge, Clin. Chim. Acta, *12*, 605 (1965).
67. P. Knapstein and J. C. Touchstone, J. Chromatogr., *37*, 83
    (1968).
68. P. Knapstein, L. Treiber, and J. C. Touchstone, Steroids, *11*,
    915 (1968).
69. J. C. Touchstone, A. K. Balin, and P. Knapstein, Steroids,
    13, 199 (1969).
70. J. C. Touchstone, T. Murawec, M. Kasparow, and A. K. Balin,
    J. Chromatogr. Sci., *8*, 81 (1970.
71. M. Gillio-Tos, S. A. Previtera, and A. Vimercati, J.
    Chromatogr., *13*, 571 (1964).
72. A. H. Der Marderosian, H. V. Pinkley, and M. F. Dobbins, Am.
    J. Pharm., *140*, 137 (1968).
73. M. F. Dobbins and J. C. Touchstone, in *Quantitative Thin
    Layer Chromatography*, J. C. Touchstone, Ed., Wiley-Inter-
    science, New York, 1973, p. 293.
74. E. Sawicki, Chemist-Analyst, *53*, 64 (1964).
75. T. Wieland, G. Luben, and H. Determann, Experientia, *18*, 430
    (1962).
76. G. M. Badger, J. K. Donnelly, and T. M. Spotswood, J.
    Chromatogr., *10*, 397 (1963).
77. T. H. Simpson and R. S. Wright, Anal. Biochem., *5*, 313 (1963).
78. C. Michalec, M. Sulc, and J. Mestan, Nature, *193*, 63 (1962).
79. J. Battailie, R. L. Dunning, and W. D. Looms, Biochem.
    Biophys. Acta, *51*, 538 (1961).
80. J. W. Copius-Peereboom and H. W. Beekes, J. Chromatogr., *14*,
    417 (1964).
81. S. Hunek, J. Chromatogr., *7*, 561 (1962).
82. C. B. Scrignar, J. Chromatogr., *14*, 189 (1964).
83. S. Udenfriend, S. Stein, P. Bohlen, W. Dairman, and M.
    Weigle, Science, *178*, 871 (1972).
84. A. M. Felix and M. H. Jimenez, J. Chromatogr., *89*, 361 (1974).
85. J. Sherma and J. C. Touchstone, Anal. Lett., *7*, 279 (1974).
86. K. Imai, P. Bohlen, S. Stein, and S. Udenfriend, Arch.
    Biochem. Biophys., *161*, 161 (1974).
87. M. Brenner and A. Niederwieser, Experientia, *16*, 378 (1960).
88. R. C. Hignett, J. Chromatogr., *31*, 571 (1967).
89. L. D. Hunter, J. Chromatogr., *20*, 595 (1965).

CHAPTER 8
Documenting the Chromatogram

8.1 EVALUATING THE CHROMATOGRAM

The process of documenting a developed chromatogram consists of two major steps: (a) evaluating the chromatogram and (b) recording the chromatogram. Good and complete evaluation aids recording.

When the developed chromatogram is removed from the developing chamber, it is first observed in visible light and then in short-wave and long-wave ultraviolet light. If the sorbent layer is one of the harder types, such as cellulose, notations may be made with a pencil right on the layer, if desired, after each observation. If the layer is one of the softer types, notations may be conveniently made with a dissecting or disposable syringe needle. The solvent front, and any irregularities in it, should be marked immediately as the plate is being removed from the developing chamber.

When any visible or UV-visible areas have been noted, the chromatogram may be further visualized by chemical reaction by spraying or dipping the plate (if feasible) with suitable reagents as determined from the literature. Any positive areas that appear immediately are noted, and the plate is heated if desired. Once again, any positive areas are noted, including location, color, intensity, etc. The notation of color is often difficult because color assessment is subjective, and shading and intensity have an influence. If this is found to be a problem, a standardized color chart, as used in the paint industry or in philately, may partially alleviate the problem. Nybom (1) devised a color key system for marking fluorescent areas obtained from plant compounds. His system is reproduced here, as it may prove of practical usefulness to other workers. See Figure 8.1.

After noting spot locations and colors, R_f values of hR_f values ($R_f \times 100$) may be measured. Spots that identify with known standard substances may be identified by name. Areas that are not immediately identified may be characterized by R_f values, color reactions, chemical tests, etc.

8.2 RECORDING THE CHROMATOGRAM

Published procedures for chromatographic separations are, from the experience of many workers, often difficult to reproduce (2). The major reason for this is the lack of experimental data recording in the publication. The fault for this lies in two places, either with the authors who fail to include such pertinent data in the manuscript, or with editors who screen such data out before publication to "simplify" the report. All authors should make a definite effort to report all details.

The following recommendations are presented here as a guide to the reporting of chromatographic data. The list covers most situations; a summary of major items of data that should be included in every report follows the detailed list.

Guide To Reporting Chromatographic Data

Tank or chamber

1. Name of manufacturer; internal dimensions.
2. How used--for example, in special chromatography room or on open bench.
3. Tank sealed by glass contact, grease, etc.
4. Lined with paper or cloth, etc., for atmosphere saturation.
5. Saturated or not, and if so, how and for how long before chromatography.
6. Temperature of room or chromatography cabinet.
7. Humidity and relative humidity--this is of extreme importance in many separations.

Paper or thin layer

1. Manufacturer of paper or thin layer; whether home-made, etc.

General rules

(1) Spots are outlined with solid or stippled lines according to their intensity and distinctness:

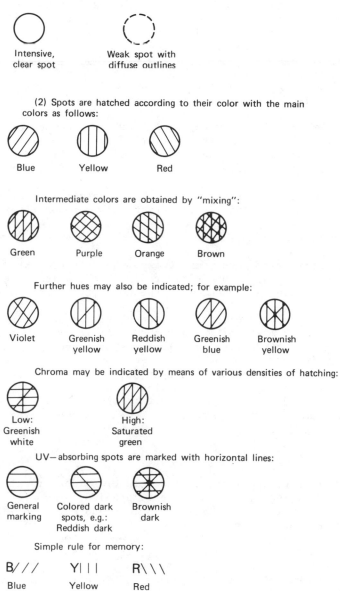

Intensive, clear spot   Weak spot with diffuse outlines

(2) Spots are hatched according to their color with the main colors as follows:

Blue   Yellow   Red

Intermediate colors are obtained by "mixing":

Green   Purple   Orange   Brown

Further hues may also be indicated; for example:

Violet   Greenish yellow   Reddish yellow   Greenish blue   Brownish yellow

Chroma may be indicated by means of various densities of hatching:

Low: Greenish white   High: Saturated green

UV—absorbing spots are marked with horizontal lines:

General marking   Colored dark spots, e.g.: Reddish dark   Brownish dark

Simple rule for memory:

B/ / /   Y| | | R\ \ \

Blue   Yellow   Red

Figure 8.1. Key for marking the color of fluorescent spots on chromatograms, according to Nybom (1). Courtesy of Journal of Chromatography.

2. Type of paper or thin layer, including whether channeled
or serrated; dimensions used; foil, plastic, or glass plates;
thickness of layer; wet or dry.
3. Any additives such as fluorescent indicators, buffers,
binders, etc.
4. Ascending or descending technique, etc.; pretreated by
user prior to chromatography; distance of origin line from edge;
length of time of run; whether rerun using same or different conditions.
5. Method of supporting paper or layer; number of plates
in the tank -- this can have a major effect on separation, and
frequently many workers keep blank plates to fill the tank
so that this factor remains constant (blanks can be dried and re-
used).
6. Method of activation, if any, together with subsequent
method of storage. This is important also for quantitative work.

Mobile phase

1. Details of each component of the solvent mixture, e.g.,
petroleum ether (b.p. 100-200°), redistilled, dried or not. If
aqueous solutions are used, then the molarity or concentration
should be stated.
2. Composition of solvent; monophasic or upper or lower layer of two-phase mixture; whether the lower layer was used for saturation or rejected; give total volumes of solvents used as well
as proportions.
3. Equilibration using one or both phases, wetting of internal lining of tank, time of equilibration.
4. Method of drying plate after solvent run.
5. Number of times mobile phase can be used before replace-
ment is necessary.

Sample application

1. Type of applicator; manufacturer's identification.
2. Spot, band, or streak, etc. applied.
3. Solvent used to dissolve solute and method of drying
spots or streaks after application.
4. Was the plate handled, or was a "no-touch" method used?
Was the whole paper or layer open to the atmosphere, or totally
covered with only the origin exposed?
5. Amount of solute and volume of solution applied, and size
of spot or streak; analytical or preparative run.

Location reagents and methods

1. Spray, dip, radiation monitoring, etc.; if viewed in UV, then wavelength used; if a photodensitometer was used, then quote wavelengths used, filters, speed, type of record, etc. (give de tails of color and stability of colored product, fluorogen, etc.); time-lapse between visualization and analysis.

2. Composition of chemical or biological reagent used; stability of reagent and storage conditions; order of mixing of components of the reagent if important, together with any time factors involved; room temperature or oven heating.

3. Details of color and other changes that occur with time.

4. Quote R_f values or, better, relative R_f values, i.e., R_x. For sugars, glucose is often used; for amino acids glycine or proline, etc. If identity is claimed with a known compound, was the known run in parallel or by applying both to the same spot (isographic), and in how many different mobile phases?

Additional details to be reported would include details of derivatives formed, if any. If the mobile phase reacts with the visualization reagent, the drying procedure should be carefully and completely described.

The above condenses to the following items that should be permanently recorded about every chromatogram, usually in the research notebook and with the chromatogram itself, accompanied by a cross reference:

> Date
> Chromatogram number
> Substances applied to plate; e.g., nature of extract, name
>   and concentration of known substances
> Amount of substances applied
> Sorbent layer, including brand name, indicator, special prep-
>   aration, thickness
> Plate size
> Activation procedure, if any
> Mobile phase systems used
> Length and type of development
> Temperature
> Developing chamber used; type, size
> Drying procedure
> Detection and visualization
> Identifications made and evaluation

It is often convenient to prepare a standard or printed form with space for this information so that it is easily filled in and filed or stored. A card index of filed or punched cards may also

be useful in that the chromatograms can be categorized in many
ways.

The actual chromatogram can be preserved or documented in a
number of ways.

One of the simplest ways to preserve a chromatogram or a portion of one is to press cellophane tape or clear contact paper on
top of the layer such that some excess hangs out from either side
of the plate. A smooth surface, such as a beaker bottom, is rub-
bed on top of the tape so that the adhesive comes into uniform contact with the layer. As the tape is carefully peeled away, the
upper portion of the layer should be attached, and the whole may be
fastened in a notebook with the excess tape at the edges.

Berlet (3) recorded chromatograms on transparent paper. After
the developed plate has been observed and marked under UV light or
sprayed with a visualization reagent, it is covered with a clean
uncoated plate of equal size. This plate will protect the layer.
A sheet of transparent tracing paper is clamped in place on top of
this with four small spring clips. The spots are traced onto the
paper with any desired notations. A light box may be located
underneath to aid visibility and the location of weakly reacted
spots. This procedure has the advantage that it leaves the layer
intact for any additional analytical procedures such as *in situ*
quantitation or additional spraying.

8.3 DOCUMENTING BY PHOTOGRAPHY

Photography is certainly the preferred method for the permanent
recording of chromatograms. Both standard color (4-6) and Polaroid
(7-9) have been used for photography in visible light, and color
(4,7,10,11) and black and white (8,9) have been used for photography in UV light.

A complete synopsis of photographic documentation of TLC
plates, including details on filters, exposure monitoring, and experimental photography, with actual photographs of TLC plates for
illustrations has been prepared by Heinz and Vitek (12). The worker interested in such documentation should consult this article.

Caldwell et al. (5) have devised two data sheets for use in
photographing a chromatogram; the first of these is shown in Figure 8.2. Data are entered on the sheet prior to development, with
any additional data or remarks entered during visualization and
evaluation. The sheet is placed adjacent to the chromatogram and
photographed. The sheet is filed in a folder, loose-leaf book, or
notebook for subsequent filing of the photograph.

Figure 8.3 illustrates a small (2.75×8 in.) legend sheet for
use with 20×20 cm plates that have been developed for a distance
of 10 cm or less. Data are placed on the sheet prior to chroma-

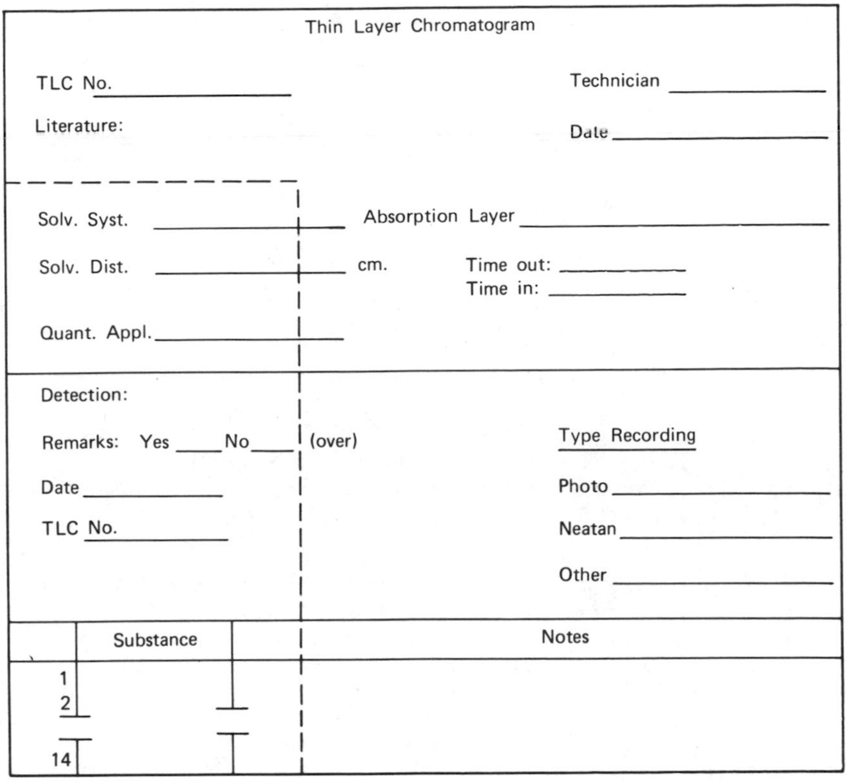

Thin Layer Chromatogram

TLC No. _____   Technician _____

Literature: Date _____

- - - - - - - - - - - - - - |

Solv. Syst. _____|____   Absorption Layer _____

Solv. Dist. _____|____ cm.   Time out: _____
                          | Time in: _____

Quant. Appl._____|____

Detection: |

Remarks: Yes ____ No ____| (over)   Type Recording

Date _____ | Photo _____

TLC No. _____ | Neatan _____

                           | Other _____

	Substance		Notes
1			
2			
14			

Figure 8.2. Legend sheet (8.5×11 in.) for use when photographing a 20×20 cm chromatogram. Portion to left of dashed line is included in photograph of plate (5). Courtesy of Journal of Chromatography.

tography, and after development and visualization additional comments may be entered before it is placed on the plate to be photographed with it. Caldwell et al. (5) reported the following photography conditions: visible light: Kodacolor X film, f5.6 for 1/50 sec, two 15-W fluorescent bulbs placed 18 in. above the plate. Ultraviolet light: High Speed Ektachrome film, f8 for 12 sec, two 15-W UV bulbs (General Electric F15T8 BLB, 3200-4000 H) placed 18 in. above the plate.

Xerox copying can also be used to record chromatograms. Felici et al (13) published a note explaining the conditions under which they obtained satisfactory results. The Xerox model 422 was found to give the best results compared to the other five copiers

```
┌─────────────────────────────────────────────────────────────────┐
│                        TLC No. _____                        │
│                                                                   │
│   1│              8│                    Date _____      │
│   2│              9│                    Solvent _____      │
│   3│             10│                    Detection _____      │
│   4│             11│                    Solv. Dist. _____      │
│   5│             12│                    Remarks:                   │
│   6│             13│                                               │
│   7│             14│                                               │
│                                                                   │
└─────────────────────────────────────────────────────────────────┘
```

Figure 8.3 Small legend sheet (2.75 8 in) for use when photographing a 20 20 cm chromatogram. When the solvent front is 10 cm
or less, this sheet will fit above it on the plate for recording
date (5). Courtesy of Journal of Chromatography.

tested. It was also preferred over Xerox models 3600 and 7000
because of their curved windows. Felici et al. note that Xerox
copiers involve reflection of blue/green light and they will not
satisfactorily reproduce blue and green colors. The intensity
setting on the copier apparently had no influence on the copies,
but it was found that intensity could be improved by copying the
copy. However, the background is also intensified, so there is a
limit to this.

Brinkmann Instruments, Inc. offers a binding liquid that dries
to form a flexible plastic coating. This liquid, known as Neatan,
is sprayed on the chromatogram and allowed to dry. After it hard-
ens, transparent tape is placed on top of the layer to peel it off.
It is not suggested for use with home-made coated plates.

Applied Science Laboratories offers a similar fluid called
Strip-Mix. The film is porous, allowing the extraction from the
sorbent of any substances that are soluble in chloroform or water.
The film may be placed, along with any attached radioactive substances, into scintillation fluid and counted without causing
significant quenching. The following steps are suggested for its
use:

1. Set the plate on a paper towel with the sorbent layer up.
2. Place spacers on opposite sides of the plate upon the
plate. These may be any narrow cardboard or metal strips approximately 0.5 mm thick and as long as the plate side.
3. Pour 20-25 mL of Strip-Mix onto one end of the plate, and

spread it evenly across the surface with a glass rod resting on the spacers.

4. Push the excess off one end onto the paper towel.

5. Allow the layer to dry 10-15 min.

6. Using a razor blade, cut out individual areas as desired or the entire layer. These areas should easily peel away.

The layer thus obtained may be used as a permanent record, or substances of interest may be extracted for further analysis, chromatography, radio-counting, etc.

REFERENCES

1. N. Nybom, J. Chromatogr., *26*, 520 (1967).
2. I. Smith, A. D. Baitsholts, A. A. Boulton, and K. Randerath, J. Chromatogr., *82*, 159 (1973).
3. H. H. Berlet, J. Chromatogr., *21*, 485 (1966).
4. R. Jackson, J. Chromatogr., *20*, 410 (1965).
5. R. L. Caldwell, J. K. Shields, and L. S. Stith, J. Chromatogr., *36*, 372 (1968).
6. J. F. Gonnet, J. Chromatogr., *86*, 192 (1973).
7. A. S. Milton, J. Chromatogr., *8*, 417 (1962).
8. E. Hansbury, J. Langham, and D. G. Ott, J. Chromatogr., *9*, 393 (1962).
9. K. Randerath and E. Randerath, J. Chromatogr., *16*, 111 (1964).
10. I. D. Jones, L. S. Bennett, and R. C. White, J. Chromatogr., *30*, 622 (1967).
11. G.PP. McSweeney, J. Chromatogr., *33*, 548 (1968).
12. D. E. Heinz and R. K. Vitek, J. Chromatogr. Sci., *13*, 570 (1975).
13. R. Felici, E. Franco, and M. Cristalli, J. Chromatogr., *90*, 208 (1974).

CHAPTER 9
Quantitation

9.1 INTRODUCTION

Concomitant with the availability of spectrodensitometers in recent years, quantitation by evaluation *in situ* is becoming common.
However, there are a number of ways in which quantitation of the
amount of a substance separated on a thin layer chromatogram can
be performed. Densitometry *in situ*, and other methods will be discussed in detail in this chapter.

Quantitative methods will not be successful unless the sample
and the reference standard are properly handled. The samples must
be kept in closed vials, and the solvents used should preferably
be of low volatility, since evaporation during application procedures can change the concentration. This is particularly true in
routine work when a standard solution must be opened repeatedly
for application of repetitive amounts on many chromatograms. A
Reacti-Vial (Pierce Chemical Co.) with a Teflon stopper through
which the needle of a syringe can be forced for sample withdrawal
seems ideal for overcoming this problem. This is not suitable for
the disposable glass sampling capillaries. Standard solutions
should not be kept too long, as they may become concentrated or may
decompose or develop extraneous contaminants.

9.2 VISUAL TECHNIQUES

From estimation by eye many years ato to *in situ* scanning in a
spectrodensitometer, quantitation of substances separated by TLC
has come a long way. Visual evaluation of thin layers can be carried out for quantitative purposes in a number of ways.

In the earliest attempts to quantitate the amount of material
separated on a plate the spot size was visually compared with the
size of a known amount of the equivalent substance separated on
the same chromatogram. Spot areas appear to be a logical first
attempt at quantitation using visual means. The relationship between the spot size and the amount of substance therein is influenced by the sorbent, the mobile phase, and the conditions of the
chamber. For this technique to succeed, good separation is needed, as well as the largest possible change of spot size per unit
of change in the amount of substance separated. At lower concentrations per spot area it is possible to obtain a linear correlation between weight and area. The area of the spot compared with
the density within the area may be the limiting factor here. It
is easier to evaluate spot size than small changes in the density
when visual evaluations are being made. Planimetry has been used
to measure the spot size, but this is difficult to do on friable
thin layers of sorbents.

The visual method has seen little use in recent reported ex-
perimentation. However, in some applications visual evaluation
could be all that is necessary. For example, in many quality control procedures a visual scan can show the presence or absence of
impurities in a given preparation.

There are two reasons for using the method:

1. To determine that impurities are within certain limits.
2. To determine the actual number and quantity of the impurities that may be present.

One example of the first use is to apply serial amounts of the
sample in question to the plate and after chromatography evaluate
the results visually. Ideally there would be only one spot present
if the solution contained a pure solute. This is the simplest
approach but has many shortcomings. At a low solute concentration
only one spot may be seen, whereas if a larger amount is applied
other minor constituents may show up in the sample. The results
may vary from laboratory to laboratory and between two individual
operators in the same laboratory, particularly if quantitation is
desired. Various conditions of plate development and spot travel
may affect the ability to see some solutes, particularly if they
are present in small amounts. The decision to apply a larger sample or to slightly vary the development conditions may depend upon

different situations. Generally, this approach should not be used
if reliable quantitative results are desired.

The other consideration in this type of study is the reactivity of the detection reagent with the solute of interest and the
impurities that may be present. It is not valid to assume that all
substances within a class will react to the same degree with the
reagent. This is also true when considering *in situ* scanning methods. Calibration curves for quantitation must be prepared for each
compound under consideration.

Probably the best way, and the method most often used, is to
apply standard quantities of the solutes in question and compare
the corresponding spot sizes after development. The use of visual techniques for quantitative purposes requires a knowledge of
the nature of the impurities and a source of the pure substance
for reference. Also required are well-separated spots. The R_f
should be high enough that the spot is not too concentrated. Spots
of R_f between 0.3 and 0.7 are more readily evaluated, since the
area per unit of solute is more uniform than those of higher or
lower R_f's. Spots of low R_f are too concentrated, and visual evaluation of area or intensity are erroneous. Likewise, the areas per
unit of mass for spots of high R_f are too diffuse and are therefore
difficult to estimate.

There are a number of ways that the spots can be visualized
other than by the inherent color of the solute discussed above.
The zones can be seen as dark spots against a fluorescent background on the F-254 sorbents. These sorbents contain a phosphor
that shows an intense fluorescent background when activated with a
short-wave ultraviolet lamp. If it is absorbing near the wavelength of activation of the phosphor, the solute will show up as a
dark spot against the brilliant background. The assessment of spot
quantity by unaided visual comparison is poor. It is difficult to
grade the spot area. The "measles effect" can be a disadvantage.
This refers to the phenomenon of seeing extra spots after looking
away from the brilliant plate several times or after blinking.

Fluorescent compounds can be detected on thin layers at the
nanogram and subnanogram level. Here again, because of visualization in dark rooms under ultraviolet light, problems arise. Pons
et al (1) and Fishbein (2), have studied the precision of quantitation of fluorescent aflatoxins separated by TLC. In quantitative measurements of fluorescence intensity, a possible error of
30-50% can occur when an unknown is visually compared to one of
two standards of different concentration. An error of 15-25% can
occur when an interpolation is made between two such standards, as
reported by Pons and Goldblatt in earlier work (3).

Johnson (4) has described the results of visual quantitation
of 4-chloroacetanilide in phenacetin. The quantity of sample that
amounted to 0.2 µg of phenacetin contained 0.01% 4-chloroacetin as

impurity. The results of the collaborative study of the use of
visual examination of the layers after separation is shown in
Table 9.1. As seen in the table, there was only fair agreement
among the different laboratories. This type of assessment of thin
layer chromatograms is subject to errors of 20% or more. As a
general evaluation technique, the procedure may be suitable for
routine quality control, but for precise quantitative purposes
visual examination is not always applicable.

TABLE 9.1   PERCENTAGE OF 4-CHLOROACETANILIDE FOUND IN
SAMPLE OF PHENACETIN QUANTITATED VISUALLY (4)

Sample No.	Laboratory				
	A	B	C	D	E
1	<0.01	<0.01	<0.01	0.005	--
2	0.01	--	0.01	--	0.02
3	0.02	0.04	0.02	0.03	--
	0.03	0.03			
4	0.07	0.05	0.07	0.07	0.06
5	0.04	0.08	0.07	0.06	--
	0.05				

9.3   EVALUATION BY MEASURING SPOT AREAS

After separation of the solute or solutes in question, the quantitation can be performed by measuring the spot area. This method
has been applied intensively in paper chromatography. It is probably easier to adapt the technique to paper chromatograms than to
TLC, since the paper is more uniform than thin layers and the spots
may in general be more discrete. However, the method has been extended to thin layers by many investigators. A number of variations of the general concept have been used; no one method seems
to be universally used.

Whether the relationship of area to concentration can be used
for quantitation depends on a number of properties of the chromatogram. Among these are:

1. Reproducible sorbent characteristics which define activity and characteristics of the sorbent that is used in the layer.
Modern commercial layers seem to fulfill this requirement. They
generally are uniform in thickness and give reproducible chromatograms and spot characteristics if the conditions can be reproduced.

2. Reproducibility of the application of the sample and the standard. Since the standards are applied at the same time as the sample, this is usually possible.

3. Use of a developing system that will separate the solutes as discrete and well-separated zones with no tailing.

4. Completeness of reaction of the solutes with the detection reagent. Dipping in the reagent is preferred to spraying, because it is easier to reproduce. However, the dipping procedures depend on the solvent-solute characteristics. Also, if aqueous sprays are used, care must be taken not to spray too heavily or the layer may be removed from its support. (See the section on application of detection reagents in Chapter 7.)

If the zones are distinctly outlined, they can be circumscribed with a sharp instrument so that there is a permanent marking of the area. This should be done at the same time for each chromatogram in order to obtain better reproducibility, since time, atmospheric conditions, and other factors can cause fading or reduction of the spots. After they are marked, the areas can be measured in a number of ways. The easiest way is to use a planimeter, as was done by Oswald and Flück (5). Another way to measure spot size or evaluate the results is to compare to areas of the standards visually, as described by Morrison and Chatten (6) and as noted earlier in this chapter.

Gänshirt and Polerman (7) described a procedure in which the spot areas were traced on writing paper, following which the spots were cut out and weighed. An adaptation of this is to trace the areas on squared paper (graph paper) and count the number of square millimeters. Since the areas are usually small, considerable error is present in these measurements.

Still another method of measuring area is to photograph the spots and then cut out the zones on the photograph and weigh them, as described by Schierf and Word (8).

To avoid preparing a calibration curve for each plate, a function expressing a relation between the amount of the solute applied and the spot area can be determined. Brenner et al. (9) have discussed application of this method to data obtained from thin layer chromatograms. Spot areas should be proportional to the logarithm of the solute amount under defined conditions. Petrowitz (10) determined tar oils, DDT, and γ-BHC within narrow limits using these relationships.

According to Purdy and Truter (11), the average deviations obtained in their work and others was 3.1% for 600 separations based on adsorption methods and 3.6% for 980 partition chromatograms. The systematic error is small, since the comparison is always made with the values of the standard solute, which should always be chromatographed on the same plate. Random errors are high because of variations in spot shape and the difficulty of defining the spot area. They can be pronounced in comparisons of results between plates or results obtained on different days.

In spite of the drawbacks mentioned, it is possible to use these simple methods to quantitate if the limitations are understood. With the uniform plates commercially available, it is easier to reproduce chromatographic results if ambient conditions can be controlled. Prerequisite for success in these methods is the use of minimal amounts of solute. Generally it has been found that practitioners of thin layer chromatography overload the plate with solute. Linear calibration curves can be more readily obtained when the spot areas are not overloaded. Solute zones that are too concentrated generally prevent complete reaction with the detection reagent. Furthermore, dark zones are more difficult to assess visually than a series of zones with varying degrees of intensity commensurate with the spot size.

9.4   ELUTION TECHNIQUES

The most widely used of all methods of quantitation of substances separated on thin layers is that of elution. The method of choice in these procedures depends almost entirely upon the nature of the solute separated as well as the sorbent used. A knowledge of the interplay between the two is very important for success in applying elution techniques. The basic problem is removal of the solute from the active sites on the sorbent surfaces. Often low recoveries are reported because these factors have not been considered. The selection of the solvent for elution is also important. After elution is accomplished the identification or quantitation of the solute can be performed by a variety of means. Depending on the nature of the compound separated, there are a number of advantages and disadvantages in detection for elution as well as the means of the actual quantitation. The focus in this section will be the chromatographic and elution problems. The compounds of interest will determine the mode of quantitation.

Before the solute can be eluted it must be located. This is simple if the compound is colored or fluoresces in ultraviolet light. However, this location procedure is not possible with the majority of compounds. They must be located by use of a reference compound separated on the same plate, which is visualized. Preferably the reference compound is spotted on the outer zones and in the middle zone of the chromatogram. This enables one to correct for uneven mobile phase flow. If reference compounds are not available, the solute in question can be applied to the chromatogram and then located by use of the appropriate detecting reagent, acting as its own reference for locating the zones to be eluted.

It is good practice to mark the solvent front at the end of a development. Usually the line (zones) of separated solute will be parallel to the solvent line. This can aid in the marking of the zones to be scraped from the support. After the chromatogram has been developed, several procedures can be used to locate the reference zones (spot where solute separated).

Some compounds, particularly if they absorb strongly in the ultraviolet regions, can readily be detected by exposing the chromatogram in the dark to an ultraviolet lamp. Often the amount of the compound must be rather high before the zones can be visualized. When marking the region to be eluted it is recommended that an area larger than the visualized zone itself be scribed into the sorbent to outline the area to be scraped off. This method also has an advantage as a preliminary screening, since the absorption in the ultraviolet region can be characteristic of the compound of interest. However, advantage can be taken of the absorptive characteristics of the compound when the chromatogram is developed on a sorbent containing phosphor. Under ultraviolet light the compounds that absorb in the ultraviolet region will appear as dark zones against the fluorescent background of the layer. The zones then can be marked into areas to be removed from the backing by scraping. The use of this technique is limited to those instances where traces of the phosphor will not interfere with the analytical technique used for the quantitation. This method is a nondestructive technique and is often used, since commercially prepared plates are available with the phosphor incorporated in the sorbent layer.

The methods described below are applicable to the majority of thin layer chromatograms. Probably the easiest method for locating a zone, particularly in the case of lipids or lipophilic substances, is to spray the chromatogram with water. The detection depends on the differential wetting properties of the sorbent and solute. The detection is easiest in oblique or transmitted light. It is generally an insensitive method, but for the reference compound this is not a matter of concern. It has been widely used in lipid work and occasionally to detect steroids (11,12). Varon et al. (13) have found the method to be sensitive for the location of estrogen acetates. This method is described in more detail in Chapter 7 under nondestructive detection methods. The technique can also be used on the solute itself, since it is nondestructive, and a reference compound may not be necessary except for use in location of the other compounds of interest.

A common nondestructive procedure is the exposure of the chromatogram to iodine vapors (14,15). The mechanism of this test is not known. It may be due to differential solubility of the iodine in the solute, formation of adducts, or the addition of iodine to double bonds present in the molecule. These reactions depend

on the type of compound present. It is good practice to test this
procedure on a chromatogram before using it in quantitative work
to be sure there is no permanent reaction of the reagent with the
compounds.

Iodine crystals are placed in the bottom of a tank, and after
some equilibration the chromatogram is placed in the tank. Usually the reaction is immediate. Yellow to brown spots are observed. After the spots are visualized, the zones are immediately
marked by scribing the area with a sharp instrument. If reaction
with the areas to be scraped must be avoided, the section of the
plate containing the reference can be cut off and exposed to the
iodine. As the iodine evaporates from the layer, the color eventually disappears.

Quantitation of zones may also be accomplished by spraying
the TLC with a color reagent, eluting the colored product, and determining it spectrophotometrically. This method, although successful in some cases, is beset with difficulties. Spraying is
difficult to reproduce, as the amount of reagent varies in different areas of the plate. The ratio of reagent to amount of solute
in the spot will also vary among different areas of the spot.
Sometimes it may be more difficult to elute the colored compound
than the original substance. Conversely, it is sometimes easier,
as described in the next paragraph.

An example of elution of a zone after spraying with a reagent
for color development was reported by Haefelfinger (16). Quantitation of diallyl-bis-nortoxiferine-dichloride was carried out by
color development with bromothymol blue. Apparently in this case
it was easier to elute the alkaloid from the plate after chromatography and after complexing with the dye. A twofold result was
obtained: easier elution of the solute as well as a means of
quantitating it. After elution, spectrophotometry or, as in this
case, optical rotation, can be used to assess the quantitative aspects.

As pointed out by Court (17), the boundary effect must be
avoided. Figure 9.1 illustrates this phenomenon. In spite of the
fact that the solvent front had moved in a uniform horizontal line
up the plate, the row of spots or line of the separated solute may
not be parallel to the solvent front. Sometimes it is possible,
particularly with biological samples, to see other solutes, either
fluorescent or absorbing under an ultraviolet lamp. These solutes
will form a line that can serve as a guide to mark the zone to be
scraped off. One can also overcome the boundary effect by including a reference in the middle region of the chromatogram, as described before.

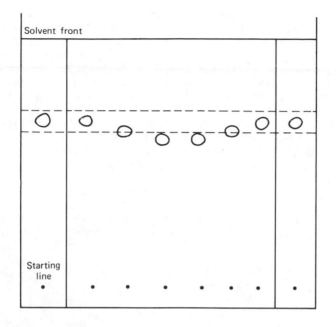

Figure 9.1.   Illustration of the boundary effect.   Outer lanes
contain the reference compounds to denote the zone to be removed.
Note that if parallel lines scribed by position of the reference
are followed, the zones in the middle of the chromatogram will be
missed.

9.4.1   Removal of the Sorbent from the Plate

After marking the zone containing the compound to be assayed, the
next step is to remove the sorbent and transfer it to a suitable
vessel for elution with the proper solvent.   The areas can be
scraped off with a razor blade or straight-edge spatula.   The sorbent can be collected on a piece of weighing paper or a similar
nonabsorbent sheet, such as cellophane.   From this the sorbent is
shaken into the tube or vessel for elution.

Many problems are related to this part of the elution process.
Air currents must be absent or small particles will be blown away
and result in loss of sample.   Some layers have a hard surface,
and care must be taken to prevent loss of the solute due to flaking.   When many samples are to be processed, this technique can be
tedious and subject to mechanical loss.

A number of suction devices on the market are designed to
avoid the problems of scraping the sorbent from the layers.   Typi-

cal of these as the all-glass suction apparatus pictured in Figure 9.2. The inlet tube of the apparatus is positioned close to
the layer within the marked zone of interest. The vacuum takes in
the sorbent and deposits it on the sintered glass disk. Two or
three small drops of water are added to deactivate the sorbent,
particularly if it is silica gel or alumina. Pure distilled solvent for elution is then drawn through the nozzle to elute the
solute out of the sorbent and into the attached collection flask.
Further operations, such as concentration, evaporation, rechromatography, and so forth may then be performed. This type of apparatus avoids mechanical losses found in scraping. The vacuum
extraction of Matthews et al. (18) is an example of this.

9.4.2 Elution of the Solute from the Sorbent

After addition of the appropriate solvent, the solute may be eluted by simple agitation of the test tube or container, followed by

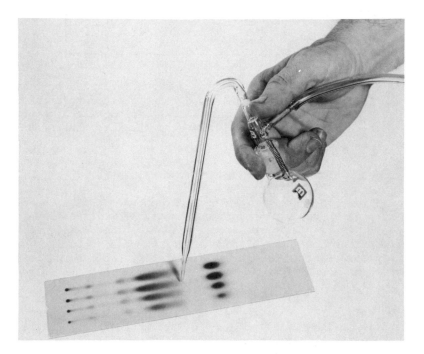

Figure 9.2. Typical suction device used to remove sorbents from
the support for elution. Courtesy of Brinkmann Instruments, Inc.

centrifugation to remove the sorbent. The solvent is then de-
canted, and the procedure is repeated as many times as necessary
to obtain complete recovery of the solute. The elution can sometimes be accomplished with a single extraction of the sorbent. A
small column containing the sorbent can also be used. The solvent
can be added to the top of the column and the effluent collected
in the appropriate vessel for the quantitative step. Lehmann et
al. (19) described a microcolumn to be used under pressure. Attal
et al. (20) used a 3-mL syringe with a 25-gauge needle. A plug of
glass wool in the bottom of the syringe served as support for the
sorbent. The syringe acting as a column is very practical. Elution methods are given in detail in Chapters 12 and 16.

The selection of the eluting solvent itself is important, especially when quantitation is desired. Completeness of elution
will be determined by the solvent. The selection of the solvent
is governed by the nature of the solute and of the sorbent. Constant percentage or completeness of recovery is desirable. Recovery can be assessed by using radioactively labeled reference
compounds. This is particularly important when working with biological samples when the expected amount is not known. Otherwise,
recovery can be determined by the amount of a known substance that
was eluted from the sorbent.

Aqueous solvents are not suitable for some elutions, since
calcium sulfate present in many TLC sorbents is slightly soluble
in water. Sometimes polar solvents such as ethanol extract other
constituents of the layers. If the R_f of the solute is over 0.8,
the developing solvent or its constituents can generally be used
for elution.

Deactivation of the sorbent by addition of enough water to
wet the small amount present has been found to increase recovery
of solutes using nonpolar solvents.

Selection of the solvent also must be considered from the
standpoint of the quantitative method to be subsequently used. If
spectrometry in the ultraviolet is to be used, the solvent must
absorb only weakly; otherwise the solvent must be evaporated and
the solute dissolved in a more appropriate solvent.

Fine particles of sorbent that can scatter light must be removed by centrifugation or filtration. Filtering through micro-
pore or membrane filters has been found useful for this purpose.
Clear infrared spectra and mass spectra have been obtained after
filtration of solutes removed from silica gel through 4-μm Milli-
pore membrane filters (21).

That deactivation of the sorbent is an important factor in
completeness of elution by solvents is illustrated by the experiments from the authors' laboratory summarized in Table 9.2. Radioactive deoxycorticosterone in equal amounts was applied to four
zones of a silica gel GF-254 plate. The chromatogram was not de-

TABLE 9.2   RECOVERY OF DEOXYCORTIOCOSTERONE-14 FROM
SILICA GEL LAYERS

1. Scraped dry into vial, scintillator then added	37%
2. Scraped wet into vial, scintillator then added	92%
3. Eluted (dry) 2 × with methanol	63%
4. Eluted (wet) 2 × with methanol	64%

veloped.   The zones were located under an ultraviolet lamp and
marked.   One zone was scraped directly into a scintillation vial;
the second was wetted with water and scraped into another vial.
One of the other two was scraped dry into a test tube and eluted
twice with methanol.   The fourth zone was wetted with water,
scraped into a test tube, and also extracted twice with methanol
and the extract transferred to the counting vial.   Scintillation
fluid was added to all four and counting was done.   In the first
two, the gel remained at the bottom of the vial and the scintillation fluid extracted the solute.   Recoveries stated in the table
were different.   Ganjam et al. (22) also point out some of the
problems related to elution of steroid from TLC plates.   They went
through considerable investigation to find a solvent that would
elute cortisol and corticosterone from silica gel.   Table 9.3
summarizes the results and also points out the well-known ineffi-
ciency in using the same solvent for eluting a variety of compounds
with varying polarities and chemical structures even in the same
class of compounds.

TABLE 9.3   COMPARISON OF DIFFERENT TECHNIQUES FOR ELUTING
CORTISOL AND CORTICOSTERONE FROM SILICA GEL

	Percent recovered*	
Elution method	Cortisol	Corticosterone
Partition of silica gel between benzene and water	1 ± 0.2	96 ± 1.3
Dichloromethane:methanol (9:1)	90 ± 1.3	92 ± 1.4
Cold methanol	16 ± 1.0	62 ± 1.0
Hot methanol	25 ± 1.0	65 ± 0.8
Cold ethanol	15 ± 1.0	55 ± 1.8
Hot ethanol	15 ± 1.3	60 ± 1.0
Chloroform:methanol (1:1)	33 ± 1.4	50 ± 1.0
Column chromatography	25 ± 1.3	71 ± 1.2

Impurities extracted from the sorbent probably are one of the largest sources of error when spectrophotometry is to be used for quantitation of eluted solutes. Silica gel and alumina contain inorganic impurities such as iron and chloride. Depending on how long the plates have been stored, many organic impurities can be absorbed from air pollutants, especially in an active organic laboratory. Impurities may also be absorbed from the packing material of the containers in which the commercial plates are supplied. Some of these problems can be avoided by using preliminary purification procedures. If the plates are prepared in the laboratory, the gel can be washed with dilute sulfuric acid or hydrochloric acid (23). Ethanolic hydrochloric acid has been used for washing the silica gel (24). Problems arise here, since these washing procedures may remove unknown amounts of binding material. Fine sorbent particles that are necessary for adhesion of the layer are removed, and new impurities may be added.

Sometimes commercial layers can be rid of extraneous material by development with the solvent to be used for elution. Methanol has been used in this way. If the solvent is allowed to overrun the plate in the developing tank, impurities are washed to the top of the plate. It is not unusual to see a yellow line at the solvent front and eventually a band of yellow at the top of the plate.

A unique way to avoid removal of the sorbent from the chromatogram and still allow elution *in situ* has recently been described by Falk and Krummen (25). The method requires a special instrument designed for the purpose and is entirely automatic. The use of this instrument, available commercially, is described below.

One to six samples are applied, evenly spaced, on TLC plates, and developed in the usual manner. A milling device is used to remove the sorbent layer around the spot to be recovered. This permits a tight seal between the backing of the plate and the Teflon elution head. The elution head is then placed over the milled-out area and pressed firmly into place with a clamping bar.

The syringe reservoirs are filled with the same solvent or with different solvents. The precise amount of solvent to be used for elution is preset, and the rest is automatic. From 1 to 5 mL of solvent flows from the syringe reservoirs through the elution head at a preselected rate of speed and into the cuvet. As the solvent flows through the elution head it flows through the sorbent and lifts the solute from the layer. When the preset volume is reached, the instrument automatically turns off. No intervention is required during the elution process, unless a different volume is desired for elution for each zone.

When the elution is complete, a pump forces air through the elution head to drive out the remaining solvent and to dry the layer under the elution head. The clamping bar is released and the

elution heads are removed. The resulting solution collected in
the cuvet or volumetric flask is free of sorbent and ready for
analysis.

The minimum amount of solvent that can be used is approximately 0.75 mL. This, of course, depends on the amount of solute
present and the elution power of the solvent and/or solubilities.
Also, the speed of elution is important. The slower the flow of
solvent, the better the elution. If set at a fast speed, larger
amounts of solvent are required because it then becomes more of a
washing effect, rather than the desired adsorption/desorption process. At the slowest possible elution speed it takes 10 min for
each 1 mL of solvent.

The utility of this apparatus and its applicability are dependent on the quantitative removal of the solute from the sorbent.
It is not always possible to obtain quantitative removal of solutes. Therefore, this type of instrumentation will have some limitations.

It presents one solution to the ever-present elution problem.
The instrument is limited to six zones. The eluting heads are
very wide, and it is necessary to have the desired solute well
separated from other materials or else scrape away the unwanted
material. It is also not adaptable to eluting solutes that have
been streaked the length of the origin of the chromatogram. This
instrument is described in more detail in Chapter 16 in connection
with other preparative techniques. Table 9.4 shows the recoveries
obtained on eluting different compounds using this procedure.

9.5 METHODS FOR MEASUREMENT AFTER ELUTION

The ultimate goal in quantitative analysis is to determine the
amount of a particular substance in a sample. When the procedure
is based on quantitation by some characteristic other than weight,
known amounts of pure reference material must be processed in the
same way to obtain calibration curves for interpolation of the
amount in the unknown. These observations also give information
on the statistical accuracy of the method.

Quantitation by gravimetric determination, in spite of its
poor accuracy, is the only technique that does not require calibration when enough solute or unidentified substance is available.
This method can be advantageous. It has been used for determination of the components of cabbage wax by Purdy and Truter (11).
Here again, one is faced with the problem of quantitative removal
of the solute from the sorbent.

Gravimetric analysis is usually included as part of the process of preparative thin layer chromatography. Sometimes a number
of chromatograms are processed together, and the desired solute is

TABLE 9.4 RECOVERIES OBTAINED BY AUTOMATIC ELUTION

Substance	Sorbent	Amount Applied (mg)	Solvent	Percentage Recovery	
				1st Fraction 0.5 mL	2nd Fraction 0.25 mL
Caffeine	Silica gel 60 F-254	36	Chloroform 90 Ethanol 10	100.5	0.25
Benzocaine	Silica gel F-254, basic	13.5	Chloroform 95 Ethanol 5	100.6	0.25
Phenacetone	Silica gel GF-254	18.4	Chloroform 95 Methanol 5 Conc. ammonia 0.15	99.2	0.25

collected from the layers and eluted in a single operation. To
find the area where the desired component has migrated, reference
substances must be applied to the chromatograms and used for guides
to locate the zones. This procedure is described in detail in preceding sections of this chapter. After elution the solvents are
evaporated, and after careful drying the sample is weighed. Since
the majority of work in TLC involves micro samples, the use of
gravimetric procedures has not been widespread.

9.5.1 Spectrophotometric Methods

Most of the quantitative evaluation of samples eluted from TLC
plates has been done by spectrophotometry. The extract is made up
to a standard volume, and spectrophotometry at the desired wavelength is performed. Many compounds will absorb in the ultraviolet region, and advantage can be taken of these absorption properties; others are colored, and a determination can be made using
the maximal wavelength of the chromophore. If the solute has none
of these properties and is colorless, it must be submitted to a
chromogenic reaction. Any of a great number of reagents may be
used to form a colored derivative after reaction with the sample
in question. Many suitable reagents for this purpose are discussed in Chapter 7. Often some of these reactions will result in
the determination becoming more sensitive, as in the case of the
reaction of dansyl chloride (26) or Fluram (27), which form fluorescent derivatives. With amino acids and other amines, the formation of fluorescent derivatives provides a method of increasing
the sensitivity as much as a hundredfold, as well as increasing
the specificity of the determination.

Some difficulties are encountered in spectrophotometry of
material eluted from thin layer chromatograms. These are inherent
in the sorbent. Some substances that absorb very strongly in the
regions below 250 nm are eluted from the sorbent, particularly from
some silica gels and aluminas. These impurities show a nonspecific
absorption spectrum. The intensity falls steadily as the wavelength increases and becomes negligible as the wavelength reaches
400 nm. Some of the impurities may be those washed to the upper
levels of the chromatogram during development and can be seen by
viewing the chromatogram in a dark room under a short-wave ultraviolet lamp.

One of the ways to minimize these impurities is to use as the
blank or control the extract of a blank portion of the sorbent
equal in area to that of the sample and from a zone of equivalent
R_f. This must be processed exactly as the sample and then used as
a blank in the reference beam of the spectrophotometer. The contribution of the blank can be a major proportion of the absorbance.
It varies widely with the source of the sorbent and the mobile

phase used for development of the chromatogram as well as the solvent used for the extraction of the solute.

Other analytical methods are available for quantitation and
identification of the solutes eluted from plates. Among these are
gas chromatography, infrared spectrometry, and nuclear magnetic
resonance. These techniques are outside the scope of this book,
although some of them are described in Chapter 16.

9.6 IN SITU QUANTITATION BY SCANNING

Quantitative TLC analysis involving elution before measurement is
time-consuming and laborious. Although measurement in situ as a
simpler method has been generally accepted, the theoretical basis
of such measurements is apparently little known. On the other
hand, recent progress in measuring techniques make direct quantitative TLC evaluation attractive and applicable to many areas of
research and routine work, if the same precision is maintained as
with elution methods that involve spectrophotometry of solutions
of the eluted solutes. Many in situ methods using densitometers
and spectrodensitometers are now available, and these can be very
precise, having a relative standard deviation of less than 5%.
Frequent reports indicate that it is possible to obtain precision
with standard deviations below 1%.

Since the early work of Privett and Black (28), Jork (29), and
Barrett et al. (30), using the densitometers then existing, many
instruments have been designed for in situ evaluation of thin
layer chromatograms. Some of these instruments represent accessories to existing spectrophotometers Others are spectrodensitometers designed for just this type of in situ scanning. These
instruments also have the capability of scanning gels from electrophoretic separations.

Basically, the optical systems of all of these instruments
can be categorized into the designs shown in Figure 9.3. They all
have a light source, condensing and focusing systems, and a photo-
sensing detector. Some of them have monochromators, whereas others
rely on filters for selection of proper wavelengths for operation.
Few of them have monochromators for both the exciting as well as
the emitted light in fluorescence.

Probably the simplest densitometer can be described as shown
in Figure 9.3(a). This design is embodied in some of the earlier
instruments that were originally designed to evaluate separated
zones in gels from electrophoretic separations. Some of these
have been used with limited success in scanning thin layer chromatograms.

In the true sense these are densitometers, since they are
rarely equipped with the adaptations to provide monochromatic

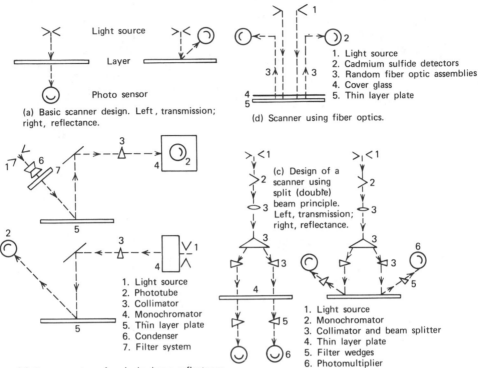

(a) Basic scanner design. Left, transmission; right, reflectance.

1. Light source
2. Cadmium sulfide detectors
3. Random fiber optic assemblies
4. Cover glass
5. Thin layer plate

(d) Scanner using fiber optics.

1. Light source
2. Phototube
3. Collimator
4. Monochromator
5. Thin layer plate
6. Condenser
7. Filter system

(b) Scanner set up for single—beam reflectance. Top, fluorescence scanning; bottom, absorbance.

(c) Design of a scanner using split (double) beam principle. Left, transmission; right, reflectance.

1. Light source
2. Monochromator
3. Collimator and beam splitter
4. Thin layer plate
5. Filter wedges
6. Photomultiplier

Figure 9.3.   How a densitometer works.

light and are usually constructed to scan in the transmission mode. The light source can be either above or below the thin layer, with the photosensor on the opposite side for transmission or on the same side for reflectance. The light source can be a hydrogen, mercury, high pressure mercury-xenon, or xenon lamp for ultraviolet light or an incandescent lamp for visible light.

Figure 9.3(b) is a simple block diagram of an arrangement that can be used in accessories to be attached to available spectrophotometers. In the lower diagram the light from source 1 passes through a momochromator 4 where the desired wavelength is selected and condensed in lens 3 to be reflected from the mirror and directed to the thin layer 5. The reflected light is then viewed by the phototube at position 2 stationed at a 45° angle to the plate. Alternatively in this system, the lamp can be positioned at 1, with the light passing through condenser 6 and filter system 7 before striking the layer at a 45° angle. The perpendicularly reflected light is caught by a deviating mirror and directed through a collimator, 3. The light then passes through the monochromator, 4, to isolate the desired wavelength and detection is by the photomultiplier, 2, which has been moved behind the monochromator. This is shown by the upper diagram in Figure 9.3(b). This instrument was designed to follow reflectance determinations.

Neither of the designs described in Figure 9.3(a) and (b) has provisions for scanning in the double-beam mode. A number of authors have pointed out that this mode is necessary for precise quantitation in scanning of thin layers because of background noise. The instrumental diagrams in Figure 9.3(c) and (d) were designed to provide double-beam operation.

Figure 9.3(c) is a diagram of an instrument designed for both double- and single-beam operation. It can be used in the transmission or reflectance modes as denoted in the diagrams. It has a beam-dividing system, which provides one beam for the reference part of the layer and another for the sample scanning.

Figure 9.3(d) describes the basic design of an instrument that presents a departure from the usual method of carrying the light to the layer. Glass fiber optics are used to transmit the light from the source directly to the plate.

Two randomly gathered fiber optic light pipes with two 1 cm heads are mounted side by side so that the beams of light from the source will be carried through the fibers perpendicularly to the layer. The reflected light is passed back through the reverse fiber and directed to the photosensors. The scanning head is close to the layer and allows it to collect most of the reflected light, since the loss of scattered light is now at a minimum.

In the discussion so far, only optical layouts have been described. Recorders to visualize the scans and integrating devices are generally included as parts of the list of equipment needed to perform quantitative work. If a monochromator is not used, filters should be available. The provisions for scanning include (1) movement of the plate over or under the light beam with a stationary viewing photomultiplier or (2) movement of the light beam along with the enclosed viewer while the plate is stationary. Most instruments have the moving plate design. Recording charts can be synchronized with the speed of the plate.

The diagram does not show how the photomultiplier response is handled. Some instruments use only an indicating meter calibrated in absorbance units. Others in double-beam operation amplify and then feed the signal into logarithmic converters. At the output the reference signal is subtracted from the sample signal and fed into an integrator. From there the signal goes to a conventional line recorder, which records the ratio of the two beam signals.

There are a number of modes of operation of these instruments, but not all modes are available on all instruments. All the instruments have some of the operating parameters, and selection of an instrument should be based on the use intended for it. The most versatile instruments have the following capabilities:

1. Scanning in the transmission mode.
2. Scanning in the reflectance mode.
3. Fluorescence scanning.
    a. Transmission mode
    b. Reflectance mode
4. Fluorescence quenching evaluation.

The choice of mode depends on the compound in question and the nature of the sorbent.

Before one can determine the capabilities of the instrument
on hand, a few basic principles of the theory and instrument variables should be understood. Calibration of the instrument in the
different modes is an important prerequisite. Novacek has reviewed the theoretical and practical aspects of densitometry as it
is practiced today (31).

Present methods are frequently based upon empirical calibration. To achieve linearity, absorbance of various functions of
transmission and reflectance are plotted versus weight or function
of weight of the light-absorbing compounds. Therefore, the question often arises whether to prefer measurements using transmission or reflectance techniques.

The following is a review of the underlying theory for transmission and reflection of light in highly scattering media. Theoretical conclusions, for example, that transmission is to be preferred to reflectance, are discussed with regard to measuring
methods used with presently available densitometric instrument
systems.

The effects of instrument variables on measurement accuracy
for the determination of colorless UV-absorbing compounds by fluorescence quenching and reflectance and the quantitation of colored compounds are discussed in terms of response-peak areas versus amount spotted.

The theory of radiative transfer in scattering media is due
to Chandrasekhar (32). However, the transport equation has no analytical solution, and therefore simplified expressions are employed. The most widely used is that of Kubelka and Munk (33,34).
The work of Goldman and Goodall (35), based on the Kubelka-Munk
theory, has shown that on theoretical grounds measurement using
transmission is more advantageous than that using reflectance.
These workers have also proved this experimentally. After exam-
ining the conditions under which the Kubelka-Munk equations hold
true, they concluded:

1. The entering light need not be exactly parallel and perpendicular to the specimen; however, the angular distribution of
incident light should be kept reasonably constant.

2. Monochromaticity of the incident light is important.
Uniform absorption by the sample within the applied spectral band
width is necessary to comply with Beer's law.
3. The illuminating aperture must be small so as to cover a
limited range of absorptions. This range depends upon the average
absorption of the measured area and the precision required.
4. Nonuniformity of the sample spot through the depth of the
layer is likely to have a greater effect on measurements using reflectance than upon those in which transmission modes are used.
5. Variations in the scattering coefficient over the area of
illumination, through the depth of the layer and over the wavelength range, are likely to be smaller than the corresponding
variation in absorption.

Using these results and the Kubelka-Munk formulas, Goldman and
Goodall computed the values for absorbance in transmittance and in
reflectance relative to the background for a wide range of variables including coefficients of absorption in relation to layer
thickness. The results are shown in Figures 9.4 and 9.5.
From the data shown in the figures, it is obvious that:

1. Linearity of response to absorption of light by a zone on
a layer is more closely approached in the transmission mode than
in the reflectance mode.

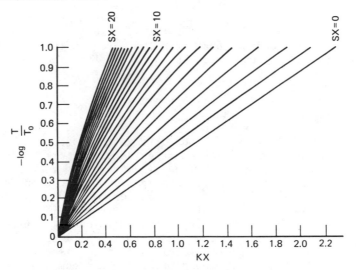

Figure 9.4. Absorbance as measured in the transmission mode.
     KX = absorption relative to layer thickness
     SX = scatter relative to layer thickness

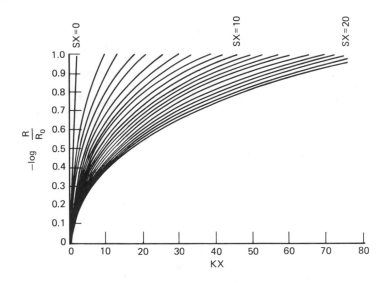

Figure 9.5.   Absorbance as measured by reflectance.
    KX = absorption relative to layer thickness
    SX = scatter relative to layer thickness

    2.   The response in reflectance is less than the Beer's law
slope (0.4343 in this case), whereas in transmission the response
is greater.   This means that for the same sample spot the scattering power of the medium substantially increases the recorded peak
area obtained in transmission and decreases the recorded peak area
in reflectance.

9.7   DENSITOMETER USE

There are, as we have seen, four different ways by which a chromatogram can be scanned with existing densitometers for solutes
separated on thin layers.   The densitometers are operated either
in the transmission or reflectance mode.   Colored compounds or
colorless non-UV absorbing compounds (derivatized) can be quantitated by the determination of absorption in the visible wavelengths, either by reflectance or transmission.   Colorless materials that absorb UV light can be determined by measuring the absorption of the incident UV radiation in the reflectance mode.
When the sorbent contains phosphor, colorless compounds that absorb in the UV regions also can be quantitated using measurements
based upon fluorescence quenching.   This technique can be carried
out in either the transmission or reflectance mode.   Fluorescence

can also be used for the quantitation of minute amounts of substance that are fluorescent or have been made fluorescent.

9.7.1 Instrument Variables

Some understanding of instrument variables is necessary for precision work in scanning thin layers. Touchstone et al. (36) have made a critical evaluation of various operating modes of a double-beam densitometer.

The properties of the light source in the instrument should be considered and the lines of highest emission energy should be known. This information is usually provided with the instrument. Different lamps have different spectral characteristics. The response curves of the detecting element, whether a photomultiplier or a cadmium sulfide diode, should also be known.

The spectral characteristics of a xenon-mercury lamp (Hanovia 901.B-11) and the response curve of the photomultipliers in one instrument show the mercury bands of the xenon-mercury lamp at 312/313 and 365/366. The band at 365/366 nm is considerably more intense than at other wavelengths. The lamp has a curve of intensity rising gradually from 200 nm to a hump at approximately 500 nm and then leveling off. The photomultipliers have a maximum response to light of 380 nm. Before this wavelength there was a shoulder, then a continual decline in response. Above 380 nm the response gradually declined and become almost zero at 570 nm. As seen in Figure 9.6, the peaks of highest emission of the lamp coincided with the bands of greatest activation of the fluorescence of quinine sulfate separated on a silica gel layer. Sometimes it is possible to obtain greater sensitivity by using light of wavelength closer to the lines of greatest emission of the lamp.

Figure 9.7 illustrates the importance in quantitative work of operating at the peak of absorption of excitation bands. In the case of 2,7-dichlorofluorescein, scanning away from the maximum of excitation (315 nm) resulted in nonlinear curves. Scanning at 315 nm gave a linear calibration curve.

The importance of scanning at the wavelength of maximum absorption of the solute in the sorbent is illustrated in Table 9.5. The maximum of absorption of a compound adsorbed on silica gel particles is not the same as the maximum of the absorption spectrum of the compound in solution in alcohol. For precise and sensitive determinations the absorption maximum must be determined *in situ* on the plate. This is possible in the instruments with sources of variable wavelength or monochromators. For the instruments using filters, narrow-bandpass filters are available, and one should work as close as possible to the wavelength of maximal absorption.

In terms of the Kubelka-Munk theories as modified by Goldman and Goodall (see previous discussions), certain other conditions of the instrument must be considered.

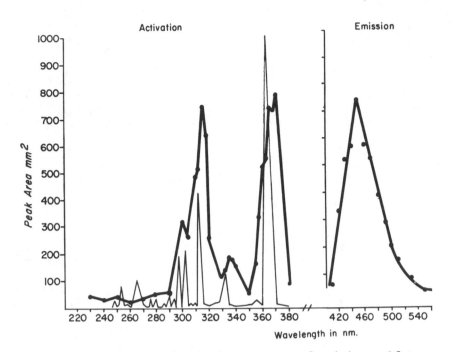

Figure 9.6. Activation and emission spectra of quinine sulfate
(dark line). Also shown are the peak of maximal energy of the
light source (light line).

1. Angular distribution of the incident light must be kept
constant. This condition is met in most instruments now available.
In spectrodensitometers the light is almost completely collimated.
2. Monochromatic light is desirable. This condition is best
fulfilled in instruments equipped with monochromators.
narrow-bandpass filters or wedge monochromators can be used.
3. The aperture through which the incident light passes
should be small enough to illuminate a limited range of absorption.

Ideally the "flying spot" design of an instrument meets these conditions best. In addition to being rather expensive in terms of
system complexity and unnecessary for most chromatography applications, this scanning technique is for single-beam operation. Sub-
stantial variations in background signal are obtained in its use,
due to a small illuminated area, because of nonuniformity of the
layer itself, and various "impurities" caused by development of
the plate.
For these reasons, the high noise levels and strongly changing
absorption profiles of the layer along the plate make the single-

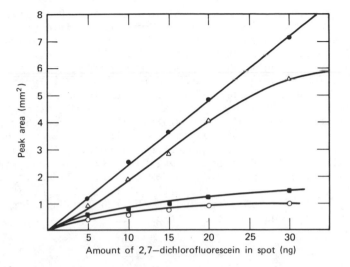

Figure 9.7 Regression lines resulting from exciting 2,7-dichloro-
fluorescein separated by TLC at different wavelengths. ● 310 nm,
Δ 300 nm, ■ 320 nm, 0 290 nm.

TABLE 9.5   EFFECT OF MEDIUM ON SPECTRAL CHARACTERISTICS
OF QUININE SULFATE

Medium	Optical Maxima of Spectra (nm)		
	Absorption	Excitation	Emission
0.1N H$_2$SO$_4$	250, 315, 350	340, 390	458, 466
Glycerol	315, 363	340, 390	450
Alcohol	280, 333	337	370
Silica gel		315, 365	450

beam spot measurements very inaccurate. For double-beam measurements there is little similarity between the measured zone and the
reference zone. There is an even noisier baseline. Thus, for
most applications using currently available sorbents, the flying
spot technique may not justify its additional cost. Therefore,
double-beam one-dimensional scanning with a longer slit may be the
method of choice for quantitative methodology. Fortunately, most
of the instruments use slits, although they do not all provide
double-beam capabilities.

For the range of absorption "seen" by a slit of given width, under the approximation that the absorbance profile of the chromatographed band in the direction of development is an isosceles triangle, Goldman and Goodall found (35) that the difference between the actual and measured absorbance could be measured. Furthermore, for a known spot band width and the anticipated peak density, a suitable scanning slit can usually be found. It is assumed that there is a negligible absorption gradient in the transverse direction along the band width. This is true only for a short distance. However, even assuming that the areas where the spot starts and ends do not contribute linearly, the peak area obtained will be in most cases a reliable basis for quantitation. In transmission the peak area (under the optical density curve) is, within certain limits, proportional to the weight of the absorbing substances, as expressed by the Kubelka-Munk formula.

With the signal-to-noise ratio taken as

$$\frac{S}{N} = \frac{\text{peak height}}{\text{average noise}}$$

lower noise should be expected with wider slits. However, in double-beam operations the noise remains fairly constant and the signal does not change appreciably with changing slit width, so that S/N may also be considered to be almost constant.

Single-beam recordings (Fig. 9.8) show typical background variations of the plate and increased noise level in both transmission and reflectance measurements. The figure shows the same background smoothed out by a double-beam system. Although for shorter distances near the center of the plate the background variations and noise obtained by reflectance in single-beam operation seem to be acceptable, one has to realize that in reflectance the peak response for the same spot is substantially lower than that in transmission mode, and consequently the S/N value (i.e., precision) should be expected to decrease.

Figure 9.9 shows recordings of the same sample strip scanned by transmission and reflectance. (The background noise due to impurities in the plate was relatively high.) The peak signals obtained in transmission are about five times as high as the corresponding signals in reflectance, the noise level being just twice as high.

Colorless UV-absorbing compounds are determined by reflectance, or in the transmission mode by fluorescence quenching. With the latter method the average noise may be higher because of uneven phosphor distribution in some layers. However, this method gives sufficiently large peak areas if the wavelength of maximum absorption in the compound is close to the peak of wavelength giving maximum fluorescence. Figure 9.10 shows an average noise level of a fluorescent plate (layer containing phosphor).

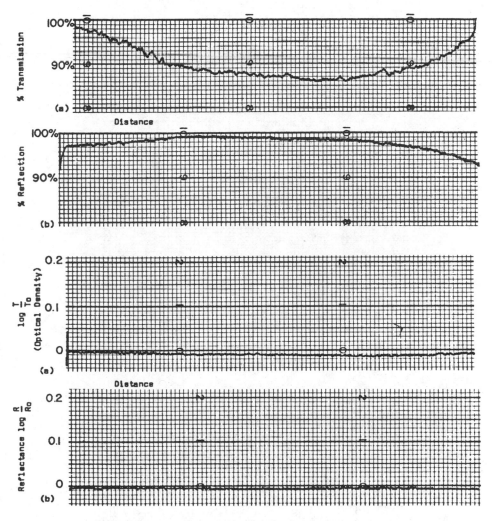

Figure 9.8. Upper curves: Scans of background in a TLC in the single-beam mode. (a) Transmission; (b) reflection, both at × = 420 nm, with slit width 0.7 mm. Lower curves: Scans of background in the same TLC as above in the double-beam mode. Conditions are the same.

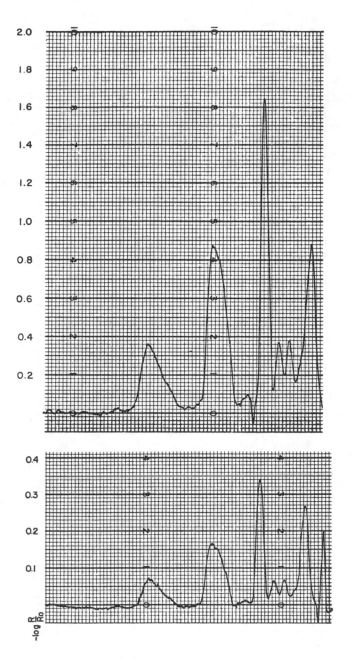

Figure 9.9. Comparison of double-beam transmission (upper curve) and double-beam reflectance (lower curve) of six components separated on a TLC.

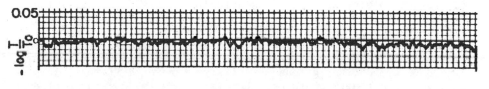

Figure 9.10. Average noise in a TLC containing phosphor scanned in the double-beam mode.

In general, the desired linear relationship is not always obtained because the technique of spot application involves factors that affect linear response. For instance, varying areas per unit weight were obtained when samples were spotted with pipets of different volumes. Plates spotted with varying volumes of one solution yielded linear relationships. Attempts to repeat this work with varying concentrations and constant volume resulted in curved lines.

According to Stahl and Jork (37), the peak area does not change for the same weight spotted with varying volume if the starting zone remains constant. For increasing weight spotted with constant volume and a constant starting zone diameter, they obtained a linear relationship. In other words, repetition of the spotting technique under controlled conditions will yield repeatable results.

Touchstone et al. (38) reported a linear response for steroids in the 0.02-10.0 µg range using a double-beam system. Similar results were obtained by others. The relative standard deviation caused by instrumental variations was found to be less than 1%. It seems reasonable to expect that if proper conditions of measurement are maintained, a reproducibility of quantitative determination better than ±2% can be achieved.

There are a number of sources of error in any quantitative determination involving thin layer chromatograms. Systematic errors are usually caused by application of an incorrect method. These include unsuitable illumination and detection geometry of the instruments, insufficient monochromaticity of light, and too small a scanning aperture, as discussed before. Operational errors may have the largest effect on the accuracy of a quantitative determination. A number of these sources will be discussed, since generally the instrumental results only indicate poor technique in the sample preparation and application, to say nothing of development of the chromatogram itself.

Spotting errors are among the most significant sources of inaccuracy. The technique of spotting may affect the delivery of volumes between 1 and 5 µL. A constant starting zone size should be maintained so that a given amount of substance spotted with

different volumes will yield identical peak areas, and different
amounts spotted with constant volume will exhibit a linear relationship in a plot of peak area of recordings of densitometer
scans versus weight. During movement of the solute through the
sorbent layer, errors occur because of the lateral diffusion during development, variations in the structure of different plates,
and variations between tanks. Lateral diffusion depends on the
relative rates of diffusion of the solute and the mobile phase.
To avoid errors when substantially different amounts of solute applied next to each other diffuse out of the straight pathway, plates
can be scored into strips about 1 cm wide before spotting. In this
way two dissimilar samples can safely be developed side by side.
The variation in the structure of different layers leads to an
incorrect extrapolation of the data obtained for a known quantity
of standard compared to the quantity of the unknown. Since
the ratio between the quantity in the final spot and the quantity
spotted varies for different layers, the use of a standard and a
test solution on each plate is recommended. In this connection,
factors such as the spot size before and after development and R_f
values become important. Variations in results due to the use of
different chambers are sometimes significant, and identical conditions should be maintained.

As discussed in Chapter 4 and pointed out by Touchstone et al.
(36), ideally, the sample or component to be quantitated should
travel in the layer to a position of R_f between 0.3 and 0.7 for
reproducible quantitative results. Shellard (39) also discusses
this variable in quantitative thin layer chromatography. An R_f of
0.3-0.7 assures that there will be a more even distribution of the
solute in the zone on the chromatogram to which it has traveled.
Solute zones that are too concentrated militate against linear
calibration curves. Zones with low R_f travel are more concentrated than those with higher R_f values, and the sharp narrow peaks
obtained on recordings give areas not truly indicative of the concentration in the zone. This effect is also seen in the spectrometry of solutions. Solutions that are too concentrated do not
follow the Beer-Lambert law. Consequently, if one considers that
for the most part quantitative TLC is a micro technique, it is
recommended that in most cases amounts of sample containing no
more than 1 µg of the solute be quantified.

The above, however, may also be construed as a general rule
in thin layer chromatography; "start small." Overloaded chromatograms rarely present the separations that are possible when
the correct amount of sample is applied to the chromatogram. The
literature abounds with resports of nonlinear calibration curves
that may be due to use of overly concentrated samples.

A more recent development in densitometry is the appearance
of a densitometer using soft laser scanning (40,41). A number of

features in this instrument indicate a promising future for it.
The width of the beam is controlled internally rather than by a
slit. Thus, the beam is not diffracted as it would be if it passed
through a slit. The beam's intensity can be adjusted so that either TLC or gel layers can be scanned. With these features the
results of scans of high sensitivity show stable baselines. Quantitation will be based on interpolation from standard curves.

A new type of quantitative method for TLC, based on consecu-
tive vaporization of the separated zones, requires special but relatively low-cost instrumentation. Usually the products of the
vaporization are either monitored by a vapor-phase detector, as
commonly used in gas chromatography, or collected and analyzed by
conventional means. The zones separated on a standard flat chromatoplate cannot be vaporized easily. Consequently, the technique
was originally carried out with layers coated on rods (42), strips
(43), or disks (44). More recently, narrow tubes coated on the
inner walls (45) were used, and instruments are now available for
scanning tubular chromatograms. An extension of this technique is
the Thermal Analytical System (TAS) whereby the zones are scraped
from the layer and vaporized to be collected and analyzed by conventional techniques (46-48). The vapors can also be collected by
condensation on the starting line of a TLC.

The scanner described by Mukherjee et al. (48,49) permits the
detection and quantitative analysis of organic substances in tubular thin layer chromatograms. The flow diagram of the scanner
equipped with a flame ionization detector is shown in Figure 9.11.
Samples to be analyzed are chromatographed in glass or quartz

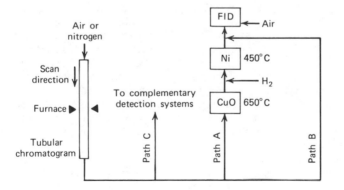

Figure 9.11. Flow diagram of a scanner for tubular thin layer
chromatograms. Path A, detection techniques PCRD, PCD, CD and CRD;
path B, detection technique PD; path C, complementary techniques,
such as coupling with a gas chromatograph.

tubes coated internally with a thin layer of silica gel or a mixture of silica gel and cupric oxide. After removal of the developing solvent, chromatograms are scanned. The substances in the various chromatographic zones are vaporized consecutively, either by pyrolysis or by combustion, and the resulting products are monitored by a flame ionization detector or a thermal conductivity detector. Vaporization is accomplished by moving the tubular thin layer chromatogram gradually through a ring-shaped furnace while a carrier gas, nitrogen or helium, flowing through the tube delivers the products of pyrolysis or combustion to the detector. The products of pyrolysis are monitored by the flame ionization detector, either directly (path B) or after their conversion to methane (path A). The carbon dioxide formed by *in situ* combustion can be reduced to methane and monitored. If a thermal conductivity detector is used (50), the products of combustion are passed through a water trap, and the carbon dioxide in the effluent is monitored. With air or a mixture of nitrogen and oxygen as carrier gas (51), the organic substances are vaporized mainly by combustion, from layers of either silica gel or silica gel containing cupric oxide. In the recording of the detector signals, both from a flame ionization detector and a thermal conductivity detector, the various zones of the tubular thin layer chromatogram appear as peaks whose areas correspond to the total carbon present in the respective zones.

For radioassay of tubular thin layer chromatograms, a proportional flow counter is employed for monitoring radioactivity. Low levels of radioactivity in various zones of a tubular thin layer chromatogram are determined accurately when the scanner is coupled with a trapping device. The labeled compounds separated in tubular thin layer chromatograms are quantitatively converted to $^{14}CO_2$ and/or $^{3}H_2O$, either by pyrolysis and subsequent combustion over cupric oxide by combustion *in situ* on absorbent layers containing cupric oxide. The products of combustion are collected serially or continuously in Hyamine solution, and the radioactivity of the fractions is measured by liquid scintillation counting.

Both radioactivity and mass in various zones of a tubular thin layer chromatogram can be monitored simultaneously by means of a stream splitter. Both radiochemical and chemical purity of labeled compounds can then be assessed and their specific activities determined.

Mangold and Mukherjee (53) have summarized their experiences with lipids using these techniques. For quantitative use a calibration curve must be set up for the compound to be assayed. The compound, of course, must be separated on the tubular TLC, vaporized, and if detected by a flame ionization detector, it can be quantitated. This is done on the basis of interpolation of the curve by plotting areas of the corresponding peaks from the recorder against the starting amounts. Essentially, the method represents an *in situ* quantitation of the TLC and can be very sensitive.

9.7.2 Calculation of the Results of Scans

As in any quantitative method based on interpretation of peaks from the responses recorded from the detectors, there must be some way to calculate the results. The recordings from densitometer and radioisotope scans show peaks corresponding to the compounds of interest. The use of present-day densitometers capable of reproducibility in scanning has made TLC a technique of high accuracy and reproducibility. Peak factors also are the basis of quantitation in radiochromatogram scanning.

There are several possible sources of error in quantitating the results of densitometer or radiochromatogram scannings. These include:

1. Sampling methods
2. The separation variables
3. Variation in detector responses
4. Techniques of calibration and measuring

The discussions on sampling methodology in Chapter 3 provide the beginning for good quantitative work. Production of reproducible quantitative data is highly dependent on sampling techniques. This involves proper preparation of the solutions for either the densitometric or radiochromatogram scanning and the method of transferring an aliquot to the plate. Good analytical techniques are important here as in any quantitative method. Precision will be determined primarily by these two factors

Most errors in quantitation originate in the application of sample to the layer. The techniques used to apply the sample will affect the shape of the spot after chromatogram development. The process of the chromatography will determine whether there is tailing or overlap of the spots. High R_f values result in wide peaks that are not amenable to reproducible results by scanning *in situ*.

As recommended before, it is important for precise scanning that the R_f of the components to be measured should be between 0.30 and 0.70. R_f values that are too high result in diffuse spots, which give broad peaks that do not give true densitometric results. Likewise, too low an R_f results in a very sharp, narrow peak in the recording that does not represent the true concentration in the zone scanned. Once again it must be pointed out that the nonlinear calibration curves often seen in the literature are generally due to overloading of the chromatographic system. Start with 1 μg of material and work to the lower levels. Most TLC procedures are capable of quantitation in the nonogram levels.

Picogram levels are possible, especially with fluorometric procedures.

Another source of error is sample decomposition during the chromatography. This can be tested by developing chromatograms with standards of known serial dilutions. If a linear increase in peak area is not obtained as a function of concentration, sample decomposition in the zone should be suspected. Errors due to tailing and overlap are often seen and can be eliminated by changing the mobile phase to obtain better separation.

The response of the detector in the instrument can be a source of error. Sensitivity, noise, and linearity are important in quantitative work. Furthermore, detector response is not the same for all wavelengths, nor do all substances show the same relative absorptive properties. The calibration must be carried out with the individual compound being measured.

Recorder tracings are generally used for evaluating quantitative results of densitometric and many radiochromatogram scans. The recorder should be properly adjusted. The methods used for calculating the peak-quantity relationship of the recordings are also subject to error.

There are a number of ways in which the area of the peaks of recordings of scans can be determined. Peak area may be determined with a planimeter, a device that computes the area of the peak as it mechanically traces the peak's outline. The precision and accuracy of planimetry depend on the skill of the operator. Precision can be improved with this technique by making repetitive tracings; however, this makes the time-consuming technique even more tedious.

A simple but very effective technique is to measure areas by the product of peak height, H, and the width, W, at one-half the peak height. This technique should only be used with symmetrical peaks or peaks that have similar shapes. The precision of this measurement can be improved by increasing recorder chart speed, so that the width of the peak can be determined more precisely.

The peak-area calibration procedure gives calibration based on area. Usually, the precision of this method is less influenced by changes in instrumental parameters. In addition, improved results are obtained with peaks whose shapes are not Gaussian. However, results from the peak-area quantitative method are more affected by neighboring peaks. Peak area should be used for better precision, and peak height (as discussed below) for maximum freedom from possible interferences.

A simple method of quantitation involves the measurement of peak heights. Calibration is obtained by chromatographing serial dilutions of the compounds of interest and measuring the heights of the peaks on the recordings. Peak height is measured as a distance from the baseline to the peak maximum. Baseline drift is

compensated for by interpolation of the baseline between the start
and finish of the peak. A standard calibration curve is obtained
by plotting the concentrations of the compound of interest versus
the peak heights obtained. Unknows are analyzed by chromatographing the same size aliquot of an unknown solution and determining the peak height of the materials for which the calibration has
been prepared. By interpolation on the standard curve, peak
heights can then be translated into the concentration of material
in the unknown sample.

The peak-height method should be used only when peak heights
change linearly with sample size. Serious errors will occur if
the peak-height method is employed when peaks are distorted or the
plate is overloaded. The peak height method is usually no more
precise than about ±5%, relative, since the results depend greatly
on the precision with which the sample has been applied. Reproducibility is improved by using frequent calibrations to compensate for changes in instrumental parameters. The peak-height
method is valid only when the R_f values are reproducible. Low R_f
values give narrow peak widths on recordings of TLC scans. Likewise, zones with higher R_f values show broad peaks. Thus, if the
retention time in GC or the R_f in TLC is not reproduced within
narrow limits, the peak-height method is not valid. To overcome
this, the peak-area procedure should be used.

Peak-height measurements can be more accurate than peak-area
measurements because peak heights are less interfered with by
neighboring, overlapping peaks. The peak-height method of quantitation also involves the least effort. It is often desirable to
initially determine the approximate concentrations of components
in a mixture by a simple peak-height method, before attempting to
set up a more rigorous high-precision procedure for routine use.

Another technique is to cut the peak out of the recorder
trace and weigh it. This method destroys the recorder trace. If
it is desired to retain the tracings, the recordings can be photo-
copied and the peak cut from the copy. The accuracy of this approach depends on the constancy of the weight of the paper and the
care used in cutting out the peak. By keeping the ratio of the
peak height to the width at half-height in the range of 1 to 10,
the accuracy of the cutting process can be improved. This method
can be used satisfactorily for irregularly shaped peaks.

Some recorders are equipped with ball-and-disk integrators
that automatically produce tracings that indicate the area of the
peak. The accuracy is still limited by the performance of the
recording potentiometer. The ball-and-disk integrator can often
be used with good accuracy for irregularly shaped peaks.

The electronic integrators used in gas chromatography generally are also satisfactory for measuring peak areas in TLC scans.
The electronic digital integrator automatically measures peak

areas and converts them into a numerical form, which is then
printed out. Sophisticated versions of these devices also correct
for baseline drift.

TLC peak areas from scans can also be conveniently measured
with the computer systems that have been developed for gas chromatography. Many different types of computer systems can be used
without change for TLC densitometry. With appropriate programming, most of these devices print out a complete report, including
names of the compounds, retention times, peak areas, area correction factors, and the weight-percent of the various sample components. Snyder and Kirkland (54) state that the maximum precision
of the various integration methods is as given in Table 9.6.
Since the type of data generated by TLC densitometry is very similar to that found for gas chromatography, these area precision
values are representative for both techniques.

TABLE 9.6   AVERAGE PRECISION OF PEAK AREA MEASUREMENT TECHNIQUE
             (NO INTERNAL STANDARD)

Method	Relative Precision, $1\sigma(\%)$
Planimeter	3
Triangulation	3
Cut and weigh	2
$H \times \frac{1}{2}W$	2
Ball-and-disk integrator	1
Electronic digital integrator	0.5
Computer	0.25

9.8   INTERNAL STANDARD TECHNIQUE

An internal standard will compensate for errors in the analytical
measurement. A known compound is added to the unknown mixture at
a fixed concentration. The added compound is also used as the
marker. The precision of the analysis does not depend on reproduc-
ing the size of the sample applied to the plate. Both peak-height
and peak-area ratio measurement can be used here. The compound to
internal standard peak-height (or peak-area) *ratio* is used for the
calibration.

Peak-height ratio calibrations are constructed by chromatographing aliquots of mixtures containing the compound of interest
in various concentrations together with a constant concentration
of the internal standard. The peak heights of the components of
interest are determined, and the compound/internal standard peak-

height ratios are plotted against concentration. This calibration
plot is linear. Measurements by the peak-height ratio method can
be reproduced with a precision of 0.7-1%.

The peak-area ratio measurement technique is most precise,
since it compensates for changes in instrumental parameters and
variation in technique. The approach requires completely separated peaks in the chromatogram. The method of calibration is identical to the peak-height ratio method, except that measurements
are based on peak area.

The selection of the internal standard is critical for both
the peak-height and peak-area ratio methods. First, the internal
standard zone should be separated from the other zones in the mixture to be analyzed and located in a "vacant" spot in the chromatogram of the unknown. To determine the most feasible position
for the internal standard, an appropriate range of unknown samples
must first be chromatographically assessed. Optimum results are
obtained when the concentration of the internal standard is adjusted so that the peak-height or peak-area ratio is approximately
unity. The internal standard must not be present in the original
sample, and it should be a compound that is stable and not reactive with sample components, sorbents, or mobile phase. If possible, the internal standard should be commercially available in
high purity.

9.9 ADDITIONAL LITERATURE

Three books are available that deal with the separation and quan-
titaion of various compound classes (55-57). Such classes as
aflatoxins, nitrosamines, phospholipids, hydrocarbons, biogenic
amines, and lipids are covered.

REFERENCES

1. W. Q. Pons, Jr., A. F. Cuculla, S. L. Lee, J. A. Robertson,
    A. O. Franz, and L. A. Goldblatt, J. Assoc. Off. Anal. Chem.,
    49, 554 (1966).
2. L. Fishbein, Chromatography of Environmental Hazards, Vol. 9,
    Elsevier, New York, 1972, p. 398.
3. W. A. Pons, Jr., and L. A. Goldblatt, in Aflatoxins, L. A.
    Goldblatt, Ed., Academic Press, New York, 1964, p. 77.
4. C. A. Johnson, in Quantitative Paper and Thin Layer Chromatog-
    raphy, E. J. Shellard, Ed., Academic Press, New York, 1968,
    p. 105.
5. N. Oswald and H. Flück, Sci. Pharmacol., 32, 136 (1964).
6. J. C. Morrison and L. G. Chatten, J. Pharm. Sci., 53, 1205
    (1964).

7. H. Gänshirt and J. Poderman, J. Chromatogr., *16*, 510 (1964).

8. G. Schierf and P. Word, J. Lipid Res., *6*, 317 (1965).

9. M. Brenner, A. Niederwieser, and G. Pataki, in *Thin Layer Chromatography*, E. Stahl, Ed., Springer-Verlag, Berlin, 1965, p. 391.

10. H. J. Petrowitz, Mitt, Deut. Ges. Holzforsch., *48*, 57 (1962).

11. S. J. Purdy and E. J. Truter, Lab. Pract., 500 (1964).

12. H. O. Bang, J. Chromatogr., *14*, 500 (1964).

13. H. H. Varon, H. A. Darnbold, M. Murphey, and J. Forsythe, Steroids, *9*, 507 (1967).

14. W. Heibrink, Fette, Seifen, Anstrichm., *66*, 569 (1964).

15. R. P. A. Sims and J. A. G. Larose, J. Am. Oil Chem. Soc., *39*, 232 (1962).

16. J. Haefelfinger, J. Chromatogr., *33*, 320 (1968).

17. W. E. Court, in *Quantitative Paper and Thin-Layer Chromatography*, E. J. Shellard, Ed., Academic Press, 1968, p. 37.

18. J. S. Matthews, A. C. Perede, and A. Aguilera, J. Chromatogr., *9*, 331 (1962).

19. G. Lehmann, H. G. Hahn, and P. Martinod, Z. Anal. Chem., *227*, 81 (1967).

20. J. Attal., S. M. Hendeles, J. A. Engels, and K. B. Eik-Nes, J. Chromatogr., *21*, 167 (1967).

21. J. C. Touchstone and M. F. Dobbins, J. Steroid Biochem., *6*, 1389 (1975).

22. V. Ganjam, C. Desjardins, and L. L. Ewing, Steroids, *16*, 1227 (1970).

23. B. A. Kottke, J. Wollenweber, and C. A. Cowen, J. Chromatogr., *21*, 429 (1966).

24. J. M. Brand, J. Chromatogr., *21*, 424 (1966).

25. H. Falk and K. Krummen, J. Chromatogr., *103*, 279 (1975).

26. J. M. Varga and F. F. Richards, Anal. Biochem., *53*, 397 (1973).

27. S. Udenfriend, S. Stein, P. Bohlen, and W. Dairman, Science, *18*, 871 (1972).

28. O. S. Privett and M. L. Black, J. Lipid Res., *2*, 37 (1961).

29. H. Jork, Deut. Apotheker Ztg., *102*, 1263 (1962).

30. O. B. Barrett, M. S. J. Dallas, and F. P. Padley, J. Am. Oil Chem. Soc., *40*, 580 (1963).

31. V. Novacek, Am. Lab., *2*(12), 129 (1969).

32. Chandrasekhar, Radiative Transfer, University Press, London, 1950.

33. P. Kubelka and F. Z. Munk, Tech. Physik, *12*, 593 (1931).

34. P. Kubelka, J. Opt. Soc. Am., *38*, 448 (1948).

35. J. Goldman and R. R. Goodall, J. Chromatogr., *32*, 24 (1968).

36. J. C. Touchstone, S. S. Levin, and T. Murawec, Anal. Chem., *43*, 858 (1971).

37. E. Stahl and H. Jork, Zeiss Inf., *68*, 52 (1968).

38. J. C. Touchstone, A. K. Balin, and P. Knapstein, Steroids, *11*, 115 (1969).
39. E. J. Shellard, in *Quantitative Paper and Thin Layer Chromatography*, E. J. Shellard, Ed., Academic Press, New York, 1969.
40. R. A. Zeineh, W. P. Nijm, and F. H. Al-Azzawi, Am. Lab., *7*(2), 51 (1975).
41. R. A. Zeineh,and W. P. Nijm, Clin. Res., *22*, 428 (1974).
42. F. B. Padley, J. Chromatogr., *39*, 37 (1969).
43. J. J. Szakasitis, P. V. Peurifoy, and L. A. Woods, Anal. Chem., *42*, 351 (1970).
44. J. H. van Dijk, A. Anal. Chem., *236*, 326 (1968).
45. E. Hahti and I. Jaakonmaki, Ann. Med. Biol. Exp. Fennai, *47*, 175 (1969).
46. R. N. Rogers, Anal. Chem., *39*, 730 (1969).
47. E. Stahl, Z. Anal. Chem., *261*, 11 (1972).
48. K. D. Mukherjee, H. Spaans, and E. J. Haahti, J. Chromatogr., *61*, 317 (1971).
49. K. D. Mukherjee, H. Spaans, and E. J. Haahti, J. Chromatogr. Sci., *10*, 193 (1972).
50. E. Haahti, P. Vihko, J. Jaakonmaki, and R. S. Evans, Jr. Chromatotr. Sci., *8*, 370 (1970).
51. K. D. Mukherjee, J. Chromatogr., *96*, 242 (1974).
52. K. D. Mukherjee and H. K. Mangold, Ergebn. Exp. Med., *20* (1974).
53. H. K. Mangold and K. D. Mukherjee, J. Chromatog. Sci., *13*, 398 (1975).
54. L. R. Snyder and J. J. Kirkland, *Modern Liquid Chromatography*, Wiley-Interscience, New York, 1974, p. 439.
55. J. C. Touchstone and J. Sherma, Eds, *Densitometry in Thin Layer Chromatography*, Wiley-Interscience, New York, 1979.
56. J. C. Touchstone and D. Rogers, Eds., *Thin Layer Chromatography: Quantitative Environmental and Clinical Applications*, Wiley-Interscience, New York, 1980.
57. J. C. Touchstone, Ed., *Advances in Thin Layer Chromatography: Clinical and Environmental Applications*, Wiley-Interscience, New York, 1982.

CHAPTER 10
Procedures for Radioactivity

10.1 INTRODUCTION

Radioisotopes in TLC have wide application, particularly in metabolic studies, because of the great sensitivity that the methods
can offer. Research in the metabolism of drugs and natural substances represents a broad field in which the localization of labeled substances in various tissues plays a great role. In order
to identify and locate as well as to quantitate these substances,
the use of a radioisotope as a tracer for substances separated by
TLC is important. Often because of low detection limits, isotopic
methodology is the means of choice, offering a number of methods
by which these tagged materials may be detected or quantitated *in
situ* after separation by TLC.

10.2 PURITY ASSESSMENT

Any discussion of the methodology in TLC of labeled compounds must
define the limits within which one may work. Therefore, the purity and handling of isotopically labeled substances are a first
consideration. When short-lived isotopes are used, one must be
consistently aware that the material may not always be pure. Even
with long-lived isotopes, such as ^{14}C, purity must be assessed,

275

particularly if the samples have been stored for long times.
Snyder and Piantadosi (1) feel that a labeled substance should always be assessed for purity immediately before use. A number of
factors of contamination and degradation enter into this conclusion.

The high sensitivity of detection methods, especially in
scintillation counting, would reveal very low levels of extraneous matter. Compounds not checked before use may show signs of
instability after chromatography; even a contaminant of low radioactivity can masquerade as a metabolite. Because of this frequent
problem and because different compounds behave differently under
the various conditions (that is, storage, type of solvent, type of
sorbent, and their use in the separation methods), the entire process of the experiment must be evaluated in terms of each component that is to be assessed or quantitated.

Compounds of high specific activity can show autoradiolysis
(2). Chemical changes can take place, especially in the case of
soft beta emitters. Preparation of samples in dilute solutions
can reduce the chance of this occurring. However, if the solvent
is attacked and highly reactive, long-lived excited species such
as free radicals or ions are formed and further complications result. Water is a poor solvent in this regard, especially under
aerobic conditions. To cut down these effects, benzene is used by
many companies that supply labeled substances. When the compound
is not soluble in benzene, alcohol or dioxane diluted with benzene
is used. The samples should be stored at low temperatures to increase stability.

Chemical changes, including those caused by air and light,
can also occur during the chromatographic procedures (3). This is
true with either labeled or nonlabeled compounds. Snyder (4) indicated that the crucial stage is when the solute is present on a
dry sorbent and exposed to the atmosphere. Alumina and highly active silica gels are particularly troublesome in this respect.
Therefore, in quantitative work the results must be assessed as
soon as the chromatogram is complete. In the authors' laboratory,
compounds have completely disappeared because of contact with the
sorbent (silica gel) in the atmosphere after only 3 hr exposure.

The usual procedures for the handling of radioactive materials must be followed. In quantitative work particularly, contamination of the chromatogram or the sampling implements must be
avoided, as pointed out in Chapter 4 on sample delivery techniques.
The delivery systems are of major concern in isotope work, since
the sensitivity is so high. The use of calibrated disposable microcapillaries can save much time and effort in precise quantitative work using labeled compounds.

These basic beginnings are necessary to the success of quantitative procedures in TLC as well as other separation methods.

10.3   LOCATING THE SEPARATED SUBSTANCES

After the chromatograms have been developed, the separated solutes
can be located in a number of ways. As discussed in Chapter 9 on
quantitative methods, reference compounds can be used, or if the
compound absorbs in the ultraviolet it can be visualized under the
UV lamp. The detection methods described in Chapter 7 can also be
used. The methodology of isotope detection in TLC is the subject
of this chapter.

There are three main methods of detection when the separated
solutes in the thin layer chromatogram are radioactive. These
are:

1.   Autoradiography *in situ*.
2.   Liquid scintillation counting after elution, or in some
cases after scraping from the plate without elution.
3.   Direct chromatogram scanning.

10.4   AUTORADIOGRAPHY

Autoradiography, exposing the chromatogram to X-ray film, provides
resolution comparable to the original chromatogram. The radioactivity level will determine the exposure time, which can vary from
several hours or several days to weeks. The quantitative analysis
is done by scanning the developed film in a densitometer in ways
described in detail in the previous chapter on quantitation. The
plate must be dried well to prevent formation of artifacts from
traces of solvent left in the layer. While in contact with the
layer, organic solvents can attack the film and produce spurious
blackening.

Pseudoautoradiography should be tested by chromatographing
chemically identical but nonradioactive samples using the conditions required for the sample. If this possibility is not excluded, a thin plastic foil can be interposed between the chromatogram
and the film to prevent *chemical* reduction of the photographic
emulsion by the separated substances, solvents, or reagents. If
the foil is not sufficiently thin, sensitivity of detection may be
impaired for ^{14}C or ^{35}S. No foil can be used when the isotope is
tritium. To prevent contamination of the film, handle the film as
if it were contaminated. Do not touch it directly. In order to
ensure the cohesion of dry layers and prevent their damage, they
are often fixed by a protective spray such as clear Krylon or
Neatan (see Chapter 8). Too thick a spray, however, may increase
self-absorption, which is undesirable, especially if the spray is
not applied in a perfectly uniform fashion.

Chromatographic spots are relatively large and diffusely limited. Thus it is desirable to have photographic films of maximum

sensitivity at the expense of fine grain. Single-coated film is preferable for ^{14}C or ^{35}S, since blackening hardly takes place at all in the distant emulsion, and background fogging is doubled in two-emulsion layers. For autoradiography, place the X-ray film on top of the TLC and cover it with a soft or elastic sheet (rubber or polyurethane foam, etc.) and a solid plate (wooden board, metal or plastic plate), then weight or clamp it down. This operation must be done in darkness or under weak illumination (detailed by the instructions given by the film manufacturer). In the case of penetrating radiation, the film can be left in its black paper envelope and the work done with lights on.

When a number of chromatograms of very weak beta emitters are to be exposed for the same length of time, it is possible to repeat the layers chromatogram-film-foam. Absorbent layers, such as lead foil or 3 mm thick aluminum, must be interposed in the case of ^{32}P or ^{36}Cl. The sandwich is then wrapped in black paper (such as the envelope in which films are purchased) or placed in an opaque box and left for the time required. The time may vary in practice between hours and months. To ensure correspondence between the chromatogram and its autoradiogram, markers such as radioactive ink or heavy graphite pencil lines are useful. The film is developed by the usual technique.

In general, No-Screen Medical X-ray safety film can be used for detecting the radioactivity. The emulsion is sensative, and the resolution suffices for TLC requirements. When tritium-labeled substances are to be recorded, Kodak NTB or Nuclear-Trak emulsion films can be used.

The newest film used for detection of radioactivity is Kodak X-omat AR, which provides high sensitivity with direct recording and with fluors and intensifying screens. The double-coated, high contrast emulsion has high sensitivity to both blue light and beta emitters. It can be processed manually or by machine.

Radiation effects on a photographic emulsion result in the appearance of a latent image that appears as silver grains after development. Within a limited range of doses (radiation intensity time), blackening is proportional to the concentration of the radioactive element in the spot. Above a certain limit, blackening will not increase, and relative comparison of spot intensities becomes difficult. Overexposure must be avoided. In some parts of the chromatogram, radioactivity may be relatively high, causing merging of partially resolved spots, while other spots may be too weak. A number of autoradiograms with different exposure times should be prepared to assess the extent of these differences and to determine optimal exposure times. If different isotopes are present on one chromatogram, it is possible to differentiate between them by carrying out autoradiography with and without suitable filters. A film that is in contact with the chromatogram can act as such a filter for another photographic plate.

An adaptation of the autoradiographic procedure is to use scintillation methods to improve the sensitivity. In the case of Tritium and similar soft beta emitters, the chromatogram can be impregnated with the scintillator, as reported by Wilson (5) (see below), or the scintillator (anthracene) may be added directly to the layer before chromatography, as described by Lüthi and Waser (6). The chromatogram is then exposed to autoradiography. When a liquid scintillator is used, be sure that no bubbles are trapped under the film. The substances to be chromatographed must be soluble in the scintillator liquid used for the impregnation, or they may spread somewhat in the layer. Some TLC sorbents, especially those containing a phosphor, may act as scintillators by themselves (7).

The blackening of the film can be used to measure the solute components as well as the radioactivity. However, to do this, regression lines (radioactivity versus darkness) must be set up, and the range of linearity of the blackening must be assessed. This means that the control of a number of chromatographic factors, exposure times, and the amount of the radioactivity in the separated zones must be considered.

The need for densitometry of autoradiograms and the long exposure times make the method time consuming. The method is not generally applicable when a wide range of radioactivity or double-labeled compounds are separated on the same layer. Consequently, it is not widely used for quantitative purposes. A general autoradiographic method (8,9) that involves the apposition of a sheet chromatogram to No-Screen X-ray film, is several thousandfold less sensitive for ^{3}H than for ^{14}C. This difference is even greater if Tritium is compared with beta emitters of higher energy, like ^{32}P. Because of their low energy and short range, only a small proportion of the primary beta particles of Tritium emerge from the surface of the chromatogram (8). Further losses occur not only in the gap between the chromatogram and the film, but also in the film itself because the range of the ^{3}H beta particles in the film is shorter than the average distance between the silver halide grains.

Wilson, in 1958 (5), added a scintillator to the chromatogram to convert the energy of the ^{3}H beta particles to light, which in turn produced an image in the photographic emulsion. The sensitivity reported for procedures of Tritium autoradiography and fluorography that permit the subsequent recovery of the radioactive compounds is 25-300 nCi/cm^2-day. If the chromatogram itself is impregnated with an X-ray emulsion, a treatment that precludes the subsequent recovery of the compounds, the lower limit of detection is reported to be 30 nCi/cm^2-day, according to Chamberlain et al. (8).

Randerath (9) made a systematic study of the factors that influence the speed of Tritium detection on chromatograms with added scintillator. The method makes possible the visualization on thin layer chromatograms of 2-3 nCi ^{3}H per square centimeter and of 0.05-0.06 nCi ^{14}C per square centimeter after an exposure for 24 hr. Scintillation radioautography can increase the sensitivity of detection by a factor of 25-50.

New England Nuclear, a major supplier of radioactive substances, offers a surface autoradiography enhancer named En hance. Available either as a spray or a liquid, it can be applied to the surface of a TLC plate to aid detection of ^{3}H levels as low as 60 dpm/mm^2 in 24 hrs.

10.4.1 Fluorography

For fluorography, a solution of the scintillator is either sprayed or poured over the dry chromatogram. A 7% (w/v) solution of 2,5-diphenyloxazole (PPO) (scintillation grade) in diethyl ether is poured from a small beaker over the entire chromatographic area as rapidly as possible. The scintillator is then distributed evenly by tilting. This operation should take only 1-3 sec, depending on the size of the layer. The treated chromatogram is then immediately brought into a vertical position and agitated until the ether has evaporated completely. The volume of the PPO solution used for this treatment is 35-40 μL/cm^2, that is, 14-16 mL for a 20×20 cm layer. All further operations are carried out in a dark room under proper lighting conditions (7.5 W bulb, Wratten 6B filter, distance 120 cm from the film). The layer is placed in contact with the emulsion of Kodak RB-54 Royal Blue Medical X-ray film (found to be the most sensitive) and kept between two glass plates in the dark in an insulated container over dry ice (-78.5°) for the time necessary to visualize the labeled compounds. The usual exposure time is 24 hr. The X-ray film is subsequently developed using the manufacturer's prescribed procedure. Randerath also developed a procedure modified from the early work of Muehler and Crabtree (10) for intensification of the final image on the film as described below.

For preparing the four solutions needed, the chemicals specified are first dissolved in 400 mL distilled water and the solutions so obtained diluted to 500 mL with distilled water. The following chemicals are required: *For solution* A, 5 mL 37% aqueous formaldehyde and 3 g Na$_2$CO$_3$; *for solution* B. 15 mL concentrated H$_2$SO$_4$ and 11.2 g K$_2$Cr$_2$O$_7$; *for solution* C, 1.9 g NaHSO$_3$, 7.5 g hydroquinone, and 1.9 mL Kodak Photo Flo 200 solution; *for solution* D, 11.2 g Na$_2$S$_2$O$_3$·5H$_2$O. For making the intensifier, 2 parts solution C is added, with stirring, to 1 part solution B. While the stirring is continued, 2 parts solution D and finally 1 part

solution B are added. The intensifier is made up fresh before use.
The stock solutions can be kept for several months. The technique
itself is as follows: After fixation the film is washed for 10
min in running water. It is allowed to harden for 5 min in solution A, washed for 5 min in running water, and briefly rinsed under distilled water. The hardened film is then immersed in the
intensifier for 10-15 min, washed 10-15 min under running water,
and air dried. Only one film should be treated at a time; it is
important not to damage the emulsion mechanically in the areas to
be intensified.

For locating radioactive compounds on the original chromatograms, radioactive ink is applied as a marker to a few points on
the layer after chromatography. Following exposure, these spots
serve as reference points with which to line up the film and the
chromatogram. The compounds are marked by perforating the super-
imposed film around the darkened areas with a needle.

These procedures were developed for use with the "standard"
layers of 0.25 mm thickness. The results may be somewhat lower if
the layers are thicker. It should be noted that the spots on the
layer should not be so diffuse that a solute is of low concentration, since this would decrease the sensitivity of the autoradiography. A good rule to follow is to obtain migration of the solutes of interest to the middle third of the chromatogram to keep
diffusion to a minimum.

The method will enable one to visualize 4-6 nCi ^{3}H per square
centimeter per day. With one-dimensional TLC (spot size 0.3 cm^2),
1.5-3 nCi ^{3}H can be located after a 24 hr exposure to film. When
expressed in terms of nci/cm^2-day or dpm/cm^2-day, the sensitivity
is practically independent of the layer material. Layers of cellulose, cellulose ion exchangers, polyamide, silica gel, and silica
gel-kieselguhr can be used. The green phosphor of Merck F-254
layers does not increase the sensitivity of the film detection
method, regardless of whether or not PPO is present and whether
daylight or X-ray film is used. However, the radioactivity detectable in a given spot is lower on cellulose, PEI-cellulose, or
silica gel layers than on polyamide or silica gel-kieselguhr
layers because spots are more compact in the former layers.

Prydz et al. (11) critically evaluated and improved the procedures of Randerath for a low-temperature solid scintillation
fluorography of TLC. The Eastman Chromagram sheet 6067 with silica gel is used. Anthracene is added by repeated dipping in a
benzene solution, or PPO is added with ether solution according to
the method of Randerath. The optimum amount of scintillator was
found from control measures in which the emitter luminescence is
measured by means of a sensitive photomultiplier. The dipping is
repeated until no further increase in beta-radioluminescence
(beta-RL) is observed. An additional scintillator amount would

have decreased the light output by self-absorption. This resulting decrease in the film blackening obtained was observed by
Randerath after extensive scintillator impregnation. The fastest
film available, Kodak RP X-Omat Estar Medical X-ray film, is used.
Two exposure temperatures, +20° and -78°, are compared. Prydz et
al. used exposure periods of 5 hr, whereas Randerath used 24 hr.
Lüthi and Waser (6) exposed their films for longer periods of time
at room temperature, at -30°, and at -70°. The actual time for
any particular application should be determined by trial and error.

The optical densities obtained on the films can be evaluated
with a photoelectric densitometer. Curves showing the density as
a function of the exposure intensity (for a fixed exposure time)
can then be obtained.

In most investigations of the scintillation fluorography
technique, spots containing known amounts of ratiotracer must be
used to evaluate the sensitivity. Randerath performs a complete
separation procedure and analyzes the results obtained for the
separated spots. To determine the spot activities, he uses liquid
scintillation counting of the eluted spots. The solid scintillator luminescence output of the various spots is measured directly with a photomultiplier. To further test the behavior of the
photographic film at low temperature and for low exposure intensities he made some exposures using an electroluminescent lamp
that gave, approximately, a Poisson distribution of photons and
for which the light intensity could be varied stepwise with a series of neutral density filters.

In thin layer chromatography the spots obtained are more concentrated than on paper. This is important for a low-level detection limit. The resulting number of grains per unit are, for a
first approximation, inversely proportional to the spot area.
However, the number of grains may increase when the area containing the activity is decreased, due to more effective production of
nucleation centers. What is important for their establishment is
the concentration of photons both in time and space.

The film can also be exposed with a mirror placed below the
chromatogram to increase the photon flux reaching the film. For
photographic plates with emulsion on both sides, the one not being
employed in the detection should preferably be covered watertight
during the development. Following fixation of both sides the fog
background will be reduced by 50%.

The optimum amount of scintillator should be used, and this
is more important for radiocarbon than for tritium. Because of
overlap between the absorption and emission spectra of the scintillators, they can exhibit self-absorption when applied in too
large an amount. Optimum amounts must be obtained by trial and
error. This is a well-known phenomenon for anthracene and has
been shown by Randerath to occur for PPO (9).

In connection with the temperature effect, it should be stated that different films show different behavior and have their own particular optimum temperatures (9). For most scintillators, the scintillator efficiencies vary only very slightly, and the cooling of the pertinent solid scintillators should not be expected to produce more than about a 5-10% increase. On the other hand, frozen benzene shows a drastic increase in efficiency upon cooling. The use of a high-efficiency scintillator, although important, is not in itself sufficient. One should also attempt to choose one whose emission spectrum best fits the spectral sensitivity range of the film material to be used. This can be determined from known characteristics of the scintillator and the specifications of the film.

Adequate discrimination between the isotopes, usually ^{3}H and ^{14}C, has been difficult to obtain with those techniques that generally involve the application of multiple photographic emulsions to the preparation. However, Gruenstein and Smith (12) described a procedure for distinguishing autoradiographically between ^{3}H and ^{14}C on a single TLC plate. Discrimination between the two isotopes is good and the procedure is nondestructive, so that samples may be recovered for further analysis. The method is based on the technique of Randerath (9) in which the sensitivity of ^{3}H detection was increased approximately 25-fold following addition of a scintillation fluor to the chromatogram. They observed that ^{14}C sensitivity is not enhanced by the scintillator and suggested that it might be possible to distinguish between ^{3}H and ^{14}C on the same plate by adjusting the ratio of ^{3}H/^{14}C radioactivities and the exposure times so that only ^{14}C would be detected before, and only ^{3}H after, application of the scintillator.

The method is illustrated in work with amino acids separated on cellulose layers. ^{3}H-lysine and ^{14}C-leucine were applied on an individual cellulose layer, and after chromatography an initial autoradiographic exposure was made for 24 hr at -78°C. Following this, the plate was covered briefly with 7 mL of diethyl ether containing 7% (w/v) of 2,5-diphenyloxazole (PPO) and dried in air at room temperature. The solution of PPO in ether was applied rapidly with a pipet to the upper edge of the plate, which was tilted approximately 10° from the horizontal. A second autoradiographic film was then exposed in a manner identical to the first, except that the exposure time was shortened to 4.5 hr. As described, this method should be applicable to all chromatograms in which the compounds being chromatographed are unreactive with the scintillator. If any of the compounds resolved by the chromatogram are soluble in ether, care must be taken that these compounds do not have time to diffuse or dissolve in the ether.

Application of this technique to experimental situations requires consideration of certain additional points. If the material to be analyzed chromatographically contains several components,

or if a single component is treated to yield several products,
then, while the ratio of ^{3}H to ^{14}C may remain quite constant for
all products, the absolute amounts of radioactivity would in some
instances vary widely among the different chromatographic spots.
In order to effectively analyze such a chromatogram, several exposures for different lengths of time, both before and after treatment of the plate with scintillator, are available. A wide range
of absolute values of radioactivity can be analyzed, provided that
very faint spots do not overlap very strong ones.

10.5   COUNTING BY LIQUID SCINTILLATION

Counting by liquid scintillation is probably the most commonly
used counting technique. Sections of the chromatogram containing
the sorbent are scraped off the plate. Then the solute is eluted
or the sorbent is transferred directly into the counting vial.
This can be done in a number of ways. Liquid scintillation is the
most sensitive to counting technique, as will be discussed in the
next section; counting directly from the layer is much less efficient than counting by the scintillation in solution.

The elution of the solute from the sorbent is subject to the
problems discussed in detail in Chapter 9. The scintillation fluid, however, can be used as the eluting agent in a procedure in
which the sorbent is transferred directly to the counting vial.
Location of the zone to be scraped off for measurement of radioactivity is dependent on a reference marker. If the compounds are
visible or absorb in ultraviolet light they can be located. Resolution is limited by the width of the zone removed from the layer. To increase resolution, a large number of small sections can
be taken. The procedure of transferring the samples to the vials
is slow. Snyder and Kimble (13) describe an apparatus that automatically scraps a predetermined width of zone into individual
counting vials. This apparatus is now available commercially.
The procedure is tedious, and when a large number of samples are
to be analyzed, it is subject to more sources of error. Generally,
however, most work with labeled compounds will involve assessment
of one, two, or three major components separated by TLC. As with
other quantitative methods, the standard must be carried through
the procedure so that regression lines and correction factors can
be determined for quantitative results.

Transverse zones on the plate are usually scraped off with a
spatula or blade. Suction devices run the risk of cross-contamination due to the passage of the powder through those parts of the
device that are used for removal of several zones. The chromatogram can be sectioned according to previous detection (physical,
chemical, by *in situ* scanning, or autoradiography) or at regular

intervals, preferably intervals short enough to retain the resolution of the chromatogram.

Individual sections or scrapings may then be eluted and the eluate subjected to radioassay. This is seldom done, since part of the radioactivity may remain uneluted and since elution involves a considerable amount of labor. In many cases, the sections or scrapings are counted directly. Counting of radioactive materials after absorbing their solutions on cellulose or glass fiber disks and evaporating is one of the standard techniques, even outside the chromatographic field. It is possible to place the section on a planchet and carry out counting, but liquid scintillation counting is more prevalent.

Snyder (14) showed that most detection reagents, including iodine vapor, do not cause appreciable quenching of the scintillation. Silver nitrate quenches appreciably if it is used for impregnation in a high concentration. Charring with sulfuric acid causes very strong quenching in eluted zones and is not recommended in scintillation procedures.

10.6 ZONAL PROFILE SCANS

Snyder (15) claims that for quantitation, the liquid scintillation radioassay of minute zones (zonal profile analysis) facilitated by automatic scraping instruments provides the best resolution and sensitivity for the detection of biochemical metabolites. Most sorbents, visual indicators, and related materials that are used in thin layer chromatography have no influence on the quantitative aspects. Furthermore, low-level samples can be counted for long periods of time so as to gain statistically good quantitative results. Zonal profile scans are especially helpful in exploring labeling patterns of compounds in systems such as metabolic profiles in urine that are being studied for the first time, as the scans can reveal peak areas of isotopic distribution that are not associated with reference compounds. It is in this manner that metabolic intermediates have been found. Details of the method are given below.

After chromatography of the sample, the dried chromatoplate is exposed to iodine vapor (or dichlorofluorescein) so that areas of sufficient mass can be noted in relation to standard compounds resolved on an adjacent lane. Next, the chromatogram is placed in the scraping device that is capable of automatically collecting small zones of the adsorbent layer along the entire chromatographic strip in counting vials used for liquid scintillation radioassay. A scintillation solution is then added, and the activity in each vial is determined in a liquid scintillation spectrometer. The vial is shaken for a predetermined time, and the gel is then allowed to settle. Plotting of the data obtained from sequential

vials along the entire chromatographic lane results in a zonal
profile scan. The entire procedure of scraping, collection, dis-
pensing of the scintillation solution, plotting, and analysis of
the zones can be carried out automatically. Figure 10.1 shows examples of zonal profile scans.

An important concern in the application of this procedure is
to ascertain that the total radioactivity in the labeled compounds
applied to the chromatogram is accounted for by the integral of
all the zones collected from the origin through the solvent front.
Quantitative recovery is possible only when self-absorption (due
to adsorption of the labeled component on adsorbent particles and
the glass surface of the vial) and other quenching phenomena are
absent. Snyder has found that relatively polar scintillation solutions containing a constant proportion of water permit the quantitative recovery of ^{14}C and tritium associated with both nonpolar
and polar lipids on silica gel layers.

Dual isotope zonal profile scans show the problem of isotopic
cross-contamination caused by tailing of radioactivity from one
compound into an adjacent zone containing a different compound.
Techniques for the detection of radioactivity are extremely sensitive compared with the detection of mass. Therefore, the tailing of components is sometimes not obvious when the separated components on a chromatogram are made visible with spray reagents or

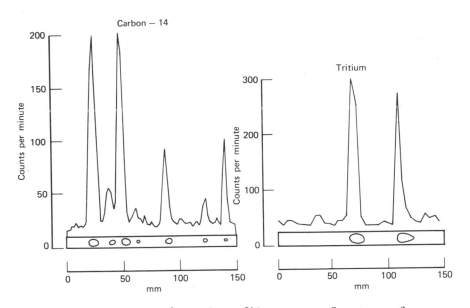

Figure 10.1 Examples of zonal profile scans. Courtesy of
Analabs, Inc.

iodine vapor. On the other hand, zonal profile analysis of the
isotopic distribution along the lane will clearly show gross contamination of unlabeled components in the system. If the tailing
of a radioactive peak is not recognized, it will be mistakenly assumed that the radioactivity associated with a particular "spot"
on a chromatogram is derived from that component made visible with
the mass-detecting reagent. The shape of the radioactive zonal
profile scan reveals maximal information about cross-contamination
on the chromatogram.

The capabilities of the technique for obtaining zonal profile
scans can be extended to the analysis of even smaller zones, for
example, 1 mm, when extremely high resolution is required. This
approach has already been shown to be effective in demonstrating
the isotopic fractionation effect of Tritium- and ^{14}C-labeled molecules. These experimental data also indicate the high resolving
power of adsorption chromatography carried out on thin layers,
even though under most circumstances such a high degree of resolution is undetectable.

If the radioactive material is soluble in the scintillation
liquid and totally desorbed by it, no complications arise from the
presence of sorbent in the vial. It is therefore worthwhile looking for a suitable scintillator mixture that would elute the material quantitatively. This is relatively easy with lipid-soluble
substances. The addition of a polar organic solvent and a certain
amount of water can even assist desorption, as reported by Turner
(16). Experience has shown that the presence at the bottom of the
vial of a white or transparent sorbent to which no radioactivity
is attached exerts little, if any, influence on the counting efficiency. Black or highly colored material on the bottom of the vial
reduces the reflecting surface and consequently lowers the energy
of the pulses. Large amounts of sorbent on the bottom of the vial
may reduce the energy of pulses generated by the external standard
source, if the latter approaches from below the vial. Thus the
plot of the ratio of efficiency to external standard channels is
slightly shifted for ^{3}H and highly quenched ^{14}C samples; this
shift is not apparent for slightly quenched ^{14}C.

Radioactivity that remains adsorbed on the sorbent may be
counted with lower efficiency for several reasons. One is the
less convenient geometry if the thin layer material lies on the
bottom. The external standard cannot completely account for this
effect, since it produces secondery beta particles and scintillations evenly in the whole volume of the liquid. The sample channel ratio method of correction, on the other hand, would be applicable if no self-absorption occurred.

Another way homogeneous dispersal of thin layer material
throughout the whole vial can be ensured is by shaking with the
thixotropic silica gel Cab-0-Sil. Particles of this preparation

are so small that ^{14}C and ^{35}S, even if they were adsorbed, do not show any detectable self-absorption losses. Caution is necessary for ^{3}H.

Self-absorption in the fibers or grains of the sorbent must be taken into account in the case of ^{14}C, ^{35}S, and ^{3}H. Some of the beta particles emitted from the adsorbed substances in the direction of the sorbent grains are completely absorbed. This is not the case for ^{14}C if the silica gel grains are less than 10 μm in diameter. Normal silica gel for TLC purposes has grains 5-25 μm in diameter; thus, part of weak beta radiation may not be counted. This loss is not very great for ^{14}C. According to Snyder and Kimble (13), a ^{14}C-labeled compound that remains on the silica gel particles (10-25 μm) will give counting results approximately 8% less when it is homogeneously suspended in a scintillation gel than when it is in solution. For ^{3}H the respective figure is 25%.

Neither the external standard nor the sample channels ratio method will correct for the complete loss of scintillations. Internal standardization by the addition of a known amount of radioactivity to the samples after they have first been counted may not be entirely satisfactory for sample self-absorption, even if the radioactivity did not differ chemically from that in the sample, since the added radioactive compound need not become absorbed to the same degree as that already retained by the sorbent. Careful calibration by graded amounts of substance subjected to chromatography, scraping, and counting in exactly the same way as the sample should eliminate the systematic error but can hardly be accurate, as the degree of sorption may not be reproducible and pipetting errors may affect the results. Quenching, including that by dissolved oxygen, would have to be corrected for individual samples in addition to the assessment of self-absorption losses.

If the labeled compound is completely dissolved, or if the sorbent containing beta emitters of sufficient energy (such as ^{32}P) is evenly distributed throughout the volume of the scintillating mixture, there is no serious objection to using external standardization, the most convenient way of correction. If the sorbent containing a sufficiently energetic emitter sediments, the sample channels ratio method is indicated, unless it is shown experimentally that the error committed by using external standardization is acceptable for the particular type of analysis. The sample channels ratio is less convenient, since it requires very long counting times in the case of low activities.

If the scintillation mixture does not dissolve all of a weak beta emitter, a reasonable compromise would be to use the external standard (or sample) channels ratio and to derive, by internal standardization with samples of the particular substance subjected to chromatography, the difference between the disintegration rates calculated from the known radioactivity added and from the external standard or channels ratio procedures. This difference, if not

very great, may then be used for correcting the results of the
analysis after the usual correction has been applied. The chemical composition of the sample of known radioactivity used for the
calculation of self-absorption losses should, of course, be as
close as possible to that of the sample.

Lacko et al. (17) described a method that eliminates some of
the difficulties involved in scraping the sorbent from the TLC
plate. In a method limited to plastic supports or possibly the
plates with the foil backing, a punch is used to remove the zone
from the chromatogram rather than scraping the sorbent from the
support. In the method as described, TLC is carried out on plastic-based plates (Eastman Chromagram #6061), and the selected
zones are cut out by scissors or with a simple manual punch. The
punch is preferred when large spots are revealed by the scanning
procedure (iodine vapor), since scissors cutting results in some
flaking at the edges and thus causes loss of material. This punch
can be made by converting an inexpensive press (Manhattan Supply
Co., 171 Ames Street, Plainview, New York 11803) to produce uniform disks with a clear edge. The disks can then be transferred
directly to vials for scintillation counting.

^{3}H-cholesterol (Mann/Schwarz) was found to be 95% radiochemi-
cally pure by TLC and was used without further purification. The
remaining 5% invariably stayed at the origin during chromatography
and did not interfere with the radioassay. Recovery of the isotope was tested by spotting identical amounts diluted with plasma
lipid extract on silica gel, and the disks were punched out without subjecting the plate to chromatography or visualization. The
same amount of ^{3}H-cholesterol was introduced directly to vials for
counting and taken as 100% for reference purposes. All three gave
essentially quantitative recovery. Toluene was the most efficient
solvent with the least amount of quenching. In all scintillation
fluids tested, the radioactivity was completely extracted from the
silica gel, for when the disks were removed from the vials the
counts did not change.

Since the extraction of lipids by the scintillation fluid is
such a crucial step in the counting process, another experiment
was carried out to test the counting efficiency and isotope recovery following TLC and visualization. ^{3}H-cholesterol was diluted
with a lipid extract of human plasma prepared by the method of
Leffler et al. (18), and identical amounts were spotted on plates.
The recovery of radioactive samples can be substantially accelerated without loss of accuracy when plastic-based plates are used
in combination with a manual punch. This technique is a substan-
tial improvement over other procedures for the TLC of ^{3}H-choles-
terol. No extraction of the samples is required before the radioactive counting, and self-absorption is avoided since the material
settles completely to the bottom of the scintillation vials. The

method should have general applicability in a wide variety of systems provided that careful testing is carried out to find the
proper experimental conditions for each compound. See Figure 10.2.

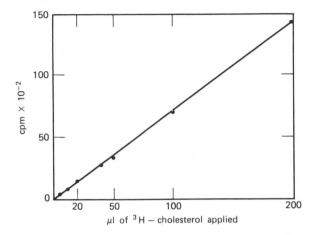

μl of ^{3}H — cholesterol applied

Figure 10.2. Relationship between amounts of isotope applied and
radioactive counts obtained.

10.7 *IN SITU* EVALUATION OF TLC PLATES

Quantitative evaluation on thin layer chromatograms of radioactive
substances by direct scanning has not seen the attention that is
deserves in spite of the fact that as late as 1968 Snyder (14)
published a comparative table listing 11 scanners available from a
number of companies. The radioactivity may be measured directly
on the plate, with either a Geiger counter with a thin end window
(19,20) or a gas flow counter (21,22). The gas flow counter is
more sensitive for the low-energy beta emitting isotopes, especially ^{3}H. In biochemical experiments in which traces are used to
follow the course of metabolism, the technique of direct scanning
of the radioactivity can be particularly useful.

10.7.1 Dot Printers and Spark Source Chambers

Pullan (23) introduced the spark chamber for rapid scanning of radioactivity on thin layer chromatograms. The procedure can also
be used for paper chromatograms.

When beta particles flow between two high-voltage electrodes placed above the chromatogram, a spark occurs that is recorded photographically with a camera positioned in the top of the chamber. In the cross-wire spark chamber, the cathode (facing the underlying chromatogram) and the anode, which is about 2.5 mm above it, are mutually perpendicular systems of parallel wires. In the sturdier design of the coil spark chamber, the cathode is in the form of several steel coils wound around thin anode wires situated in their axes. Argon with methane or methylal is flushed through the electrode system. The resulting photograph resembles an autoradiograph but is obtained with an exposure 1000 or more times shorter. The resolution is such that the edges of two adjacent spots must be separated by more than 6 mm in the case of ^{14}C and 3 mm in the case of ^3H. The pattern can be greatly improved by "scanning," that is, by moving the electrode system either manually or mechanically during the exposure. Decontamination of the cross-wire electrodes becomes a major problem unless the chromatogram has been covered with a thin plastic sheet; this is, of course, not feasible in the case of tritium. The advantages of the methodology here are that scanning times are short and quantitative results can be obtained from measurements of peak area (24). The method is nondestructive, and thus the chromatogram can be subjected to other analytical procedures.

More recently, Hesselbo (25) described a modified design of Pullan's spark chamber. The work of Smith and co-worders (26) has extended the use of this type of TLC scanning. In principle, the spark chamber (as used by Smith) consists of a set of coiled copper cathodes connected in parallel, each with a central stainless steel wire anode, these also connected in parallel. About 50-60 coils, each approximately 25 cm long, are cemented to a plate of glass such that the overall useful area of the plate is 25×25 cm. Hence, the usual 20-cm-square TLC plate can be examined. Each cathode plus central anode constitutes an independent condenser, and when any short is applied across the system a current will flow. As used here, a radiodisintegration will occur at the position of any one of the separated radioactive substances. This will be evidenced by a spark jumping the electrode gap, and these sparks can be photographed to form a picture of the two-dimensional separation. The gap between the electrode plate and the chromatogram lying on the base of the apparatus is a few millimeters, and the space is flushed with 10% methane in argon for some moments before use. Photography is by means of a Polaroid camera fixed in position at the top of a black chimney placed above the electrode plate.

The apparatus is suitable for use with any isotope, functioning extremely well with ^{14}C and ^{32}P and less well, but quite adequately, with ^3H. Photographs can usually be obtained within 10 min, but exposures of up to 1 hr can yield good pictures, as can such short exposures as 5 sec with chromatograms having zones of high activity.

To locate the accurate position of the separated spots on the original paper or layer, two procedures can be used. First, a radioactive grid is included with every chromatogram scanned to enable the picture to be realigned with the original. The grid is a series of radioactive ink dots in two dimensions, one set of dots running along the origin line and two sets of differently spaced dots following the mobile-phase flow direction. The grid and chromatogram are always placed in the apparatus in a standard way so that mixed realigning cannot occur.

Second, the small Polaroid picture is placed face upwards and covered with a sheet of glass or plastic film, and the spots are carefully ringed. This sheet is then placed on an overhead projector, beamed onto a wall on which the chromatogram is pinned, and enlarged until the grid spots from the beam become exactly coincident with those of the original grid. Then the other spots from the separated compounds must also be coincident with their beamed spots, so these can be carefully encircled to mark their positions permanently. Substances with very similar mobilities can be readily delineated in this way.

Berthold supplies a commercially available form of this apparatus called the Beta Camera LB292, shown in Figure 10.3. The designation "beta camera" comes from the fact that, contrary to the thin layer scanner, the chromatogram is not scanned and measured point by point, but rather the entire chromatogram is recorded all at once. During the recording, the spark chamber of the apparatus is located directly above the chromatogram. Sparks are produced between the anode and cathode of the spark chamber by the beta particles emitted from the chromatogram. These sparks are generated exactly above the source of the radiation. In practice, spots that contain no more than 100 dpm ^{3}H or 20 dpm ^{14}C on an area up to 3 mm in diameter can clearly be demonstrated on the film after an exposure time of 15 min. This is between 100 and 1000 times shorter than measurement by autoradiography. The usual exposure times are between 2 and 10 min.

It was originally hoped that quantitative data would also be obtained via extensive computer hookup. Unfortunately, this is not possible. Zones of very high activity act as spark sinks and appear even hotter, and less active spots appear weaker than they really are. Nevertheless, quantitation is readily achieved by cutting out the respective areas, placing them in one of the usual POPOP mixtures, and counting in a liquid scintillation counter; differently sized zones have practically no effect on counting background. In one sense, therefore, the visual picture can be misleading, and it must always be remembered that the picture observed is entirely a qualitative one within very wide limits.

Figure 10.3. Berthold Beta Camera LB292. Courtesy of Beta
Analytical, Inc.

An example of the use of the apparatus follows. A thin layer
chromatogram on which nucleotides were separated was scanned. As
Tritium decay particle energy is approximately one-tenth that of
the ^{14}C particle energy, the plastic sheet covering the spark chamber had to be removed and the exposure time increased appreciably.
Radioactivity of the different spots varied over a factor of about
60 times, the lowest being 0.3% of the total of 3-4 µCi applied.
Exposure of the whole chromatogram for 15 min at f22 showed only
four major spots, whereas exposure for 60 at the smaller aperture
of f32 resulted in the appearance of at least five more zones.
When the first four major spots were masked out by six thicknesses

of Whatman No. 1 filter paper (1 mm thick) and reexposed for 1.5
hr with a camera aperture of f32, several more spots appeared.

The masking-off procedure prevents the major spots from acting as electron sinks, thus allowing the most minor spots to be
visualized, including five not seen on the original autoradiograph,
which was obtained after a 4-day exposure. Hence, within a much
smaller time it is possible to take a number of photographs with
the spark chamber, to mask off and/or excise the major radioactive
spots, and to locate specifically the quantitatively minor spots
or compounds present. The method has time-saving advantages when
compared with radioautography.

10.7.2 Direct Scanning for Radioassay

Several instruments are available for direct scanning of a plate
for radioisotope detection of labeled compounds. These are applicable to both paper and thin layer chromatograms. They also provide non-destructible first analysis. However, they have probably
the lowest sensitivity of the radiochemical methods discussed in
this chapter, mainly because of the effects of layer thickness.
Some of them are limited to narrow-width paper or TLC "strips." The
more advanced instruments can handle the standard 20×20 cm TLC
chromatograms. In contrast to the detection by "spark chamber" or
autoradiography, which "photographs" the entire plate at once,
scanners with a Geiger-Muller (GM) rate meter scan the plate as it
moves at a constant speed under the window.

For one-dimensional chromatograms, the scanners might be the
most convenient method of detection. The use of the film technique described previously is probably more widespread.

For scanning, the TLC plate is moved along under the detector
at a constant speed and the count rate is recorded. After a suitable calibration, radioactivity present in a spot (or zone) can be
calculated from the peak area on the record. GM tubes are available with or without a thin end window. An efficiency of 1.5% for
^{3}H may be attained. Tubes provided with a lateral aperture that
serves as the slit through which the chromatogram is viewed are
rather efficient (about 37%) for ^{14}C and ^{35}S.

If the chromatogram is in the form of sheets, rather than
strips, it can be cut. There is the danger that in the case of
oblique or irregular flow, individual chromatograms will be wrongly
identified. Autoradiography or an informative GM scanning then
becomes necessary; this may cause delay with low activities. In
the case of glass-supported thin layer chromatograms, cutting is
obviously inconvenient, if not impossible, but in some cases the
layer can be removed. For thin chromatograms, one detector is
generally used and the plates are scanned in a horizontal position.
Resolution of adjacent peaks is given by the slit width, yet in

the case of the Desaga scanner it was shown that, for a 2 mm nominal slit width, a detector situated 1.5 mm above the chromatogram yields peaks as broad as if the slit width were 12 mm. Decreasing the detector height above the plate improves the resolution.

A formula relating scan speed to percentage error and confidence level for a given length and radioactivity of a spot as well as the nominal slit width was given by Wood (27). Irregular oscillations of the recorder are due to statistical fluctuations in the disintegration rate. Noise can be reduced by increasing the time constant in the recorder circuit, but this broadens the peak and displaces its maximum toward later time intervals. Thus the lowest time constant compatible with acceptable noise level, the slowest scan speed compatible with the time available, the lowest detector height compatible with contamination risk, and a reasonably narrow slit will give optimal results.

In addition to the steady movement of the chromatogram through the counter, discontinuous movement is also possible in steps whose duration is given by either a preset time or a preset number of counts. Recording is either graphical or digital. With a preselected number of counts, the program must ensure that impulse rates below a certain limit are skipped rapidly (low-background reject system) to prevent excessively long scanning times.

More recent improvements in instrumentation permit scanning of as many as 12 individual lanes on a 20×20 plate. This can be done in either the X mode or an X-Y mode. In the first, after one lane is scanned, the plate is returned to the original position and then switched over to the next lane for a scan of that lane. This is repeated until all the lanes on the chromatogram are scanned or a preset number of lanes on the plate are covered. The detector does not count during the rapid right-to-left retrace. In the X-Y mode the plate moves in such a way that it zigzags left to right, right to left, down the Y direction. The Y increment depends on the width of the lane to be scanned. The instrument automatically shuts off when the detector reaches the end of the plate (see Figure 10.4).

The detector is usually a windowless flow-through Geiger chamber. A variety of slit widths, adjustable height, and windowless or Mylar windows or changeable diaphragms are also available. Also available are dot printers, which give two-dimensional radioactivity distribution. The dot intensity is proportional to the radioactivity present on the plate. The dot printer record is similar to that obtained by autoradiography but is obtained in a short time.

Berthold also manufacturers the TLC Linear Analyzer, which counts the entire chromatogram simultaneously using a position sensitive detector. The major advantage of this instrument compared with the TLC scanner is speed. The analyzer is approximately

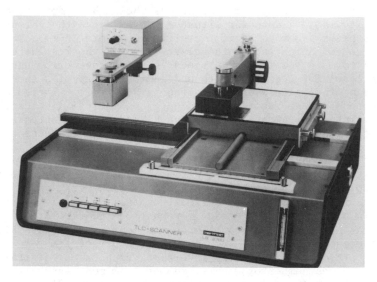

Figure 10.4. Berthold instrument for scanning radioactivity on a
TLC plate. Courtesy of Beta Analytical, Inc.

100 times faster than the TLC scanner, which only counts that part
of the chromatogram which is directly below the slit diaphragm with
its typical width of 2 mm.

    Figure 10.5 illustrates the data presentation of the same TLC
plate from three of the Berthold instruments: (a) the TLC scanner
with recorder, (b) the Beta camera, and (c) the linear analyzer.

    The use of these scanners for any given application will be
determined by the level of activity present in the chromatogram
scanned. Scanning speed, height of the Geiger tube, slit widths,
windowless or windowed, and the resolution of the zones on the
chromatogram are variables that must be considered. The uniformity
of the layer thickness is also important. Under proper conditions,
as little as 100 dpm ^{14}C, 50 dpm ^{32}P, or 1000 dpm ^{3}H can be detected in a single spot. The counting yield, depending on the
layer thickness, is claimed to be as high as 15-30% for ^{14}C, 40-50%
for ^{32}P, and 0.7-3% for ^{3}H. Good reproducibility was attained in
direct measurements of tritiated glutamate and 20-reduced desoxy-
corticosterone as shown by Wenzel and Brüchmüller (28). Table
10.1 shows that quantitative recovery (14%) was obtained when the
same amount of radioactivity was applied to several chromatograms.
Recovery was good in spite of pipeting errors. However, chromatography of a radioactive reference compound is recommended in
assays of absolute activity because counting efficiency varies inversely with the thickness of the layer.

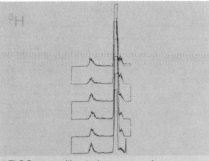

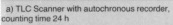
a) TLC Scanner with autochronous recorder, counting time 24 h

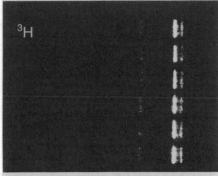

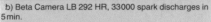
b) Beta Camera LB 292 HR, 33000 spark discharges in 5 min.

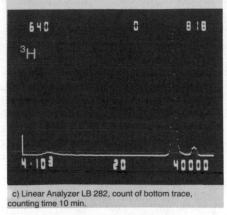
c) Linear Analyzer LB 282, count of bottom trace, counting time 10 min.

Figure 10.5. Data presentation from 3 different Berthold instruments. Courtesy of Beta Analytical, Inc.

297

TABLE 10.1   REPRODUCIBILITY IN DIRECT SCANNING OF
RADIOCHROMATOGRAMS

Radioactive substance	Activity (cpm) after 4 hr incubation*		
	0 mole-% NAD	5 mole-% NAD	500 mole-% NAD
Glutamate	37,900	13,900	13,600
20β-Hydroxysteroid	--	22,700	26,000
Total activity, cpm	37,900	36,600	35,200
%	104	100	96.5

* Activity of individual substances recovered at the indicated
concentration of NAD (expressed as mole-% of the glutamate substrate).

The activity of a doubly labeled sample may be assayed directly, although the accuracy is reduced. Wenzel (29) described
the principle of the method, which depends upon measuring collec-
tively the activity from ^{3}H and ^{14}C with the aid of a windowless
thin layer or 4π counter. The activity from the ^{14}C is then measured separately using the attenuated efficiency of the end window (0.6 mg/cm^2) counter. When a windowless scan is carried out,
both the ^{3}H- and ^{14}C-labeled zones on the chromatogram are detected. When the end window is inserted into the GM tube, only the
^{14}C will be detected, thus giving a differentiated chromatogram
scanning as seen in Figure 10.6.

In situ scanning of thin layer chromatograms was used for detecting tritiated deoxycorticosterone and corticosterone after incubation of the former with rat liver microsomes by Levin et al.
(30). this has proved very useful in locating metabolites. After
extraction of the medium with methylene chloride, the metabolites
were separated on silica gel GF thin layers using the mobile phase
chloroform:methanol:water (188:12:1). The layer was scanned in
the fluorescence quenching mode at a wavelength of 250 nm on a
Schoeffel Model 3000 double-beam scanner. Then the chromatogram
was scanned on a Berthold "Dunnschicht-Scanner" using the windowless GM tube. As seen in Figure 10.7, a differentiation could be
made between the substrate, metabolites, and the extraneous unknown material separated from the extracts. From interpolation of
the standard curves of the deoxycorticosterone (DOC) and corticosterone (b), the amounts of these substances could be determined.
Furthermore, the conversion in the metabolism could be determined
as well as the specific activity. This type of information is
becoming important as more research is being done in toxicology
and drug metabolism. The direct scanning both with the densitom-

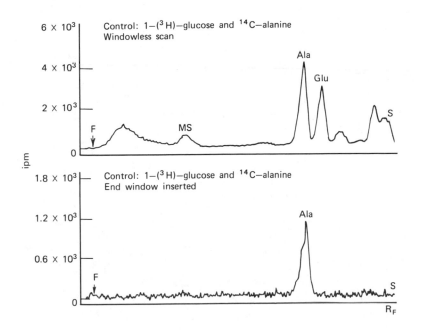

Figure 10.6. Use of end-window and windowless scans to differentiate ^{14}C and ^{3}H components separated by TLC. Ala: alamine; Glu:
glucose. Courtesy of Shandon Southern Instruments.

eter and the radioscanner facilitates the location of compounds of
interest. Thus, the *in situ* radiochromatogram scanner can be a
very useful tool in many areas.

Thin layer chromatography is a very rapid and effective method for purifying labeled compounds. In this respect it is also
useful for assessing the presence or absence of impurities of
breakdown products of the labeled substance. In most cases of organic or biochemical preparative radiochemistry it is advantageous
to be able to scan for radioactivity without destroying the chromatogram. Therefore, chromatography with direct scanning of the
separated radioactive compounds is the best way for quality control
as well as purification of substrates. Burger (31) used the direct
scanning method for assessing purity of a ^{14}C- and ^{3}H-labeled
herbicide. The limit of detection for ^{14}C in thin layer plates
was about 5×10^{-4} mCi per spot.

Although not as sensitive as scintillation counting, direct
scanning of radioactive compounds separated on thin layers offers
some advantages, as pointed out in the previous discussions. With

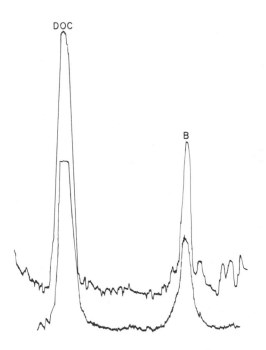

Figure 10.7. Scan of a TLC plate representing a separation of corticosterone (B) and 11-deoxycorticosterone (DOC). Lower curve: radioactive scan; upper curve: densitometer curve of fluorescence quenching scan.

newer and improved models of the radioscanners now available, the technique of direct scanning of thin layer separations should and will receive more attention.

REFERENCES

1. F. Snyder and C. Piantadosi, Advan. Lipid Res., *4*, 253 (1966).
2. R. S. Bayly and E. A. Evans, Storage and Stability of Compounds Labelled with Radioisotopes (Review 7), Radiochemical Center, Amersham, Netherlands, 1968.
3. K. Macek, J. Chromatogr., *33*, 332 (1968).
4. F. Snyder, Advan. Tracer Methodology, *4*, 81 (1968).
5. A. T. Wilson, Nature, *182*, 524 (1958).
6. U. Lüthi and P. G. Waser, Nature, *205*, 1190 (1965).
7. S. Prydz and K. S. Skammelsrud, J. Chromatogr., *32*, 732 (1968).

8. J. Chamberlain, A. Hughes, A. W. Rogers, and G. H. Thomas, Nature, *201*, 774 (1964).
9. K. Randerath, Anal. Biochem., *34*, 188 (1970).
10. L. E. Muehler and J. J. Crabtree, Phot. J., *86B*, 32 (1946).
11. S. Prydz, T. B. Melstand, and J. F. Karen, Anal. Chem., *45*, 2106 (1973).
12. E. Gruenstein and T. W. Smith, Anal. Biochem., *61*, 429 (1974).
13. F. Snyder and H. Kimble, Anal. Biochem., *11*, 510 (1965).
14. F. Snyder, Advan. Tracer Methodology, *4*, 81 (1968).
15. F. Snyder, J. Chromatogr., *78*, 141 (1973).
16. J. C. Turner, Intern. J. Appl. Radiation Isotopes, *19*, 557 (1968).
17. A. G. Lacko, H. L. Rutenberg, and L. A. Soloff, Clin. Chim. Acta, *33*, 506 (1972).
18. H. H. Leffler, F. W. Sunderman, and F. S. Sunderman, Jr., in *Lipids and Steroid Hormones in Clinical Medicine*, J. B. Lippincott Co., Philadelphia, 1960, p. 18.
19. A. Breccia and F. S. Spalletti, Nature, *198*, 756 (1963).
20. J. Rozenberg and M. Bolgar, Anal. Chem., *35*, 1559 (1963).
21. P. E. Schulze and M. Wenzel, Angew. Chem., Intern. Ed. Engl., *1*, 580 (1962).
22. P. Karlson, R. Mauer, and M. Wenzel, Z. Naturforsch., *18*, 219 (1963).
23. B. R. Pullan, in *Quantitative Paper and Thin Layer Chromatography*, E. J. Shellard, Ed., Academic Press, New York, 1968.
24. J. R. Ravenhill and A. T. James, J. Chromatogr., *26*, 89 (1967).
25. T. Hesselbo in *Chromatographic and Electrophoretic Techniques*, Vol. 1, 3rd ed., I. Smith, Ed., Wiley, New York, 1969, p. 693.
26. I. Smith, S. E. March, P. E. Mullen, and P. D. Mitchell, J. Chromatogr., *28*, 75 (1972).
27. B. A. Wood, in *Quantitative Paper and Thin Layer Chromatography*, E. J. Shellard, Ed., Academic Press, New York, 1968, p. 107.
28. M. Wenzel and M. Brüchmüller, Z. Naturforsch., *21b*, 1242 (1966).
29. M. Wenzel, Naturwissenschaften, *52*, 129 (1965).
30. S. S. Levin, H. Schleger, D. Y. Cooper, and O. Rosenthal, Fed. Proc., *30*, 308 (1971).
31. T. Burger, J. Labelled Comp., *4*, 262 (1968).

CHAPTER 11
Reproducibility

11.1 INTRODUCTION

The field of chromatography has grown considerably since its introduction. However, even now there are many problems with the factors affecting reproducibility. Many practitioners of chromatography look down on TLC as being a nonscientific, nonreproducible
method. This, of course, is not true. Thin layer chromatography
is often used as a qualitative method from which no reproducibility is generally expected. For example, the standard is chromatographed along with the sample, and the purpose is served. With
the use of internal standards, TLC can be used for quantitative
purposes with results as good as and often better than other forms
of chromatography. Often, also, TLC is the method of choice in
many analytical procedures.

There are a number of reasons for poor reproducibility in TLC.
Probably the most important one is inherent in the conditions under
which the chromatogram is developed--the atmosphere in the developing chamber. This in essence calls for control of the environmental factors during development of the chromatogram as well as during the sample application and handling. Reproducibility concerns
R_f as well as quantitation.

11.2   EFFECT OF LAYER THICKNESS

Early in his work with TLC, Kirchner showed that by controlling
conditions, the reproducibility of R_f values could be held to ±0.05
(1,2). The need for a uniform TLC plate is to some extent satis-
fied by the modern commercial plates, which tend to have a well-
controlled and reproducible thickness. The recent development of
high performance TLC layers (3) having highly reproducible particle
size as well as layer thickness will further advance the reproducibility of chromatographic results. Variation of R_f values with a
variation in the thickness of the plate has been shown by other
workers (4-7). Solvent flow is slowed in thick layers, and when
the layer thickness varies on an individual plate the erratic effect on R_f value can become more pronounced. Pataki and Keller
(6) show the effects of layer thickness on R_f values in a series
of experiments.
        Dallas (8) has pointed out, however, that especially with the
more polar solvents, the heat of adsorption onto the layer can be
appreciable. The dissipation of this heat should be less the
thicker the layer, and one can expect the temperature to be greater
in a thicker layer than in a thin layer. Therefore, if there is a
decrease in R_f with temperature as normally found in adsorption
chromatography, this heat effect can lead to an increased R_f with
increased layer thickness.
        Geiss et al. (9,10) showed that dry solvent vapor can displace
water from a partially activated plate and thereby increase its
activity. The displacement occurs more rapidly with a thin layer
than with a thick layer. This can result in an increase of R_f value with layer thickness.
        However, Jänchen (11) reports that when wedge layers were used
with varying development techniques, it could be demonstrated that
the influence of layer thickness on R_f in adsorption chromatography
could be attributed to the effect of the vapor-phase situation. A
comparison of a "saturated jar" and a "saturated sandwich" illustrated that certain development chambers commonly considered to
feature saturated conditions are far from being saturated. Jänchen
feels that the saturation and relationship between the layer and
the vapor phase of the developing system are the controlling factors rather than layer thickness.

11.3   EFFECT OF MOISTURE ON THE SORBENT

It is well known that the water content of a sorbent affects the
"activity" of that sorbent. The amount of moisture, directly related to activity, has a pronounced effect on R_f values as well as
on resolution. Hermanek et al. (12) showed the effect of the ac-

tivity of the sorbent on R_f values. R_f values increased with a decrease in activity. The effect of the activity of the sorbent due to varying degrees of water content is considerable. Care should be taken in handling the plates during all phases of the procedure, including storage. Herein lies one of the greater failures of reproducibility between laboratories. Some environments have consistently higher humidity and pollutant levels than others. Dallas (4) indicated that half of the total amount of moisture taken up by silica gel at equilibrium was absorbed in about 3 min. Even breathing on the plate during the application of the sample could affect R_f values.

Geiss et al. (13) showed that alumina layers exposed to an atmosphere of 65% humidity gained 2% moisture in 6 min and 3% in 20 min. They also indicated that the R_f of a substance will vary as much as 300% between an atmosphere of 1% humidity and one of 80% relative humidity. Dallas (4) also investigated the effect of humidity on R_f values.

Considerable changes in R_f values as a result of different amounts of water vapor in the ambient atmosphere was seen in studies of the TLC of hypnotics by deZeeuw (14). With increasing humidity (lower sorbent activity), a rise in R_f was noted. DeZeeuw recommended the use of a constant-humidity room and indicated that reheating the plates after spotting to reactivate the layer is dangerous with respect to reproducibility. In other work, he pointed out (15) that the vapors absorbed from the atmosphere greatly determine the separation and that partition chromatography rather than adsorption chromatography may be the true factor in obtaining the separations seen.

Many researchers advocate drying all TLC plates in an oven and storing them in a constant humidity cabinet of 50% relative humidity, for example. Solutions of sulfuric acid in water can be used for this purpose.

11.4 EFFECT OF CHAMBER SATURATION

The effect of saturation of the developing chamber with the mobile phases used for chromatography can be seen in the different R_f values. If the chamber is not saturated, the solvent evaporates from the layer, with evaporation increasing as the mobile phase travels over the layer. This can change R_f values to the higher ranges. Honegger (5), while investigating chamber saturation, temperature, and activity of the sorbent (silica gel) found that the activity of the layer influenced the separation to the greatest extent. Chamber saturation, particularly when the same mobile phase is used for several chromatograms, can be a decided factor. Each time a chromatogram is developed the saturation changes, and

the composition of the mobile phase changes because the sorbent
absorbs one component of the mobile phase mixture preferentially.
Each time the chamber is opened, the moisture content also changes.
Thus it is very difficult to control these factors in TLC as it is
now practiced. Some of the other apparatus, particularly the high-
performance TLC system described by Starker and Hampl (16), overcomes many of these shortcomings. The system is closed. The solvent delivery is in a closed system. The chamber is small, and
only small amounts of solvent are needed. The apparatus use is
limited to 5 cm plates, and circular chromatograms result.

Thus, for reproducible work, a mobile phase composed of more
than a single solvent should only be used once, and it must be allowed to remain in the chamber for a time sufficient to achieve
saturation of the vapor space.

11.5 EFFECT OF TEMPERATURE

Temperature has not been found to have too great an influence except in some cases (3,5). Increases in temperature cause an increase in R_f, probably due to increased evaporation from the layer
with the resultant increase of mobile phase flowing over the layer.
Higher temperatures will also change the solubility factors of the
solute in the mobile phase as well as the interactions between the
sorbent and mobile phase, although with small temperature changes
the effects would be small. (See also Section 11.2 above.)

11.6 EFFECT OF DEPTH OF DEVELOPING PHASE

The effect of distance from solvent level to the point of sample
application on the R_f will depend on the sorbent used, the compounds being separated, and the mobile phase used (16,17).
Furukawa (18) showed when the developing solvent is a two-component system used in separating mixtures containing compounds of
both high and low R_f's, the level of the origin will show an effect. There is a separation of the components from the mobile
phase in the sorbent, and the compounds of higher R_f, which travel
with the less polar solvent, are not affected by the distance of
the application zone above the solvent. The higher above the mobile phase the sample is applied, the greater will be the effect
of the separation of solvents in the sorbent. Most sorbents will
selectively adsorb the solvent of higher polarity. The greater
the total distance the mobile phase travels, the greater the effect it has on the R_f values, since the solvent flow through the
zone containing the solute lasts a longer time.

This is important when mixed mobile phases whose components
have different polarities are used. For these reasons the depth

of the mobile phase in the developing chamber as well as the distance of the starting line from the lower edge of the plate should be standardized.

11.7 NATURE OF THE SORBENT

One of the most important factors in reproducibility between laboratories is the particle size differences that exist in the sorbents available from the different manufacturers. The quality and activity can also vary from batch to batch from the same manufacturer. These generally affect the time of development but do not greatly affect the R_f values. Geiss (19) showed that differences of as much as 50% have been observed in the R_f values for separations using silica gel and alumina layers from several different manufacturers.

Most manufacturers activate the TLC plates before packaging. The heating varies from 105 to 110° for 30-60 min, mainly to drive off the water and activate the sorbent. However, depending on how the plates are stored, the conditions will vary from laboratory to laboratory. If plates are stored in desiccators or in the presence of desiccants, care should be taken to protect the plates during sample application. This can be done by placing another glass plate on the sorbent layer above the sample applications zone.

The activity of alumina layers is controlled to a greater extent by the temperature of activation. The greater the activity, the harder it will be to retain this property during storage and handling of the plates. Reichel (20) has shown that in the thin layer chromatography of insecticides by the normal procedure, R_f values are quite strongly affected by the humidity of the laboratory atmosphere. It is time consuming and not very convenient to bring the layer to a precise degree of activity before the development of the chromatogram. Methoxychlor, dieldrin, and heptachlor were poorly resolved when the relative humidity was 14% but were completely resolved on alumina plates developed with Skellysolve B when the relative humidity was 60%.

For this reason, Dallas (21) advocates the use of relative R_f values, which are independent of the adsorbent activity over a wide range. If this is the case, the conclusion is that the mean energy of adsorption of a solute on the sorbent surface is constant over a wide range of activity.

More recently, Halpaap (22) investigated parameters of the sorbent in specific systems used in thin layer chromatography. He found that particle size and distribution are as important as the primary characteristics such as the pore system and activity of the layer since they control the flow characteristics of the developing solution within the interspaces of the particles and thus within the pores.

Still another consideration in the nature of the sorbent is
the binder used in preparing the material before spreading it on
the plate. The binders used by different manufacturers are different. Some TLC plates have hard surfaces, while others seem to
be very fragile, with layers that brush off readily. These differences in the nature of the layer have decided influences on the
R_f values obtained. However, at the same time the separating characteristics of the layer are also changing. This means that the
resolution on some plates will be better than on others, depending
on the nature of the compounds to be separated.

Standardization of particular sorbents selected from the different types, with narrow pore distribution and particle size distribution, is prerequisite for the reproducibility of chromatographic processes. The present trend of TLC plate manufacturers
is in this direction.

A report by Ripphahn and Halpaap (3) has shown that control
of pore size can greatly improve the quality of separations and
quantitative results in TLC. The development of "silica gel 60"
(silica gel with pore size of 60 A) is based on this concept. The
particle size is also carefully controlled. Optimization of the
chromatographic procedure enabled adaptation of the method on a
reduced scale. Smaller plates can be used with shorter development times since efficiency of separation is much greater. This
means that precision in TLC methodology is now possible because
conditions can be more readily controlled.

11.8   EFFECT OF THE MEDIUM pH

Sorbents from different manufacturers apparently have differences
in pH. Although this may be due to different methods of preparation, it may also be due to the environment under which the TLC
plates are stored. Since TLC sorbents usually have high activity,
they are prone to absorption of pollutants from the atmosphere.
Depending on the nature of the solutes being separated, such contamination can have a decided effect on the R_f of the solute as
well as on the resolution obtained.

Tichy (23) has indicated that silica gel G seems to act as a
buffer. Therefore, in its presence the autoreduction of molybdate
solutions must be suppressed by higher concentrations of acid in
the molybdate solutions used for detection or quantitation.

11.9   EFFECT OF SAMPLE SIZE

The concentration of the solution and the complexity of the solute
applied to the layer can affect the R_f and the resolution. This
in turn will affect the reproducibility of the separation. The

effect will vary in different situations, and the individual case
must be evaluated by itself.

Application of too much sample tends to overload the chromatographic system, by exceeding the mobile phase capacity or the linear capacity of the sorbent. This results in zones or spots with
tailing that are hard to pinpoint for measurement of the R_f values
and to evaluate for quantitation by spot size or direct densitometry *in situ*. It is sometimes difficult to determine whether the
tail represents another component not resolved or is merely a
trailing of the major component. Beyond this, the R_f value can
increase or decrease with sample size depending on whether the
isotherm is concave or convex.

The R_f values can also be affected by the ionic bonding between different components (solutes) in the sample. Attraction or
repulsion of components of lower concentration by the component of
higher concentration can result in changes of the R_f values of
some components. Geiss (19) has shown the effect of applying over-
large quantities of solute to the TLC plate. The R_f's of sample
components that are of too high a concentration appear to decrease
in spite of the fact that the other components, not being of high
concentration, move in a normal fashion. The R_f of a sample solute can be affected, that is, increased or decreased, by the prox-
imity of other solutes as a result of attractive forces as mentioned above. For reference purposes, the R_f values of the pure
solute chromatographed singly should therefore always be quoted.

11.10   EFFECT OF SOLVENT PARAMETERS

Depending on the mobile phase used, its reuse can affect the R_f
value. With volatile mixed mobile phases, the more volatile component will continually decrease with use (23), or it can be diluted with volatile components of the mixture being separated.

As the mobile phase advances through the sorbent layer, the
ratio of mobile phase to stationary phase is not constant. It is
smaller near the solvent front than some distance behind it.
Dallas (4) found that if the plate is allowed to stand 15 min
after the mobile phase first reaches the layer limit, the R_f values, although a little higher than if measured immediately, are
more constant and reproducible. This is because the ratio of liquid to stationary phase has had more time to become constant over
the length of the chromatogram.

The R_f values measured in this way are independent of the distance between the point of sample application and the solvent
front if the mobile phase is a single pure solvent. With a mixed
solvent, however, there often is a solvent gradient up the plate
and R_f values are no longer independent of the distance the mobile

phase has traveled. The distance, therefore, is often standardized at 10 or 15 cm.

The use of the more lipophilic solvents as a rule results in
a decrease of R_f values, whereas an increase is obtained from the
more hydrophilic solvents.

The small k, the capacity factor of a substance, is defined
as the ratio of the retention time in the stationary phase to the
retention time in the mobile phase:

$$k = \frac{t_s}{t_m}$$

The viscosity of a solvent has a distinct influence on the k
value and thus on the migration time. The decline in viscosity at
higher temperatures has a favorable influence on the k values.

The velocity of the mobile phase can affect the R_f because
the rate of attainment of equilibrium is not instantaneous. In
ascending development the angle the plate makes with the vertical
can also affect R_f; a greater incline will result in increased
solvent flow. The mobile phase velocity is also affected by pore
and particle sizes.

Solvents of highest purity must be used for development in
order to obtain accurate and reproducible R_f values. The presence
of a small amount of impurity of different polarity can have a
large effect on R_f values. Few investigators realize, for example,
that chloroform usually has ethanol added as preservative. Distillation will remove this alcohol. Experimental records should
note whether or not the chloroform contained preservative. Furthermore, autooxidation of aged chloroform forms polar reactants
that grossly affect mobility of the solutes being separated.

When using mixed solvents a fresh mixture should be used for
each chromatographic development to allow for changes in composition resulting from differential evaporation or adsorption during
development or from chemical interaction between the solvent components.

REFERENCES

1. J. G. Kirchner, J. M. Miller, and G. J. Keller, Anal. Chem.,
    23, 420 (1951).
2. J. M. Miller and J. G. Kirchner, Anal. Chem., 25, 1107 (1953).
3. J. Ripphahn and H. Halpaap, J. Chromatogr., 112, 81 (1975).
4. M. S. J. Dallas, J. Chromatogr., 17, 267 (1965).
5. C. G. Honegger, Helv. Chim. Acta, 46, 1772 (1963).
6. G. Pataki and M. Keller, Helv. Chim Acta, 46, 1054 (1963).
7. L. Starka and R. Hampl, J. Chromatogr., 12, 347 (1963).

8. M. S. J. Dallas, J. Chromatogr., *33*, 193 (1968).
9. F. Geiss, H. Schlitt, and A. Klose, Z. Anal. Chem., *213*, 132 (1965).
10. F. Geiss, H. Schlitt, and A. Klose, Z. Anal. Chem., *213*, 331 (1965).
11. D. Jänchen, J. Chromatogr., *33*, 195 (1968).
12. S. Hermanek, V. Schwarz, and Z. Cekan, Pharmazie, *16*, 566 (1961).
13. F. Geiss, H. Schlitt, R. J. Ritter, and W. M. Weimar, J. Chromatogr., *12*, 469 (1963).
14. R. A. deZeeuw, J. Chromatogr., *33*, 227 (1968).
15. R. A. deZeeuw, J. Chromatogr., *32*, 43 (1968).
16. L. Starker and R. Hampl, J. Chromatogr., *12*, 347 (1963).
17. E. J. Shellard, Lab. Pract., *13*, 290 (1964).
18. T. Furukawa, J. Sci. Hiroshima Univ. Serv., *A21*, 285 (1958).
19. F. Geiss, J. Chromatogr., *33*, 9 (1968).
20. W. L. Reichel, J. Chromatogr., *26*, 304 (1967).
21. M. S. J. Dallas, J. Chromatogr., *33*, 58 (1968).
22. H. Halpaap, J. Chromatogr., *78*, 77 (1973).
23. J. Tichy, J. Chromatogr., *78*, 89 (1973).

CHAPTER 12
Preparative Thin Layer Chromatography

12.1 INTRODUCTION

Preparative thin layer chromatography (PLC) may be defined as the thin layer chromatography of relatively large amounts of material in order to prepare and isolate quantities of separated substances for further work such as additional chromatography, infrared analysis, melting point determination, or synthesis.

In general, the procedures used with analytical TLC may be used with PLC, the major difference being that the thickness of the sorbent layer is at least twice that used in analytical work. Layer thicknesses range between 500 μm (0.5 mm) and 10,000 μm (10 mm), but the commonly used thicknesses are 1000 μm (1.0 mm) and 2000 μm (2.0 mm). Commercially coated preparative layer plates are available from only a few suppliers in thicknesses of 500, 1000, 1500, and 2000 μm. Chapter 3 lists the plates available. Ritter and Meyer (1) chose 1-mm thick layers for the majority of their work because thicker layers did not consistently provide good resolution.

The resolving power of preparative plates is generally not as good as that of analytical layer plates, but compared with column chromatography, which is also commonly used for preparative separations, PLC has a number of advantages:

313

1.   Smaller particle size (5-40 µ) of the PLC layer results
in sharper, distinct separations. However, small particle size
column packings prepacked into tubes as "dry columns" are now
available from a number of suppliers.
2.   Conditions necessary for development of PLC may be experimentally determined beforehand by using rapid, analytical TLC.
3.   Separated zones may be easily removed from a PLC plate
and eluted.

12.2 PLATES AND LAYERS

Plate size for PLC is usually 20×20 cm or 20×40 cm, rather than
the smaller sizes, because of the quantities normally dealt with.
These two sizes are available commercially coated with preparative
layers. Several manufacturers provide smaller PLC plates, i.e.
5×20 cm and coated with silica gel G or silica gel H with or without fluorescent indicator. For the preparative separation of
extreme amounts, the sample may be applied to multiple plates,
which can be developed concurrently in a large tank.
    Many laboratories prefer to use commercially available (up to
2000 µm thick) PLC plates rather than making their own. Such use
will avoid preparation problems such as the need for a special
high-capacity spreader, the preparation time and mess, and the
precautions that must be taken when handling, storing, and drying
the thick layers. If layers thicker than 2000 µm are desired, it
will be necessary for the researcher to follow certain precautions
in preparing them.
    The same quality sorbents used to coat analytical (250 µm)
layer plates are not generally suitable for PLC plates. If a thick
layer is made from a sorbent without a binding agent, the layer
will be mechanically very unstable, rendering it unsuitable for
use. When an analytical sorbent containing binder is used to
overcome this problem, thick layers will most often pit, crack, or
flake and thus will usually be unacceptable for use.
    To minimize these problems, sorbent manufacturers have for-
mulated a series of sorbents especially for PLC. These are identified with the suffix letter P, as in silica gel PF-254. These
sorbents are a mixture of different size particles to improve adhesion and may or may not contain calcium sulfate. They are recommended for making layers from 500 to 2000 µm thick.
    To fill the need for sorbents that can be used for still
thicker layers, another series of sorbents has been formulated.
These are designated "P+CaSO$_4$" and are suitable for layers from 2
to 10 mm thick. Just as analytical sorbents are not recommended
for making thick preparative layers, the preparative layer sorbents
are not meant to be used for thinner analytical layers. The pre-

preparative layers 500-2000 μm thick are used most often, and for
this reason plates precoated with layers this thick are available
commercially. These layers have the necessary higher capacity,
yet still offer a good degree of resolution. Layers greater than
2000 μm thick are used primarily for coarse separation or presepa-
ration, as they do not have the resolution of the thinner layers.

Affonso (2-4) prepared 1-5 mm thick plates of calcium sulfate
that were hard but limited in their separation capabilities because of the adsorption characteristics of the calcium sulfate.

Kirchner (5) developed a method for preparing layers 3-12 mm
thick with a stainless steel frame with internal wires for support
rather than a supporting plate. The layers were formed from silica gel with 20% gypsum binder, and because there was no support
plate, both sides of the layer were available for high-capacity
sample application.

For convenience to the user, Table 12.1 lists the preparative
layer sorbents that are available commercially. It will be noted
that these are limited to the two most widely used sorbents, alumina and silica gel, and that all contain a fluorescent indicator.
A number of firms supply the various celluloses used for TLC (see
Table 2.2), and although it is not specifically mentioned in supplier literature, these powders should be suitable for preparative
layers if desired, although the sample capacity of such a cellulose layer would be much smaller than (as little as 1/50) that of
a silica gel layer of a comparable thickness. The major factor
here is the separation process taking place; partition on cellulose
does not have the capacity compared with adsorption on silica gel.

A variety of precoated preparative layer plates are commercially available. These are listed with the analytical layers in
Chapter 3. In addition to the selection of sorbents and the convenience that they offer, these plates have hard surfaces with
high mechanical stability. The hard surface has two advantages;
it is less subject to damage through handling and it allows the
sample to be applied through manual or mechanical streaking. It
is difficult to hand-streak a home-made preparative layer because
of the softness of the layer. Mechanical application is usually
used in this case.

Table 12.2 gives the procedures for preparing 2 mm thick preparative layers. Additional procedures may be found in Chapter 2.
Storage racks for 20×40 cm preparative plates are available from
Brinkmann Instruments, Inc., as is a Plexiglas desiccator large
enough to store the plates. A wooden storage cabinet that will
hold 24 20×40 cm plates is also available.

TABLE 12.1   COMMERCIALLY AVAILABLE PREPARATIVE
LAYER SORBENTS (BRINKMANN)

Sorbent	UV Indicator	Maximum Layer Thickness (mm)
Aluminum oxide 60* PF-254	254	2
Aluminum oxide 60 PF-254+366	254 and 366	2
Aluminum oxide 150† PF-254	254	
Aluminum oxide 150 PF-254+366	254 and 366	2
Silica gel 60* PF-254	254	2
Silica gel 60 PF-254+366	254 and 366	2
Silica gel 60 PF-254 with CaSO$_4$	254	10
Silica gel 60 PF-254 silanized RP-2‡	254	2

*   Previously designated type T; 60 indicates the pore diameter in angstroms.
†   Previously designated type E; 150 indicates the pore diameter in angstroms.
‡   Reversed-phase applications.

12.3   SAMPLE APPLICATION

Because the general goal of PLC is the separation of quantities of substances to be isolated, larger than normal amounts of sample are applied to the thick layer. A number of factors must be considered, including (a) the amount of sample to be applied; (b) the type of application technique to be used--whether streaking, spotting, or other; and (c) the method of sample application to be used--manual or automatic. Chapter 4 covers sample application in detail.

The maximum sample load for a 1000 µm thick silica gel layer (20×20 cm) is about 100 mg, less for cellulose and alumina.

In preparative work the majority of samples are applied by streaking, and few methods other than this have been reported. Honegger (6) and Meyer (7) cut V-shaped troughs into the layer in which to load the solid sample. The troughs of Honegger were 1-2 mm wide and half as deep as the layer. This is a tedious, time-consuming process that must be done carefully to avoid cutting the trough down to the plate, which would result in a discontinuous mobile phase flow.

Narasimhulu et al. (8) used a somewhat similar technique called "direct spot transfer." After samples were chromatographed and visualized by UV quenching, if it was desirable to further purify or chromatograph the separated zones, they were moistened with one or two drops of water, and the zone, down to the glass, was scraped off with a spatula. The scrapings were immediately transferred to a new plate in an area on the origin that had been cleared of sorbent. The cleared area had approximately half the diameter of the area from which the scrapings came. Another drop of water was added, and the scrapings were carefully pressed into place with a spatula. After suitable drying the plate was developed.

These techniques should be generally applicable to the application of small amounts of solid sample to a TLC plate.

Lehman and Field (9) developed an apparatus for transferring spot zones from one layer to another for rechromatography.

Connolly et al. (10) used a knife to cut two lines, approximately 3 mm apart, along the origin of the plate. Then they carefully blew away any sorbent dust, leaving a 3-mm wide ridge. With this method, sample solution is applied to the ridge with a pipet or syringe. The sample solvent is allowed to evaporate, and the lines on both sides of the ridge are filled with sorbent by placing a sheet of aluminum foil with a long, 1-mm wide slit cut in it over the knife cuts and using a spatula to press the sorbent in.

The generally preferred technique for sample application in PLC is automatic streaking. Sample spotting, although a simple, convenient method, can become quite tedious if a large volume of sample has to be applied manually. A spot produces a circular starting zone, which, particularly if it contains a relatively large amount of material, can interfere with the completely uniform advance of the mobile phase. Lipid and fatlike substances are very resistant to development in polar (aqueous) mobile phases, especially when they are very concentrated in a spot.

Manual streaking must be performed with care, as it can damage the layer, causing irregular development. It is often not entirely uniform, with the result that the separation suffers.

Figure 12.1 shows the separation of a multicomponent sample solution that was carefully streaked by hand compared with an identical sample that was automatically streaked. It would be difficult to isolate a desired substance from its neighbor without contamination from the hand-applied sample because of the waviness of the bands. This type of separation may be a hindrance in isolation because it is not always possible to visualize the entire

TABLE 12.2 INSTRUCTIONS FOR PREPARING PREPARATIVE TLC SORBENT LAYERS

Recommended slurry mixtures are based on the quantity needed to coat 20 20 cm plates with a 2 mm thick layer.

Sorbent	Slurry*	Drying & Activation	Comments
EM Aluminum oxide PF	300 g + 300 mL dist. H2O (shake 1 min). After 15 min standing, shake again before use.	Approx. 15 hr horizontal at ambient; then 3-4 hr at 110-120°.	Separations usually accomplished by multiple development. After solvent reaches upper edge of plate, layer must be air-dried before second development. Use a solvent that moves sample only 1-2 cm during first run.
MN silica gel P + CaSO4	150 g + 260 mL dist. H2O	Approx. 1-2 days at ambient; than 3-4 hr at 110-120°.	Separations usually accomplished by multiple development. After solvent reaches upper edge of plate, layer must be air-dried before second development. Use a solvent that moves sample only 1-2 cm during first run.
EM silica gel PF	150 g + 375 mL dist. H2O (shake 1 min).	Approx. 15 hr at ambient; then 3-4 hr at 110-120°.	Plates can be dried with an infrared lamp or heater.

318

EM silica gel PF + CaSO$_4$	800 g + 1600 mL dist. H$_2$O (shake 1 min until slurry is uniform).	Approx. 1-2 days at ambient; then 3-4 hr at 110-120°.	Glass plates should be cleaned only with water, not with organic solvents.
EM silica gel PF, silanized	150 g + 270 mL dist. H$_2$O/methanol (2:1). (see comments)	Approx. 1-2 days at ambient; no activation.	Shake until slurry is homogeneous; then add additional 18 g sorbent. Shake again and do not use for at least 15 min.

* Figures given for water or solvent addition are in all cases approximate. Slight variations may be necessary or desirable to suit local requirements.

NOTE: Courtesy of Brinkmann Instruments, Inc.

319

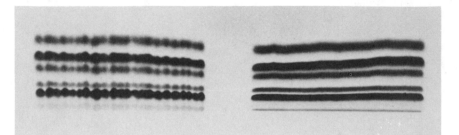

Figure 12.1.   The separation of a multicomponent solution that was
streaked by hand compared with automatic streaking.   Approximately
150 µL of a 5% solution of steroids was applied by hand-spotting
on the left and with the TLC Sample Streaker on the right.   The
plate coated with Adsorbosil-5 was run in a solvent system of
chloroform and acetone (98:2) and visualized by sulfuric-dichro-
mate charring.   From top to bottom: (1) cholesterol, (2) proges-
terone, (3) estrone, (4) dehydroepiandrosterone, and (5) testosterone.   Courtesy of Applied Science Laboratories.

separation if the substances are not fluorescent or if chemical
visualization would change the nature of the substance.   It is
sometimes necessary to assume a sharp, straight separation.
        Considering these points, the best alternative is streak-
ing application with an automatic instrument.   Other factors to be
considered are usefulness and cost.   Sophisticated sample streakers
can cost anywhere from $400 up to $2800, and it would be unwise to
spend such money for an instrument that would seldom be used.
Laboratories that do a great deal of synthesis, or isolation of
naturally occurring substances, may find it worthwhile to invest
in such an instrument.   A number of commercial sample applicators
and the characteristics of their operation are illustrated in
Chapter 4.
        Ideally, a sample streaker should have the following charac-
teristics:

        1.   It must be able to apply the sample solution to the layer
without damaging it.   A damaged layer will impede the flow of the
mobile phase, resulting in an erratic and nonreproducible separation.   This application is normally done without physical contact,
which is important when fragile home-made layers are being used.
        2.   The instrument should be capable of streaking the sample
as a narrow line uniform in concentration and size over its entire
length.   This will allow for a sharp, distinct band separation
upon development.

   3. When quantitative evaluation of sample recovery is desired, it is necessary that the volume of the sample solution to
be applied, and therefore the amount of sample solute, be accurately defined and reproduced. When applying the sample, it is
important to avoid drop formation and evaporation at the tip of
the applicator, because even a single drop may represent a relatively large percentage error. This is often considered unimpor-
tant in preparative work.
   4. When applying larger solution volumes, with the higher
solute concentrations normally encountered in preparative work,
the mechanized sample streaking should preferably be from the same
side and not back and forth. This will allow the application conditions to be as consistent as possible throughout the length of
the streak. Drying time between applications should remain consistent to keep the band width of the applied sample compact.
   5. The length of the streak to be applied should be adjustable to allow for plate size, amount of sample, and number of samples.

Several manufacturers provide preadsorbent plates with 500
µm or 1000 µm thick silica gel layers for PLC (see Chapter 3).
This preadsorbent layer allows the application of up to 5 mL of
sample, which does not have to be carefully spotted or streaked
for effective separation. Such layers save valuable time in sample application.

12.4 DETECTION

The entire width of a plate should not be streaked, because there
is a general tendency for the mobile phase to travel faster or
slower along the plate edge, which results in curved bands being
formed upon development. When streaking a 20 cm wide plate, it is
suggested that the streak be ended approximately 2-3 cm in from the
left edge and 3 cm in from the right edge.
   Before the plate is streaked, a suitably sized aliquot of
the sample is spotted at the origin approximately 1.5 cm in from
the right edge. After this aliquot is developed with the streak,
it can be visualized by any technique desired without affecting
the streak and thus serves to locate the separated components of
the sample. This portion of the plate may be cut away from the
rest of the plate, using a glass cutter. The small, narrow strip
is easily handled during visualization. It is suggested that this
be done if heating is required for visualization. Heating may
decompose the sample components, and this procedure will allow
visualization without changing the composition of the unvisualized
sample on the remainder of the plate. When a portion is to be

visualized by spraying, the large portion of the plate may be
masked off by using a clean glass plate 20 cm wide.

If it is desired to apply known compound standards to the
plate to aid identification, keep them approximately 1 cm apart
and apply them next to the aliquot of the sample so that direct
comparison may be made. Another sample aliquot spotted at the opposite end of the streak and visualized with the first one is
helpful in ascertaining exactly how the sample bands developed.
Figure 12.2 diagrams the general sample application procedure for ·
PLC.

Most precoated preparative layer plates are commonly available with 254 nm or 254+366 nm fluorescent indicators incorporated
into the layer. These plates should be used as much as possible
because they provide a good general nondestructive mode of detection, which is the most desirable mode for PLC. Other modes and
methods of detection suitable for preparative work are covered in
Chapter 7.

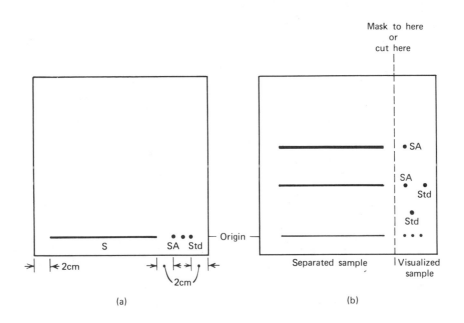

Figure 12.2. Sample application procedure for PLC. (a) Application. (b) Development and visualization. S = sample, saturated;
SA = sample aliquot; Std = known standard.

12.5   ELUTION OF DESIRED ZONES

After the preparative plate has been developed and the separated
bands have been located, the desired substances may be removed from
the plate in a number of different ways.

Camag has developed the Eluchrom automatic elution system,
primarily for quantitative analysis work. This system will elute
up to six zones that are no more than 2 cm in diameter, by passing
a known volume of eluting solvent through the *in situ* layer and
collecting it in a special vial. The eluent may then be transferred into a spectrophotometer or fluorimeter for quantitation.
Vitek et al. (11) have completely described this apparatus.

When dealing with either long wide bands or small zones of
separated substances, more manual methods must be used. These involve removing the layer containing the substance of interest completely from the plate, eluting the substance out of the sorbent
with appropriate solvents, and then filtering and concentrating
the solvent into a desired state for later use. Such procedures
are illustrated in the literature (12-15).

The most common procedure is to simply outline the desired
zone with a syringe needle or spatula and, with a spatula, scrape
off the layer down to the support onto a piece of glassine weighing
paper. The layer material is transferred into a small tube with a
Teflon-lined screw cap. A few drops of distilled water are used
to moisten the powder in order to inactivate it so that it will release the bound substance to be extracted.

In order to extract the desired substance for maximum recovery, a very polar solvent such as methanol, acteone, or water
should be used. Often, however, water cannot be used because it
would interfere with the subsequent operations to be carried out
on the sample. Water is more difficult to handle than either methanol or acetone, which are preferred.

After the solvent is added to the tube containing the sample,
the tube is capped and shaken for a given period of time, often
only 5 min, by hand or mechanical shaker. The tube is centrifuged
and the supernatant is pipeted carefully into another vessel for
the subsequent operation, which is usually evaporation or concentration. At this point, it is generally recommended to extract the
precipitated layer sorbent a second time following the same procedure.

To avoid scraping by hand with a spatula and the problems that
could ensue, such as loss by spillage or blowing, a number of different collectors for the layer material have been developed that
are based on vacuum suction. Mottier and Potterat (16) were the
first to publish the use of such a device. They applied a vacuum
to the constricted, cotton-containing end of a glass tube 8 mm in
diameter, and sucked the layer up the open end, retaining it on
the cotton which could then be extracted. Glass wool or a 200

mesh stainless steel screen supporting a Teflon mat may also be
used (14). Goldrick and Hirsch (15) used a sintered glass disk to
trap the sorbent.

A number of commercially made vacuum collectors are available.
The simplest, shown in Figure 12.3, is manufactured by Kontes Glass
Co. and is available through a number of distributors. This sample recovery tube has beveled, polished ends for use against the
layer and contains a sintered glass disk to retain the sorbent.
After the layer material has been collected, the tube is held vertically in a collection tube or flask and solvent is washed
through the layer to elute the sample. It is generally good practice to moisten the layer material in the recovery tube with one
or two micro drops of distilled water to inactivate the sorbent
before the solvent is added.

Figure 12.3.   Sample recovery tube for TLC.   Courtesy of Kontes
Glass Co.

Brinkmann Instruments, Inc. makes a similar collector that
employs a sintered disk joined into a 14/20-neck round-bottomed
flask (see Figure 12.4). The elution solvent may then be sucked
directly through the layer on the disk, and the eluted sample collected in the flask. These collectors are limited in the amount
of material they may trap at any given time and were not intended
to collect an entire band off a thick-layer preparative plate.
The general apparatus must be scaled up to accommodate larger
amounts of collected material.

Brinkmann Instruments, Inc. manufacturers a "vacuum cleaner"
type of zone collector that is designed to collect larger amounts
of material. The layer is sucked into a Soxhlet extraction thimble, which can then be placed directly into a Soxhlet extraction
apparatus for efficient removal of the sample out of the sorbent
material. Two sizes of collector are available, a micro version
that holds a thimble of 2.5 mL capacity and a macro version that
holds a 65 mL capacity thimbles. This apparatus was originally
designed by Ritter and Meyer (1).

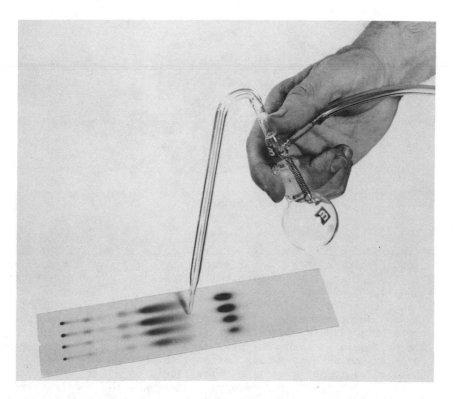

Figure 12.4. Spot collector with elution flask. Courtesy of
Brinkmann Instruments, Inc.

    When silica gel is eluted with methanol, soluble silicic acid
methyl ester may be formed and passed through the filtration medium supporting the silica gel. To avoid contamination of this sort
in the final extract, evaporate the polar solvent to dryness and
dissolve the residue in a less-polar solvent such as benzene. The
contaminant will remain undissolved and suspended, and careful
filtration will produce the desired extract.
    A useful piece of inexpensive equipment to have on hand when
eluting small areas from a TLC plate is a Swinney filter holder
for attachment to a standard syringe, shown in Figure 12.5. These
are available in the line of filtration apparatus carried by most
laboratory supply houses. They are made of chromed brass and contain a removable, perforated stainless steel disk for the support
of a 13 mm diameter membrane filter. The filtration side of the
apparatus connects to a syringe, and the effluent end will retain

Figure 12.5. Swinney filter holder.

a syringe needle for delivery. It is ideally suited for the filtration of volumes of 1-5 mL of almost any biological fluid or any
other liquid that will not chemically attack the particular filter
being used.

     This apparatus can be used to transfer a minimum amount of
the liquid being filtered. A syringe of a size suitable for the
volume of liquid being filtered is fitted with a needle. If the
volume of liquid is inconveniently excessive, the liquid may be reduced in volume by evaporation, often with minimal or no heat,
which could cause decomposition. The liquid is drawn up into the
syringe, the syringe is turned upside down, and the needle is removed. The Swinney holder, fitted with the filter and delivery
needle, is then placed on the syringe, it is righted, and the
plunger is slowly depressed to filter the liquid into the desired
receiver. Nagel and Dittmer have reported their uses of the
Swinney holder (17).

     Stutz et al. (18) have developed an apparatus for eluting the
large areas encountered on preparative layer chromatograms. It is
designed for continuous *in situ* elution of the areas in a specially
constructed chamber. Because of this feature, the procedure eliminates the mess and loss often encountered with scraping procedures and avoids carryover of fine sorbent particles into the extraction (eluting) solvent. The interested reader is referred to
the original publication for complete details.

     If only small, analytical size samples are to be eluted, it
may be convenient to use plastic- or aluminum-backed layers, which
can be cut apart with scissors. The pieces containing the desired
zones may then be eluted (extracted) with a suitable solvent, as
described earlier. The Eluquick apparatus, produced by Camag will
do this automatically with up to 12 samples at a time.

REFERENCES

1. F. J. Ritter and G. M. Meyer, Nature, *193*, 941 (1962).
2. A. Affonso, J. Chromatogr., *71*, 332 (1966).
3. A. Affonso, J. Chromatogr., *22*, 452 (1966).
4. A. Affonso, J. Chromatogr., *27*, 324 (1967).
5. J. G. Kirchner, J. Chromatogr., *63*, 45 (1971).
6. C. G. Honegger, Helv. Chim. Acta, *45*, 1409 (1962).
7. H. Meyer, Chem. Abstr., *83*, 117990c (1975).
8. S. Narasimhulu, I Keswani, and G. L. Flickinger, Steroids, *12*, 1 (1968).
9. M. C. Lehman and K. W. Field, J. Chem. Educ., *51*, 704 (1974).
10. J. P. Connolly, P. J. Flanagan, R. O. Dorchai, and J. B. Thomson, J. Chromatogr., *15*, 105 (1964).
11. R. K. Vitek, C. J. Seul, M. Baier, and E. Lau, Am. Lab., *6*(2), 109-10, 112-16 (1974).
12. M. K. Seikel, M. A. Millet, and J. F. Saeman, J. Chromatogr., *15*, 115 (1964).
13. D. R. Gilmore and A. Cortes, J. Chromatogr., *21*, 148 (1966).
14. M. A. Millet, W. E. Moore, and J. F. Saeman, Anal. Chem., *36*, 491 (1964).
15. B. Goldrick and J. Hirsch, J. Lipid Res., *4*, 482 (1963).
16. M. Mottier and M. Potterat, Anal. Chim. Acta, *13*, 46 (1955).
17. J. H. Nagle and J. C. Dittmer, J. Chromatogr., *42*, 121 (1969).
18. M. H. Stutz, W. D. Ludemann, and S. Sags, Anal. Chem., *40*, 258 (1968).

CHAPTER 13
Reversed Phase Thin Layer Chromatography

13.1 INTRODUCTION

Most TLC separations are performed in the "normal phase" mode on
silica gel using relatively nonpolar mobile phases for chromatographic development. This mode of development suffices for many
applications, and will continue to be used. Recent introduction
of chemically bonded reversed phase TLC plates by several manufacturers has made reversed-phase thin layer chromatography (RPTLC)
an important new tool. RPTLC expands the scope of separations by
thin layer chromatography to include compounds traditionally not
separable by normal phase TLC because of their high polarity.

The earliest work in liquid chromatography involved the use
of a solid or liquid stationary phase more polar than the mobile
phase; it was thus termed "normal phase." Reversed (or reverse)
phase chromatography involves the use of a nonpolar, usually hy-
drocarbonaceous stationary phase and a relatively more polar mobile phase. Classical reversed phase liquid-liquid partition systems involved a column, paper sheet, or thin layer supporting a
physically coated nonpolar stationary liquid such as paraffin,
silicone oil, or squalane. A modification of this approach has
been termed "mixed RPTLC," in which a one or two carbon silane
reagent was bonded to the silica gel surface to reduce silanol
group activity and a liquid phase was physically coated onto the
silanized silica gel.

Most layers available for RPTLC are silica gels to which
n-alkyl substituents, particularly octadecyldisilyl (C_{18}), are
chemically bonded. Most present-day high performance liquid chromatography (HPLC) is done on reversed phase columns, and the technology for preparing the RPTLC sorbents was derived from the
knowledge gained there.

Thus, reversed phase thin layer chromatography can be divided
into two categories based on whether the sorbent is (a) physically
coated or (b) chemically bonded with the alkyl groupings.

13.2 TYPES OF BONDING

Physically coated plates consist of conventionally prepared normal
phase TLC plates that are coated with a nonpolar liquid phase by
dipping the plate solutions containing the nonpolar liquid.
The nonpolar phase can also be applied by "chromatographing" the
liquid phase solution onto the TLC plate. The percentage of
loading is controlled by the concentration of the solution used
to coat the chromatographic support. A typical concentration is
10%.

Table 13.1 lists compounds and mixtures that have been used
to coat TLC plates. Most commonly used are paraffin oil (1),
silicone oil (2), octanol (3), and oleyl alcohol (4). Breyer et
al. (5) showed the effects of varying the chain length of the
coated stationary phase for a homologous series of alcohols, alcohols with chain lengths below 12 and above 16 gave unsatisfactory
separations or irreproducible results. Several naturally occurring oils, including castor oil (6), peanut oil (7), and corn oil
(8) have also been used, although the differences between these
phases has not been well evaluated. RPTLC separations have been
performed on plates coated with amines or organophosphorous compounds, particularly in extraction chromatography (a separation
technique based on complex formation or ion exchange). Amine preparations such as Aliquot 336 (8), Alamine 336S (9), Amberlite
LA-1 (7), and Primene JM-T (10) are commercially available for
coating.

Chemically bonded stationary phases are more common in HPTLC
than in TLC, and results obtained with bonded phases in HPLC are
directly applicable to TLC. A few generalizations can be presented here. Bonded alkyl phases are produced by the reaction of
alkylchlorosilane with the silica support. The alkylchlorosilane
electrophilically attacks the silanol oxygen (Si-OH) found on the
silica surface, with the subsequent elimination of water.
Bonded phases commonly prepared are either (a) monomeric
bonded phases from the reaction of silica with monofunctional
alkylsilane reagents (i.e., the reaction of dimethyloctadecyl-

TABLE 13.1.   COMPOUNDS USED FOR COATING SORBENTS
TO BE USED IN RPTLC

Alcohols	Amines	Oils	Others
C_8	Aliquot 336	Castor oil	Fatty acids
C_9	Alamine 336S	Corn oil	Pyrozol
C_{10}	Amberlite LA-1	Paraffin oil	Squalane
C_{12}	Primene JM-T	Peanut oil	Triolein
C_{14}	Amberlite LA-2	Pump oil	Hexadecane
C_{16}		Silicone oil	Undecane
C_{18}			
Oleyl alcohol			

chlorosilane with silica results in a bonded layer one molecule
thick), or (b) polymeric phases from the use of difunctional or
trifunctional alkylsilane reagents. Trace amounts of water present
in the reaction media cause some self-polymerization of the silane
reagent in addition to previously discussed electrophilic attack
on the silanol groups. Truly monomeric phases can be produced
from polyfunctional alkylsilanes if water is rigorously excluded
from the reaction media.

The reproducibility of production of polymeric phases is erratic, due to the high dependence of the reaction on trace amounts
of water present. The reproducibility of the monomeric bonded surface is excellent because the reaction is complete, and is only
limited by steric hindrance of the bonded phase towards itself and
unreacted silanol groups. Monofunctional silanes are hydrolyzed by
water, but as long as an excess of the reagent is present, this
side reaction generally does not affect the bonding efficiency.

Bonded reversed phase TLC plates are prepared in two ways:
(a) Silanized silica gel is mixed with a binder, slurried in a nonpolar solvent, and spread on glass plates to dry. When dry, the
plates are ready for use.   (b) Commercially prepared silica TLC
plates are silanized by an *in situ* reaction. Gilpin and Sisco (11)
have described in detail the apparatus and steps used in this process. Stated simply, the TLC plates are immersed in a silanizing
reaction mixture, from which water has been rigorously excluded.
In this way, highly reproducible bonded phases are prepared.

Whatman KC_{18} layers consist of absorbent treated by bonding
octadecylsilane to the surface of a special $10\mu m$ silica gel via stable Si-O-Si bonds. The layers are designed to correlate to certain
C_{18} bonded (10-12% carbon loading) HPLC high performance reversed
phase columns, specifically the Whatman Partisil-5-ODS column. The
same chemical bonding techniques used to form C_{18} columns are modified to produce KC_{18} precoated layers. The surface hydroxides,
however, are fully covered (capped) by reacting further with

hexamethyldisilazane. The resulting particles are fully silanized, with a total carbon load of 12%. They possess the resolving power and efficiency typical of Whatman high performance silica gel layers and also have a high sample capacity due to the optimized carbon load.

Although other commercial plates contain a higher nominal carbon percentage (up to 16%), the high retention of hydrophobic solutes on Whatman C_{18} layers is due to the greater availability of the C_{18} groups. The groups do not interact with each other in aqueous solvents, leading, therefore, to a high effective capacity and increased retention as water content increases. The lower percentage of carbon also leads to better wettability of the layer and relatively short development times with aqueous mobile phases.

13.3 THEORIES OF SEPARATION IN REVERSED PHASE TLC

Theories of solute retention in reversed phase chromatography are a much debated subject and many papers have appeared concerning the coated and bonded reversed phases. Theories of solute retention in physically coated TLC plates are the least controversial. The major contribution to retention in coated plates RPTLC is from partitioning of the solute between two immiscible liquids. The concentration of solute in each phase is described by the distribution coefficient, K_p, which in turn can be related to the free energy change for the solution process where $\Delta G^\circ = RT \ln(K_p)$. This observed partitioning mechanism for coated RPTLC makes this technique valuable for the chromatographic measurement of K_p values for certain eluent systems. For example, several workers have derived relationships between R_f values (from coated RPTLC) and K_p values (obtained by the "shake" method) (2,4,12). Measurement of K_p values for solvent systems is one of the largest current applications of coated plate RPTLC, because this is a quick and easy technique for determining relative degrees of lipophilicity. Although partitioning is probably the major factor in retention, active silanol groups are also present on the silica support, and in some systems, adsorption may occur (2).

Solute retention on chemically bonded phases is difficult to explain, and no real theory is available. Most of the work in the literature is concerned with HPLC, but it is applicable to bonded phase RPTLC. It is thought that the nature of a bonded phase is somewhat different from that of the corresponding physically coated liquid. Since bonded phase molecules are attached at one end, the resulting phase is more ordered than a liquid; therefore, bonded phases have fewer degrees of freedom than liquids, and in the case of monomeric phases, are only one molecule thick. Thus the liquid partitioning mechanism credited to coated supports is usually discounted as an explanation of bonded phase solute retention.

Theories that retention on chemically bonded phases is governed by hydrophobic interactions have been proposed. The hydrophobic effect can be described as the tendency of a nonpolar solute molecule to reduce its surface area when exposed to water -- either through association with other nonpolar molecules, or through removal from the solution by adsorption (13).

The principal separation mechanism on KC_{18} layers may be the van der Waals attraction between the hydrocarbon part of the sample molecules and the octadecyl groups of the layer. These forces depend in part on the carbon chain length, being approximately twice as great for C_{18} groups as for C_8 groups. Nonpolar solutes have greater retention (lower R_f values) as chain length increases (14). In addition to increased retention, capacity and stability are also greater with the longer chain length (15). With regard to selectivity, there appears to be some increase as chain length of the bonded silane is increased (16), but the differences are not as great as the retention changes (13). Control of selectivity through modification of the mobile phase is preferred. Selectivity increases with increasing percentage of water in methanol-water mobile phases (17).

The solute retention mechanism on C_{18} reversed phase layers (or HPLC columns) is not clearly understood. Because of the differences in the nature of traditional coated liquids and chemically bonded phases, the simple liquid partitioning mechanism attributed to the former is considered inadequate to explain bonded phase retention. It is now recognized that the retention process is more complicated and cannot be directly placed within one of the classical modes of chromatography (18). The role of organic modifiers (e.g., acetonitrile, dimethylsulfoxide, tetrahydrofuran) that are apparently extracted into the stationary phase (19) is still unclear. The mechanism of separation of polar solutes on C_{18} bonded layers using nonaqueous mobile phases has been termed "adsorption" in certain publications (20), but this is clearly oversimplified because of the high carbon coverage and capping of the silica gel adsorption sites. Sander et al. (21) have reviewed the theory of solute retention in reversed phase TLC.

13.4 MANUFACTURERS OF RPTLC LAYERS

Several manufacturers prepare RPTLC layers. Only one of these offers the "physically coated" type; the others are all of the "chemically bonded" type.

Chain lengths range from C_2 to C_{18} and percent surface coverages are from 50 to 100% of accessible reactive sites. The RPS Uniplate of Analtech Inc. consists of a hard analytical layer impregnated with a long chain hydrocarbon, and it is designed

to correlate retention behavior observed in RPTLC with that seen
in modern HPLC. The hydrocarbon coating is not appreciably soluble
in water or alcohol, and the plates are intended for single use
only. Uniplates are available with 250, 500, and 1000 μm layer
thicknesses for analytical and preparatory applications. Furthermore, the plates may be obtained with preadsorbent (preconcentrat-
ing) zones for increased resolution of highly retained nonpolar
substances.

E. Merck produces three plates that have different chromatographic selectivities. These plates have C_2, C_8, or C_{18} bonded
hydrocarbon chains, designated RP-2, RP-8, and RP-18 respectively.
Macherey-Nagel manufactures three reversed phase plates (Nano-RP-
Plates). These use a C_{18} bonded phase but differ in their percentage of carbon loading. The Nano-RP plates are compatible with
highly aqueous eluents and are limited only by their surface wettability. For example, at high water concentrations, SIL C_{18} 50
plates (50% of reactive silanol groups are bonded) work best, and
with a 1:3 methanol-to-water eluent, development occurs at a rate
of 4.0 cm/15 min. Plates with higher carbon loading develop considerably slower with high percentages of water. The Opti UP C_{12}
plate, manufactured by Tridom, has a C_{12} bonded stationary phase
and is indicator free. It is reported to be compatible with most
reversed mobile phases including those with high percentages of
water. This is of importance not only for choice of eluent, but
also in detection, since water based visualizing reagents may be
used. Whatman, Inc. offers two types of reversed phase plates, a
bonded C_{18} plate (KC_{18}) and a dual-mode, reversed phase and normal
phase plate (Multi-K type CS5). The KC_{18} product is available in
several forms, with or without indicator. The stability of this
plate to water (stable up to 40% water) may be increased by the addition of NaCl to the mobile phase. The Multi-K type "CS5" plate
is a hybrid of reversed and normal phase TLC. The plates consist
of a 3 cm wide C_{18} bonded layer contiguous with a silica gel analytical layer on a single 20×20 cm plate. In actual use, the sample is spotted on the reversed phase layer and developed under reversed phase conditions. Additional separation is obtained by
normal phase development 90 degrees from that of the first. Many
particularly complex separations have been obtained on this plate.

13.5 USE OF RPTLC

Sample preparation methods in RPTLC are generally the same as in
conventional TLC (see Chapter 4). The sample is to be applied in
such a way that the substances of interest will travel as discrete
zones and have the same R_f values as the reference substances.
The preparation of the sample should also consider the ratio between extraneous material and the desired solute, since this ratio

will determine the success of the separation. Too high concentration of impurities will affect separation since these can interact with analyte and cause changes in the R_f.

Cleanup of samples can be facilitated by the use of preadsorbent layers now generally available on TLC plates. A crude sample can be applied to this layer, and predevelopment with suitable solvents can extract the sample and deposit it on the starting line, which is represented by the junction between the preadsorbent and the sorbent in the developing area. In some cases, application of the sample directly to the C_{18} layer is sufficient because of the high capacity inherent in these sorbents.

13.6 PLATE PREPARATION AND APPLICATION OF SAMPLES

Reversed phase plates do not normally require activation prior to use. Adjustment of water content is not critical as with silica gel, because virtually all accessible silanol groups are covered by reaction with hydrocarbon.

It is advisable, however, to wash the layers prior to use. Few if any TLC plates as obtained from the manufacturer have sorbents free of extraneous material. This material interferes with sensitive detection methods, and results in a dark background with many of the reagents used for detection and quantitation. The use of chloroform-methanol (1:1) for washing plates usually provides a clean background. Plates are developed overnight in the developing tank. Sometimes more polar solvent in the developing mobile phase can be used for washing the plates. The sorbent must be dried before sample application is carried out.

Standards and samples should be dissolved in the most polar solvent possible to minimize excessive band spreading (16). Purification of complex samples can also be achieved in this way if some highly nonpolar impurities are left undissolved in the weak solvent (selective solvation). Because of stability and wettability limitations, sample solutions should be made up in totally or predominantly organic solvent rather than in water. Methanol, which is quickly evaporated and readily wets the layer, is convenient for substances capable of hydrogen bonding. For less polar compounds, methylene chloride and acetone are convenient solvents.

13.7 MOBILE PHASE SELECTION

Separation in reversed phase thin layer chromatography requires the selection of the solvents of the mobile phase in a reversed order of strength from the conventional. This usually results in a reversed order of the R_f values for a series of compounds from

that seen in conventional TLC. It is possible to obtain separations on RPTLC using entirely organic solvents. Reverse order of
separation will generally require that water be a larger component
of the mobile phase. Mobile phases can be two-component mixtures
of water and a polar organic solvent such as alcohol, acetonitrile,
acetone, dioxane, or an ether. With adjustment of the proportions
of the components of the mobile phase the separation can, in effect, be optimized. As polarity of the sample increases, the water
content of the mobile phase should decrease. In water-alcohol mixtures, improved resolution may be achieved by changing the alcohol
to one of higher or lower chain length with a constant concentration, as well as by altering the alcohol-water ratio. Reversed
phase layers are less subject to effects of humidity, and solvent
demixing is rare.

An approach to the selection of a mobile phase involves finding a single solvent or solvent mixture that results in median R_f
values for the solutes to be separated. This can usually be obtained by consulting the literature or by trial and error. If adequate resolution is not obtained, the components are varied to
change the selectivity while maintaining the same strength.
Strength determines R_f but not selectivity. Change of the mobile
phase by use of solvent strength parameter (P) values reported by
Snyder (22) are often successful. See Chapter 5 for details and a
table of values.

The highest P values represent the strongest solute for adsorption (silica gel) TLC but the weakest for RPTLC. The strength
of a mobile phase (P') for RPTLC can be approximated from the relationship $P' = F_a P_a + F_b P_b + F_c P_c$, where F is the volume fraction of
the pure solvents a, b, and c, and P is their strength. Certain
constituents of mobile phases are added to provide specific interactions with one or more compounds to be separated, e.g., calcium
ions for the separation of bile acids (23).

A mobile phase for initial use in reversed phase TLC has
proved to be ethanol-water (80:20 v/v). Selectivity may be improved by substituting some other solvents for ethanol and
altering the ratio with water. Introduction of 1-20% of tetrahydrofuran, dimethylformamide, or dimethylsulfoxide can be used to
change the selectivity in difficult separations. Methylene chloride and other lipophilic solvents have also been used successfully, and totally organic (nonaqueous) mobile phases are becoming
very widely used on reversed phase plates. Mixtures of water with
a polar solvent do not completely wet alkyl-bonded silica gel,
leading to some problems.

Separations of certain compounds can be improved by the addition of a lipophilic ion-pairing salt to the mobile phase. The
pH of the mobile phase is maintained at a value such that the solutes are ionized and can bond with the counter ion. The mechanism

of separation is either an ion exchange type of reaction between the ionized solutes in the mobile phase and the counterion ion that becomes bound to the C_{18} layer, or liquid-liquid partition on the C_{18} layer of the neutralized ion pairs formed in the mobile phase. Ion pair chromatography on C_{18} layers has been applied to the resolution of phenols and quinones using tetrabutylammonium hydrogen sulfate counter (24) and to sulfothiazine bases and sulfoxides using heptane sulfonic acid counter ion (25).

Solvent mixtures containing more than 40% water may disrupt the binding properties of the sorbent. Adding ion-pairing salts or NaCl in concentrations of 0.1 to 0.5 M overcomes this. Addition of salt will ensure the stability of the layer without affecting most separations and may increase the migration rate of the mobile phase (17).

13.8 DETECTION IN RPTLC

The location and quantitation of substances separated by RPTLC is somewhat hampered by the fact that there is a high level of carbon content. This eliminates the popular charring reagents widely used in universal detecting methodology. However, there are a number of common reagents that are not so drastic that they may be used for RPTLC. The first consideration is with the compounds that show ultraviolet absorption near the wavelength of activation of the phosphor in the plates containing this property. Fluorescence quenching can provide sensitive methodology depending upon the compound (26,27).

Spraying with phosphomolybdic acid can be used since this is a mild reagent requiring little heat. Iodine vapor can be used as in conventional TLC. When used properly, a charring spray of 10% sulfuric acid in ethanol is very useful. The reagent is sprayed lightly on the chromatogram, which is then heated at 170° for 2-3 min. This has been used to detect bile acids and has the reproducibility necessary for use in quantitation (23). As a general detection reagent, this method can be useful provided the spray is not too heavy and heating is not too long. The fluorescamine reagent produces fluorescent derivatives of amines and amino acids separated by RPTLC, using the usual conditions. In general, the investigator should consult the literature since methodology for RPTLC is constantly appearing.

REFERENCES

1. R. J. Dailey, C. B. J. Gray, and J. R. Brown, J. Chromatogr., 76, 175 (1973).

2. M. Kuchar, V. Rejholec, M. Jelinkova, V. Ralek, and O. Nemecek, J. Chromatogr., *162*, 197 (1979).
3. A. E. Bird and A. C. Marshall, J. Chromatogr., *63*, 313 (1971).
4. A. Hulshoff and J. H. Perrin, J. Chromatogr., *120*, 65 (1976).
5. A. E. Breyer, M. Fischl, and E. J. Seltzer, J. Chromatogr., *82*, 37 (1973).
6. J. M. Pla-Delfina, J. Moreno, J. Duran, and A. delPozo, Pharmacokinet. Biopharm., *3*, 115 (1975).
7. I. D. Jones, L. S. Butler, E. Cribbs, and R. C. White, J. Chromatogr., *70*, 87 (1972).
8. U. A. Th. Brinkman, G. DeVries, R. Jochemsen, and G. J. DeJong, J. Chromatogr., *102*, 309 (1974).
9. H. R. Leene, G. DeVries, and U. A. Th. Brinkman, J. Chromatogr., *80*, 221 (1973).
10. U. A. Th. Brinkman, G. DeVries, and H. R. Leene, J. Chromatogr., *69*, 181 (1972).
11. R. K. Gilpin and W. R. Sisco, J. Chromatogr., *124*, 257 (1976).
12. G. L. Biagi, A. M. Barbaro, M. F. Gamber, and M. C. Guerra, J. Chromatogr., *41*, 371 (1969).
13. C. Horvath and S. Melander, J. Chromatogr. Sci., *15*, 393 (1977).
14. J. Halasz, Anal. Chem., *52*, 1393 (1980).
15. M. C. Hennion, C. Picard and M. Caude, J. Chromatogr., *166*, 21 (1978).
16. A. M. Siouffi, T. Wawrzynowicz, F. Bressolle, and G. Guiochon, J. Chromatogr., *186*, 563 (1979).
17. U. A. Th. Brinkman and G. DeVries, J. Chromatogr., *192*, 331 (1980).
18. B. L. Karger and R. W. Giese, Anal. Chem., *50*, 1048A (1978).
19. R. P. W. Scott and P. Kucera, J. Chromatogr., *142*, 213 (1977).
20. H. Halpaap, K. F. Krebs and H. E. Hauck, High Resolut. Chromatogr. & Chromatogr. Commun., *3*, 215 (1980).
21. L. C. Sander, R. L. Sturgen, L. R. Field, J. Chromatogr. Sci., *18*, 133 (1980).
22. L. R. Snyder, J. Chromatogr., *92*, 223 (1974).
23. J. C. Touchstone, R. E. Levitt, S. S. Levin, and R. D. Soloway, Lipids, *15*, 386 (1980).
24. M. V. Marshall, M. A. Gonzalez, T. L. McLemare, D. L. Busbee and A. C. Griffin, J. Chromatogr., *197*, 217 (1980).
25. D. J. Volkmann, J. High Resolut. Chromatogr. & Chromatogr. Commun., *2*, 729 (1979).
26. J. Sherma and M. J. Blim, J. High Resolut. Chromatogr. & Chromatogr. Commun., *1*, 309 (1978).
27. A. C. W. Po and W. J. Irwin, J. High Resolut. Chromatogr. & Chromatogr. Commun., *2*, 623 (1979).

CHAPTER 14
High Performance Thin Layer Chromatography

14.1 INTRODUCTION

Thin layer chromatography in the conventional sense is rapidly being improved with high performance thin layer chromatography
(HPTLC), separations and quantitative analysis can be carried out
in as little as five minutes. The number of samples that can be
separated on a single HPTLC layer is limited only by the ability
to apply small volumes of sample solution. As many as sixty samples have been separated on a single 10×10 cm HPTLC plate and
scanned with densitometry, all within one hour.

The basic difference between conventional TLC and HPTLC is in
the particle and pore size of the sorbents used. Recent manufacturing advances have resulted in the ability to reproduce sorbents of
uniform pore and particle size. This ability has resulted in sorbents that are very reproducible and have high resolution. This
means that results can be obtained in shorter times with less mobile phase.

The only sorbent presently used for high performance layers is
silica gel, since it is the major sorbent used in conventional TLC.
High performance TLC utilizes optimized particle size of the silica
gel, thinner (100 µm) layers, and smaller samples. Because of the
high efficiency of the HPTLC, separations can be performed with a
sample migration distance of 3-6 cm. Compared with conventional
TLC, a fivefold to tenfold reduction in analysis time is obtained

(1). The main limitation of HPTLC seems to be the small sample
volume (nanoliters) that must be applied for optimum performance.
Recently, this problem has been solved by the introduction of the
"contact spotting" technique, which makes it possible to spot up
to a volume of about 1000 μL onto the HPTLC plate while maintaining a spot diameter of 0.1 to 0.5 mm.

14.2   SAMPLE PREPARATION

Sample preparation in HPTLC follows that of conventional methods,
but it is more important to have as great a concentration of the
desired solute as possible, since less sample must be applied to
the layer. The capacity for sample in HPTLC is low, and because
of limitations in chromatographic development distance, the extraneous materials may not be sufficiently separated for precise
quantitation. This prerequisite sample preparation is a factor in
the success of the results that can be obtained by HPTLC. The
primary consideration is the smallest amount of sample that can be
applied and still permit detection of the component of interest.
This is particularly important if densitometry is to be performed.
Refer to the preparation techniques discussed in Chapter 4.

14.3   SAMPLE APPLICATION

Success in HPTLC, particularly when quantitation is desired, is
strongly dependent on the quality of sample application. Since mobile phase travel is generally in the 5-7 cm range, there is little
space for separation to occur. Resolution is therefore important.
When considering the formula for resolution, N, it is apparent that
the size of the application point is the critical factor for optimum resolution (1); $N=2D/(W_2+W_1)$; D is the distance between the
apex of two peaks, and W_1 and W_2 are the baselines of the two peaks
obtained by scanning the thin layer chromatogram by densitometry.
     The wider the baseline of peaks in the chromatogram, the lower
the resolution. A comparison of "streaking" and "spotting" sample
application methods showed "streaking" of the sample results in
sharper separations or narrower separation zones. This will result
in higher resolution, as verified by a series of studies, including the use of a five-component dye mixture. The resolution for
the separations resulting from the samples that are "streaked" was
as much as 32% greater than that from the spotted samples. Thus,
it must be emphasized that sample application is an important factor in resolution and that "streaking" is the method of choice (1).

14.3.1 Use of "Preabsorbent" Layers

The recent development of layers containing preabsorbent zones has
facilitated the use of high performance thin layer chromatograms.
It is difficult to manually apply samples so, as to take advantage
of the high efficiency of HPTLC without the use of special devices
developed for this purpose (2). The preabsorbent area gives a
means of producing a very fine line as a "streak" for the origin
of the chromatogram. It also facilitates sample application, since
the sample can be applied rapidly. The application must be made as
a "streak" or line, since spotted zones result in circular zones
after development. Streaking across lanes scored in the layer is
recommended. Application of the sample vertically in the middle
of the lane also results in a circular concentrated zone, which
can decrease the efficiency of the sorbent.

Instruments are available for sample application; they can be
used when multiple samples must be evaluated.

For controlled application of micro samples in HPTLC, Camag
offers the Nanomat, which can be used to apply samples for both
linear and circular development. This apparatus is pictured in
Figure 14.1.

Figure 14.1. Nanomat sample applicator. Courtesy of Applied
Analytical Industries, Inc.

In use, a fixed volume capillary pipet is filled with sample solution and inserted into the magnetic head of the Nanomat. The pipet is held suspended by a magnetic sleeve a few millimeters above the plate surface. Upon actuating the release button on the instrument, the pipet descends onto the layer and delivers the sample. It then returns to its original position , from which it can be removed for replacempnt or refilling. The lowering speed and the contact time of the pipet can be regulated to prevent damage to the layer. Before applying the next sample, the application arm is moved to the next position. When preparing circular chromatograms, the plate is rotated.

The Nanomat can accommodate several sample delivery devices including (a) 100 or 200 nL fixed volume nanopipets; (b) the HPTLC sample application capillary for qualitative applications only; (c) the Camag Nano-Applicator for sample volumes between 10 and 230 nL; and (d) disposable micropipets of volume 0.5-5 µL, such as those made by Drummond Scientific Company. The Nano-Applicator is shown in Figure 14.2.

Figure 14.2. The Nono-Applicator. Courtesy of Applied Analytical Industries, Inc.

Fenimore (3) reported that the size of the developed spot depends not so much on the size of the starting spot as it does on the quality of the developing mobile phase. Initial zones smaller than 1 mm do not appear to be of advantage. However, the size of the starting zone will be dependent somewhat on the nature of the solvent used to make the solutions for application. In HPTLC when lipophilic substances are applied in 0.3 µL of heptane, a zone of 0.45 mm in diameter is possible; if acetone is ued, a diameter of 1.4 mm results. Therefore, an optimization is necessary between the concentration and volume of the solution to obtain the prerequisite starting size. As described earlier, this does not appear to be a factor when layers with preadsorbent zones are used to apply the sample as a line below the adsorbent zone.

One "contact spotters" is commercially available. The Clarke Contact Spotter (3) combines what are usually the separate operations of sample concentration and application by removing the solvent prior to, rather than after, introduction of the sample onto the adsorbent. This is accomplished by evaporating a relatively large volume of sample solution, as much as 100 µL, on a specially prepared, nonwetting polymer film. Because of the large contact angle between the solution and the polymer surface, the sample forms a symmetrical droplet or bead and retains this symmetry until evaporation is complete. The residue is then transferred to the absorbent layer by simple contact with the film.

This permits precise spotting of as many as 15 samples on a 5×10 cm or 10×10 cm HPTLC plate. The transfer film is positioned over a series of depressions in a metal plate, and with the application of vacuum through orifices in the centers of the depressions, the film is made to conform to the contour of the plate surface. Sample solutions are pipetted into the depression; gentle heat and nitrogen flow evaporate the solvent; the HPTLC plate is positioned absorbent side down over the film; and with slight pressure replacing the vacuum, the spots are all transferred simultaneously.

With very pure samples, those that form crystalline or completely solid residues, there may be some difficulty in achieving complete transfer on contact spotting. This is seldom a problem with biological extracts because these residues tend to retain sufficient liquidity for transfer. In those instances where this is not the case, small amounts of appropriate nonvolatile solvent may be added to the sample solvent so that one nonoliter or less will remain after evaporation.

This spotting technique has been used with excellent results in the determination of psychopharmacological agents in blood plasma. Because of the low concentrations encountered at therapeutic levels with many of these drugs, and the limited amount of sample usually available, it is often necessary to apply as much of the sample as possible to the HPTLC plate. This is facilitated by using relatively large volumes of sample extract while retaining the small spot sizes required by HPTLC.

14.4   MOBILE PHASE SELECTION

Solvent selection in HPTLC follows much of the same procedure as that of conventional TLC. It must be considered that the results obtained with knowledge gained from conventional TLC will not give the same results on HPTLC. Slight adjustments in solvent strength and selectivity may be necessary to bring the separations within the confines of the smaller HPTLC layer.

14.5   DEVELOPMENT TECHNIQUES FOR HPTLC

14.5.1   Linear Development

HPTLC can be developed in the linear, circular, or anticircular mode. The circular technique is an extremely powerful tool for the resolution of components in the low R_f region and the anticircular technique for the separation of components in the high R_f region (4). Both techniques appear to hold considerable promise.
     Linear HPTLC development can be carried out in the usual chromatography tank. If available, museum or pathology jars of the smaller sizes are preferable. The operations are identical with those of conventional TLC. As little as 4 mL of mobile phase are sufficient for the development of a 10×10 cm HPTLC plate. In a twin-trough chamber the layer can be preconditioned by the vapors of mobile phase filled in the trough opposite to the plate; when presaturation is completed, the chamber is tilted, so that mobile phase flows over into the other trough and development starts.
     Camag supplies the HPTLC Linear Development Chamber, in which the chromatogram is developed from both sides after samples have been applied along both opposing edges. This way the number of samples per plate is doubled although the maximum separation distance is 50 mm. In this chamber the layer is arranged in a sandwich configuration horizontally. The solvent enters the layer via a capillary slit through its frontal cross section. The developing process stops automatically as both solvent fronts meet in the middle. If a separation distance longer than 50 mm is desired, the plate can be developed from only one side.
     If preconditioning of the layer is felt to be necessary, a solvent-soaked filter paper between the layer and its counter plate
          the layer for a defined length of time over a sufficient quantity of conditioning liquid before the plate is insert-

ed into the chamber. Having the developing solvent enter the layer
via its frontal cross section can accentuate the influence of con-
taminations of the layer. Contaminants are picked up during storage in the laboratory atmosphere and are concentrated in the fron
tal cross section. From there they are washed into the layer,
whereas with tank development they are diluted into the entire
solvent reservoir. It is recommended that HPTLC silica gel plates
be prewashed, particularly when they are designated for development in the Camag HPTLC Linear Development Chamber.

The storage of HPTLC plates is important, and one should "reactivate" plates in a drying oven; this must be followed by suitable intermediate storage. Because of the water adsorption hys-
teresis, a silica gel layer behaves differently if it has been
equilibrated with a given relative humidity from the more dry form
or the more wet condition (5).

14.5.2 Circular Development

Camag offers a circular development system for HPTLC, as originally proposed by R. E. Kaiser, known as the U-Chamber (6). It is
pictured in Figure 14.3. The manufacturer claims a number of ben-

Figure 14.3. The Camag U-Chamber. Courtesy of Applied Analytical
Industries, Inc.

fits with its use: (a) with circular development on silica HPTLC,
the optimum separation distance is 20-25 mm. Such a separation
could be achieved within 1-4 min; (b) chromatograms show separations with 4 to 5 times better resolution than linear ones on the
same layer material with the same mobile phase; (c) the relation
between flow rate and resolution, well known from HPLC, exists as
well in HPTLC. Only in U-Chamber circular developing techniques
can control of the flow rate be achieved with reasonable means;
(d) the mobile phase is delivered to the layer from a reservoir
isolated from the surrounding atmosphere. Therefore, the composition remains constant throughout the separation, thereby enhancing
reproducibility.

In use, the HPTLC plate is precisely positioned by the plate
holder ring with its layer facing downward on the U-Chamber body.
The apparatus must be leveled. The mobile phase is fed to the
center of the plate via a platinum-iridium capillary of 0.2 mm
internal diameter. Constant flow of the mobile phase is achieved
by a dosage syringe of 250, 500, or 1,000 µL capacity, operated by
a stepping motor drive. Flow rate and total volume can be preselected and are electronically controlled. Because of this, reproducibility of conditions is high.

The gas phase, which has been made up externally, may be
passed through the chamber via a channel and out through the center
bore. Sample application is done to the dry layer after the plate
is locked into place in order to ensure that the center of the ring
of starting points and the point of mobile phase introduction precisely coincide.

14.5.3 Anticircular Development

In anticircular development of HPTLC plates, the mobile phase en-
ters the layer in a circular line and flows toward the center. The
samples are applied on a circle slightly inwards of the solvent
feeding line. A milled out interruption of the layer on the outer
edge prevents solvent flow in the outward direction.

According to the chromatographic flow law, the linear solvent
migration speed decreases with the square of the distance. Since
the area to be wetted in this configuration also decreases with
the square of the antiradial distance, the linear speed of solvent
migration is (nearly) constant. Thus, with respect to separation
distance, this developing mode is absolutely the fastest. The
speed of migration, however, cannot be controlled; the solvent
feeding rate depends entirely on capillary effects.

Initially, the layer is held suspended in the plate holder
ring 0.5 mm above the solvent transfer ring. In this position the
layer can be preconditioned by feeding externally prepared gas
phase through the center bore where it flows outward and emerges

through the peripheral slit. Upon lowering the plate onto the
chamber, developing solvent instantaneously starts to enter the
layer from the ring channel. No further gas phase circulation is
possible while chromatographic development is in progress.

14.5.4 Continuous Development

In the linear development mode, the maximum number of components
that can be resolved in one development is approximately ten. For
more complex mixtures the resolution can be improved over that obtained in the single linear development mode if the sample is separated by multiple development (7), programmed multiple development
(8), or continuous development (9). In the multiple development
mode, the plate is developed several times and the solvent is evaporated from the plate at the end of each development stage before
putting the plate back into the mobile phase. Redevelopment of the
plate causes spot reconcentration to take place.

On the second and subsequent passes, the solvent flows over
the trailing edge of the spot before reaching the leading edge,
contracting the spot in the top-to-bottom direction. This method
of development is very flexible as quantitative measurement can be
made at the end of any development stage. HPTLC is a rapidly developing technique of great promise for use in the field of trace
analysis of complex organic mixtures. The development techniques
allow components to be separated a few at a time or further by
multiple developments as they become separated. The process of
drying the plate and then redeveloping it in the same or a different mobile phase leads to a natural spot concentration phenomenon,
which aids resolution and improves detection limits. As only part
of the sample is separated per development, migration distances
can be kept short, maintaining a compact spot profile and minimiz-
ing analysis time. The continuous development mode takes advantage
of the high mobile phase velocity through the first few centimeters
of the HPTLC plate to provide the conditions necessary for fast
development. As the mobile phase velocity is much higher, fewer
polar solvents can be used for a required separation, providing a
more selective interaction without incurring a large time penalty.
The poor resolution and low mobile phase velocity made these development techniques of little use in conventional TLC, but with
the modern HPTLC plates these disadvantages have been eliminated.

The remaining problem in HPTLC as far as trace analysis is
concerned is the plate sample capacity. Samples obtained from
biological sources after evaporation are often quite viscous, and
only a fraction of the extract can be applied to the plate. The
final extract solution must not be too viscous to wet the capillary
surface. The development of the contact spotting techniques enable
100 μL or so of extract to be evaporated to a residue and the residue transferred to the plate as a compact spot. This eliminates

the viscosity problem and allows a much more complete sample transfer from the extract to the plate to be made.

14.6 CONSIDERATIONS FOR DETECTION AND DENSITOMETRY

The spectrophotometric detection system should provide a UV-visible
absorption spectrum and fluorescence excitation and emission spectrum of the separated components on the plate. These spectral
parameters are required to optimize detector sensitivity, to select the best wavelength for trace analysis by maximizing the
signal-to-noise ratio, and to select a scanning wavelength useful
for multicomponent simultaneous determinations. Because of the
small spot size, the slit length for scanning should be in the
range 2-4 mm and function without loss of incident light intensity
(microoptics). Usually the practical detection limit is in the
nanogram range (10-15 ng) for UV-visible measurements and in the
picogram range (100 to 500 pg) for fluorescence. For some compounds with particularly good spectral properties, much lower detection limits can be obtained.
     As reported by Touchstone et al., densitometry of HPTLC plates
after charring can be a very reproducible and sensitive technique
(10). The sensitivity limits in HPTLC, as in other analytical
techniques, depends on the ability to develop sensitivity. Thus,
it is recommended to use fluorescence (inherent or induced) whenever possible.

REFERENCES

1. J. C. Touchstone and S. S. Levin, J. Liquid Chromatogr., *3*,
    1853 (1980).
2. R. E. Kaiser, A. Zlatkis, and R. E. Kaiser, Eds., J.
    Chromatogr. Lib., *9*, 88 (1977).
3. D. C. Fenimore and C. J. Meyer, J. Chromatogr., *186*, 555
    (1970).
4. R. E. Kaiser, J. High Resolut. Chromatogr & Chromatogr.
    Commun., *1*, 164 (1978).
5. K. Ch, J. Chromatogr., *97*, 137 (1974).
6. R. E. Kaiser in *HPTLC-High Performance Thin Layer Chromatog-
    raphy*, A. Zlatkis and R. Kaiser, Ed., Elsevier, 1977, p. 73.
7. J. A. Perry, T. H. Jupille and L. J. Gluck, *Separation and
    Purification Methods*, E. S. Perry and A. J. Van Oss, Eds.,
    Marcel Dekker, New York, Vol. 4, p. 105 (1976).
8. T. H. Jupille, J. Am. Oil Chem. Soc., *54*, 179 (1977).
9. J. A. Perry, J. Chromatogr., *165*, 117 (1979).
10. J. A. Touchstone, M. F. Dobbins, C. Z. Hirsch, A. Baldino, and
    D. Kritchevsky, Clin Chem., *24*, 1476 (1978).

Special Techniques for Thin Layer Chromatography

15.1 INTRODUCTION

There are a number of special techniques in TLC that are used occasionally for specific purposes. Many require expensive and
elaborate apparatus not found in the average chromatography laboratory. Some of this equipment will be discussed in this chapter.

15.2 DEVELOPING TECHNIQUES

15.2.1 Vapor-Programmed Development

Vapor-programmed (VP) development is a recent technique in which
the effect of the vapor environment on resolution is used to optimize separation. Difficult separations can often be accomplished because optimum vapor conditions have been established over the
entire plate. DeZeeuw has used this technique for sulfonamides
and barbiturates (,2) and for local anesthetics (3).

For the convenience of the researcher, there are commercially
manufactured vapor-programming chambers. One, manufactured by
Desaga is available in two sizes, one for 20×20 cm plates and one
for 20×40 cm plates. This apparatus is shown in Figure 15.1, protected by a Plexiglas hood. This VP chamber consists of a metal

Figure 15.1.   Vapor-programming chamber.   Courtesy of Brinkmann
Instruments, Inc.

block holding 21 troughs (each 20×1×2 cm or 40×1×2 cm), a solvent
reservoir, and six sets of Teflon spacers (0.3, 0.5, 0.8, 1.0,
1.2, 1.5 mm thick).   Using one set of spacers, the plate with the
samples applied is mounted with the layer side down toward the
troughs, which have been filled with suitable liquids.   The mobile
phase is fed to the TLC plate from a reservoir through a paper
wick.   The different vapors that have been absorbed by the dry
layer before development dissolve in the migrating mobile phase
during development.   This results in a controlled mobile-phase
gradient, which is specifically a reversed gradient.

In conventional gradient techniques, the solvent with the
lowest polarity is found in the solvent front; the opposite is
true with the vapor-programmed technique.   Because the upper part
of the layer contains the more polar solvent, the faster migrating
compounds advance more rapidly and are resolved from the slower
moving compounds in the lower portion of the plate.

DeZeeuw (2), using this technique, allowed the plate to remain in the chamber in contact with the solvent vapors for 10 min
prior to adding the mobile phase to the reservoir and starting

that development. He states that "the actual advantage of VP-TLC
is that close-lying spots can be pulled apart and then guided to a
position in the chromatogram which is not yet occupied by another
compound. Accordingly, the entire plate length can be utilized to
cover the spread of the spots. Furthermore, due to the fact that
any desired vapor composition can be selected and applied to any
point of the plate, the analyst can select the optimal conditions
for a particularly difficult separation of two compounds, without
disturbing the separation already obtained of the other components
present in the sample....We indeed postulated that the increase in
resolution which can be obtained in unsaturated chambers could
possibly be due to a concentration gradient of vapor in the dry
adsorbent. In order to be able to control such a gradient and/or
to improve the gradient, we thus developed the vapor-programming
chamber."

Geiss and Schlitt (4) developed another design of vapor-programming chamber called the Vario-KS chamber, which is manufactured
and distributed by Camag, Inc. This chamber is shown in Figure
15.2. It is supplied with three different troughs that are referred to as conditioning trays: one with 5 divisions, one with
10 divisions, and one with 25 square divisions.

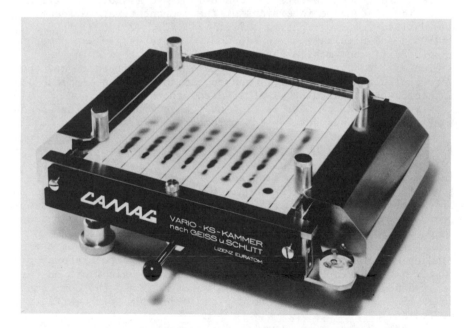

Figure 15.2. Vario-KS chamber for vapor programming TLC plates.
Courtesy of Camag, Inc.

The mobile-phase tanks are available both without divisions and with five equal divisions. Sandroni and Schlitt (5) reported the application of the latter for the concurrent development of a 20×20 cm plate with up to five different mobile phases, which are delivered to the plate with five paper wicks. Because of the different flow rates of the different mobile phases through the layer, some sorbent must be scraped from the plate so that the layer consists of five sections, each 16.5×2.5 cm. To avoid mixing of the vapor phases between sections, a sandwich slide made of six 1 mm thick polypropylene barriers glued together is inserted between the sections before development.

Figure 15.3 shows three applications for the vapor-programming chamber: (a) Preloading with five solvents for the rapid selection of the most suitable separation system. A dye mixture was developed with pure ethyl acetate on silica gel after preloading the layer with five different solvents that were less polar than or equally polar to the ethyl acetate mobile phase. (Left to right: ethyl acetate, benzene, carbon tetrachloride, cyclohexane, n-hexane.) (b) Effect of relative humidity. Preloading with water-sulfuric acid mixtures in the conditioning trough resulted in eight humidity zones for the determination of the effect of relative humidity on separation. Shown is a dye mixture on silica gel. Three sequence inversions of the components are evident. Humidities, left to right, are 72%, 65%, 47%, 42%, 32%, 18%, 14%, and 9%. (c) Simultaneous development. A dye sample was developed in five different mobile phases simultaneously to aid in the choice of the best phase for the separation. Left to right, these were cyclohexane-benzene 1:1; benzene; dichloroethane-benzene 1:1; dichloroethane; and diethyl ether-dichloroethane 1:1.

The Vario-KS chamber is also available with a heating accessory for continuous development chromatography.

15.2.2 Radial Chromatography

In radial (circular) chromatography, the sample substance to be separated is spotted in the middle of a horizontal plate. The mobile phase for development is supplied to the center of the sample zone by means of a wick or micro dropping pipet. The separated zones develop as concentric rings.

With circular development, the sample component zones are expanding continuously so that they are being drawn out into progressively narrower concentric rings. In linear development, on the other hand, there is no such expansive effect, and the component zones become progressively broader and wider due to the effects of diffusion.

A second factor in circular development is a result of the combination of a progressively expanding solvent front and a lim-

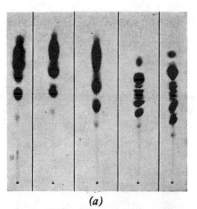

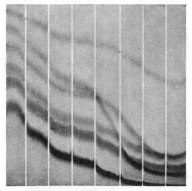

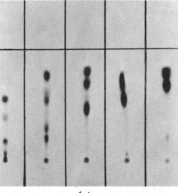

Figure 15.3. Applications of the vapor-programming chamber. (a) Choosing the most suitable vapor separation system. (b) Effect of relative humidity. (c) Choosing the best mobile phase for a separation. See text for details. Courtesy of Camag, Inc.

ited point source of mobile phase input. With this arrangement, mobile phase feed will always be faster at the trailing edge of a component zone, tending to compress it. This factor, along with a progressive drop in the sample loading, significantly suppresses the tendency of a component to trail during development.

All of these factors result in a rapid resolution of a mixture into its various components in a very short developmental distance and time.

DeThomas and Pascual (6) have designed a simple apparatus for demonstrating radial chromatography, and Rachinskii (7) has considered the basic theoretical principles involved in this action. This form of TLC is not as versatile or as convenient to use as normal development methods. For example, comparison with standard substances, elution of separated zones, and quantitation are more difficult to perform than with conventional methods. Because of these factors, radial chromatography is seldom used.

Schleicher and Schuell, Inc. is marketing a device called the SelectaSol Solvent Selector System, which employs circular development to aid in the selection of a suitable mobile phase that can then be used for normal development. The entire apparatus is shown in Figure 15.4. A template is supplied so that the sample spots may be applied in correct positions on the plate. The chamber base has 16 small wells in which up to 16 different mobile phases may

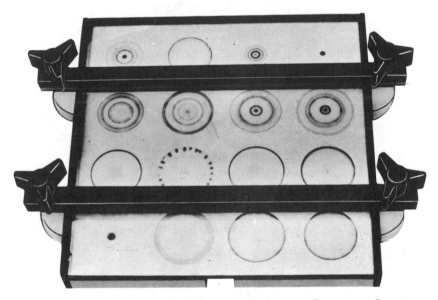

Figure 15.4. The Selecta Sol Solvent Selector System. Courtesy of Schleicher and Schuell, Inc.

be tested.   A wicking plate contains wicks that immerse in these
wells as it is sealed against the base with a gasket.   The spotted
20×20 cm plate contacts the wicks aided by the proper spacing of a
gasket.   The entire assembly is screwed together, and development
is allowed to proceed horizontally.

A cross sectional diagram of the development process is shown
in Figure 15.5.

As mentioned in Chapter 14, Camag is offering an apparatus
called a U-Chamber (see Figure 15.3) for fast screening circular
chromatography.   Separations normally require only 1-4 min, using
the 5×5 cm High Performance TLC plates (HPTLC) manufactured by E.
Merck (available from Camag and EM Laboratories).

In operation, the spotted HPTLC plate is positioned by a plate
holder ring with its layer facing downward on the U-Chamber body.
Mobile phase for development is fed to the center of the plate by
a platinum-iridium capillary.   Constant flow of the mobile phase
is achieved by a motor-driven 250 µL syringe.   Normal development
requires this volume of mobile phase or less.

Vapor phase, made up externally, may be passed through the
chamber and out a center bore before, during, and after development.   The name of the apparatus is derived from the mode of operation, with everything occurring underneath.

15.2.3   Hot Plate Chromatography

According to Turina et al. (8), warming a chromatographic plate
during the TLC process causes the solvents to evaporate from the
plate, and the chromatograms thus obtained show two advantages:
better resolution of the spots, and the possibility of detection
of trace components that are undetectable under usual conditions.

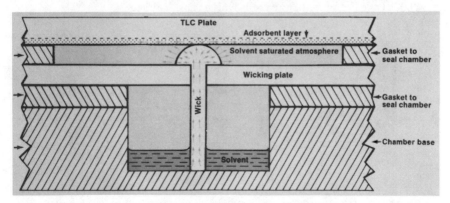

Figure 15.5.   Cross section of the Camag U-Chamber (for operation
see text).   Courtesy of Applied Science Laboratories.

These effects, which are obtained under special conditions, can be explained by the discontinuous countercurrent model of the chromatographic process (9).

Turina and Jamnicki (10) consider the theoretical mathematical principles of hot plate chromatography, which is seldom used in this form because it has little advantage for the separation of many compound types.

15.2.4 Programmed Multiple Development

Programmed multiple development (PMD) is an automated form of multiple development for TLC that was first reported in 1973 by Perry, Haag, and Glunz (11). In PMD, the thin layer plate is automatically cycled through a preset number of developments. In each succeeding development, the mobile phase is allowed to advance further. After each development, controlled evaporation by heat or inert gas causes the solvent front to recede, usually to or beyond the point of the initial zone of sample migration. After the last desired cycle, continued controlled evaporation prevents further development. PMD exploits the reconcentration that occurs each time the solvent front traverses the sample zone. As a result, spot broadening during development is counteracted. PMD is claimed to have advantages over conventional TLC in plate efficiency, speed, and sensitivity. These vary according to the operational parameters chosen.

An apparatus for PMD commercially available from Regis Chemical Co. is illustrated in Figure 15.6. It consists of two units, a programmer (left) and a developer (right). The programmer is a computerized controller that directs predetermined PMD parameters to be carried out by the developer. The developer, as shown in Figure 15.7, consists of a solvent trough, a sandwich chamber for the TLC plate, and an infrared radiating device.

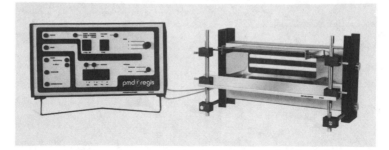

Figure 15.6. Instrumentation for programmed multiple development (PMD). Courtesy of Regis Chemical Co.

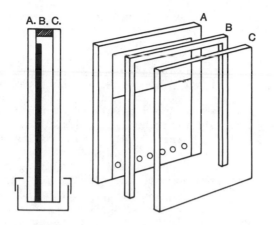

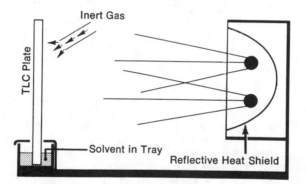

Figure 15.7. Developer unit for PMD. (a) TLC plate, (b) spacer,
(c) cover plate. The thin layer plate is about 5 in distance from
and centered with respect to the radiator. Courtesy of Regis
Chemical Co.

    Jupille and Perry (12) have reported that spot placement and
size are independent of spotting technique in PMD. The spots produced by this method are uniformly and reproducibly compact regardless of the initial top-to-bottom spread, and identical spots
arrive at the same level on the TLC plate regardless of initial
displacement of the spot origin from the plate edge. PMD recon-
centrates and aligns both high-R_f and low-R_f spots. For spots

with low R_f values, reconcentration during solvent removal is more effective than reconcentration during mobile-phase advance.

The typical spot on a PMD chromatogram is an elliptical band with the top-to-bottom width smaller than the side-to-side width.

One important consideration must be kept in mind when using PMD. Mobile phases with two or more components may change in composition during a long development and not provide reproducible separations. This would be very likely to occur with mixtures of solvents with widely differing boiling points, such as those containing diethyl ether. Using the attributes of PMD to their fullest, it may be possible to effect separations with only single-component phases. Low-polarity solvents are generally recommended. Mobile phases used for conventional TLC cannot usually be used for PMD because of such factors. A direct correlation between the separations obtained by the two procedures should therefore not be assumed; it will have to be proved experimentally.

It has also been generally found that the harder layers of silica gel plates made with polymer binders are superior for PMD development to the softer silica gel G plates with gypsum binder.

Jupille and Perry (13) have published a complete review of PMD.

15.3 PYROLYSIS AND TLC

Rogers (14) needed a simple method for studying higher molecular weight pyrolysis products as a continuous function of temperature. He devised a method that involved programming a TLC plate across the exit port of a pyrolysis cell as a function of the temperature of the sample. The TLC plate was then developed in the usual manner, and reagent sprays specific for the compounds of interest were employed for detection.

A pyrolysis cell was assembled from a 3/8-in stainless steel Swagelok T-joint, using the 180° leg of the T for sample insertion. The 90° leg of the T is used for the carrier gas inlet and must be positioned vertically. A 4.5-in section of 3/8-in stainless steel tubing was used as the pyrolysis chamber. This chamber was heated by a programmable tube furnace with an inside diameter of approximately 7/8-in. A 4-5 mm platinum or ceramic sample boat may be used for weighing and pyrolysis, but a length of platinum wire must be attached to the boat to enable sample insertion and recovery of the boat. Control and readout thermocouples are welded to the 3/8-in tubing. The sample port may be closed during a run with a brass (to prevent galling) Swagelok plug. A 3/8-in to 1/4-in adapter can be used to fit a section of 1/4-in outside diameter stainless steel capillary to the cell to increase the frontal velocity of the carrier gas downstream from the sample. This section of tubing should be kept as short as possible within the limitations imposed by the dimensions and heating capacity of the tube furnace. Rogers used a 2.5-in section of 0.1-in inside diameter. The sample boat should be as close to this reduced diameter tubing as possible.

The TLC plate is vertically held within 1 mm of the end of the stainless steel capillary and is drawn past the orifice on a trolley as a function of the sample temperature. This movement may be effected by extending the pen carriage of any suitable recorder. Rogers (14) found that a recorder with a variable span that can be set to move exactly 5 in for 20.65 mV was ideal. Such an instrument, operating from a chromel-alumel readout thermocouple at the same position versus an ice reference, can be used to push the plate 1 in per 100°.

Sample is applied along a band 1.5 in from the top of the plate, starting no less than 1 in from the side of the plate when the recorder is zeroed. The starting point is marked corresponding to recorder zero. The recorder is turned on, and the furnace is heated at the desired programmed rate. The plate should be pushed along the trolley track by the recorder at a rate proportional to the heating rate, thereby making it possible to correlate a position on the plate with the sample temperature. At the end of the pyrolysis the plate is allowed to cool; it is then inverted and developed in the chromatography chamber.

It is not necessary to use specific carrier gases for best results. However, specific gases may be used to study particular reactions and interactions during pyrolysis.

Desaga (Brinkmann Instruments, Inc., U.S. distributor) manufactures an apparatus for coupling high temperature vaporization with TLC. Called as TAS (thermomicro, application, separation) oven or Tasomat, its major use is for the separation of volatile substances from nonvolatile substances. The basic assembly is diagrammed in Figure 15.8.

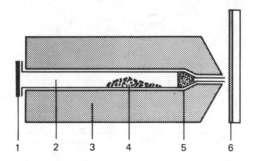

Figure 15.8. Cross-section of TAS oven with cartridge and TLC plate. 1) Seal, silicone disk. 2) Glass cartridge. 3) Heating block. 4) Sample. 5) Quartz wool. 6) TLC layer. Courtesy of Brinkmann Instruments, Inc.

The mixture to be separated is placed into a glass cartridge, one end of which forms a capillary. This tube is inserted into an oven that has been preheated to the desired temperature. The maximum is 350°. Within 15-90 sec the escaping volatile substances are transferred onto a TLC plate mounted on the oven and facing the open end of the capillary. A sideways movement of the plate may result in additional prefractionation into low- and high-boiling-point constituents.

Stahl (15-17) has used the TAS procedure in the investigation of complex substances such as plant material, detergents, and shoe polish. He calls the entire procedure "thermofractography" (TFG), and the chromatogram after development a "thermofractogram." The chromatogram shows the substances separated by temperature on the abscissa, and separated by their chromatographic behavior on the ordinate. A hypothetical thermofractogram is shown in Figure 15.9. The instrument is shown in Figure 15.10.

Because of their high polarity, many high molecular weight natural and synthetic substances cannot be separated directly by

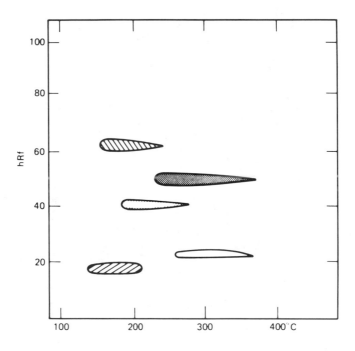

Figure 15.9. Thermofractogram. Substances separated by temperature are on the ascissa, and separated by their chromatographic behavior on the ordinate.

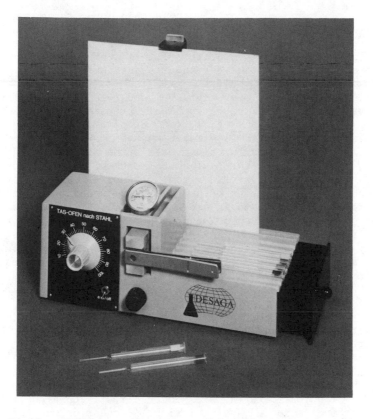

Figure 15.10.   TAS oven.   Courtesy of Brinkmann Instruments, Inc.

chromatographic methods.   By reproducibly degrading these substances to known lower molecular weight compounds, the original
substance may be identified by a "structural unit analysis."   The
best procedures are thermolysis (up to 500°) and pyrolysis (500-
1200°) (17).

15.4   BIOAUTOGRAPHY

Bioautography is often the only way to detect antibiotics that have
been separated on paper or thin layer chromatograms.   It is based
on the biological effects of the substances to be detected.   In
general, these effects can be the inhibition or promotion of growth
of the organism that is exposed to the zone that was separated on a
paper or thin layer chromatogram.   Zones of inhibition in paper

chromatographs and zones of growth in thin layer chromatograms make possible the detection of antibiotics. Several hundred antibiotics have been discovered in the past thirty years. Owing to the wide variations in chemical properties of these compounds, bioautography is the only general method for their detection on thin layer chromatograms.

The beginnings of bioautography appear in the work of Goodall and Levi (18), who applied it to paper chromatograms. Fischer and Lantner (19) and Nicolaus et al. (20) used it fifteen years later for detecting antibiotics separated on thin layers.

In the first steps in the investigation of an unknown antibiotic, when it is not yet available in pure form, bioautography becomes a method for classification of crude antibiotics by thin layer chromatography. The method was developed for identifying the antibiotics contained in a crude mixture during the early stages of isolation of these compounds. The method attempted to assess rapidly the probability that the antibiotic in question was already known. The method was used as a screening technique, and 84 known antibiotics were investigated. The procedure uses three solvent systems to yield four groups of antibiotics. These are subsequently divided into 15 subgroups by use of 11 additional solvent systems.

For these studies, Eastman Chromagram sheets with silica gel 6060 were deactivated in air at room temperature (50-65% relative humidity) for 24 hr before use. The chromatograms were developed in paper-lined chambers in the usual way after sample application. After development, the chromatograms were scanned under ultraviolet light and then subjected to bioautography.

The dried chromatogram was placed directly on a sheet of filter paper over a seeded agar gel slab. Incubation was then carried out at 37° for 18 hr. Inhibition of the growth of specific organisms then could be seen on the agar in the areas where the antibiotic had separated.

There are a number of ways by which bioautography can be performed with TLC. Nicolaus et al. (20) prepared an agar medium inoculated with the test organisms to which triphenyltetrazolium chloride had been added as a dehydrogenase inhibitor. This agar medium was poured over the TLC and incubated in the usual manner. The zones of separated antibiotic will, after a defined incubation time, show up as areas of no growth.

Kline and Golab (21) first sprayed an agar solution onto the TLC plate to solidify it. Then the inoculated agar, cooled to 48°, was poured directly over the surface of the prepared plate. Bickek et al. (22) simply pressed the plate directly on the seeded agar slab.

For chromatograms that are friable and contain no binder, a special technique was described by Meyers and Smith (23). A wet

same size as the plate. The filter paper tabs that extend beyond
each end are folded back over the glass plate, and the assembly is
placed on top of the sorbent of the plate. A sandwich results,
with the sorbent and filter paper between glass plates. The filter
paper tabs are folded back over the ends of the chromatographic
plate on the bottom. The latter is removed, leaving the filter
paper stuck to the glass support by means of the tabs folded over
it. The sandwich is then positioned on the surface of the prepared agar, which has been seeded with the specific organism.
With *Streptococcus lactis*, which grows irrespective of the presence of oxygen, results can be had overnight after incubation at
37°, with the TLC and filter paper lying on the agar surface.
Antibiotics that are active against gram-negative organisms and
antibiotics active against fungi can also be detected when suitable test organisms are inoculated in the agar.

Meyers and Erickson (24) showed that gram-positive organisms
are noninhibitory for *Streptococcus lactis*, and the latter can be
replaced by *Staphylococcus aureus* when the agar is supplemented
with triphenyltetrazolium chloride and potassium nitrate. The
potassium nitrate is an oxygen donor in terminal respiration;
Staphylococcus aureus will then grow on agar plates covered with
the TLC plate during incubation.

The methods described so far are based on inhibition or promotion of growth in the areas where the antibiotic has separated
on the plate placed on the agar. Hamilton and Cook (25) described
the use of the phytopathogenic bacterium *Xanthomonas primi* for
studies in location of antibiotics separated TLC. The organism does not reduce tetrazolium dyes and is an obligate aerobe.
It hydrolyzes gelatin and starch and produces acid from several
sugars. These characteristics can then be used as indicators of
growth when specific color reagents are used on the agar plates.

Zuidweg et al. (26) applied bioautography to layers of
Sephadex used as a sorbent for TLC. After development and drying,
the chromatogram was pressed on the seeded agar plate covered with
a sheet of lens paper. After contact for 30 min the plate was removed and the lens paper taken off carefully. The agar plate was
then incubated at the conditions optimal for the test organism.
The technique of pouring seeded agar over the developed chromatogram can result in the antimicrobial substance spreading over the
plate to produce much larger areas of inhibition. This could be
overcome, as mentioned, by spraying a base of the agar over the
sorbent before pouring the main portion of the agar on the plate.
Pressing the plate on the agar requires the use of a microorganism that can grow anaerobically.

One way to overcome some of these disadvantages uses agar
but depends on formation of a "lawn" on the surface of the agar.
The surface of the hardened but unseeded agar is inoculated with

a heavy suspension of the organisms. After incubation the conflu-
ent growth areas are raised on the surface of the agar, and zones
of inhibition appear clear and depressed. If the colonial form is
opaque or pigmented, the contrast between zones of growth and inhibition is very noticeable against the background of the silica
gel. The spreading effect that results from pouring agar on the
sorbent can be overcome by laying preformed sheets of agar on the
surface of the sorbent.

Perlman et al. (27) have developed bioautography as a means
of bioassay of cytotoxic substances. The differential agar-dif-
fusion procedure for assessing antitumor compounds is based on the
cytocidal effects of the test compound or zones separated on plates.
Eagle's KB cells, Earle's L cells, and L-1210 cells were suspended
in Waymouth's Medium GM752/1 supplemented with calf serum for use
as the assay medium. After development to separate the compounds
of interest, the plate was placed on the agar surface for 40 min
and removed. The plates were then incubated for 18 hr at 37°. For
detection, the plates were flooded with vital stain and the dye
reduction noted. The assay technique measures only cytocidal activity and is based on inhibiting cells from oxidizing glucose.

For preparative TLC, Lefemine and Hausmann (28) located colorless neutramycin on the chromatogram by making a "print" of the
chromatogram with filter paper while the layer was still wet with
the mobile phase. A sheet of glass was used to make even contact
with the support. Enough of the separated antibiotic is transferred to the paper after contact for 10 min. The paper is then
rid of the solvent and pressed against the agar plate. After the
bioautography, the desired zones can be located on the preparative
plate for further analysis or elution.

Homans and Fuchs (29) described a method for detection of fun-
gistatic substances directly on the plate. After development and
drying of the chromatogram to remove solvent, a conidial suspension of a sensitive fungus in a suitable medium is sprayed directly on the layer. The thin layer plates are then incubated in a
moist atmosphere for 2-3 days at 25°. The positions of the zones
of inhibition of growth are measured. From the quantitative
standpoint, bioautography may be helpful, but the relatively high
variability of biological quantitation must be considered. Consequently, quantitative usage has seen little practice.

Cephalosporin C and related compounds can be determined according to the method of Miller (30). The chromatogram was subjected to bioautography with B. subtilis as the test organism.
After incubation the maximum widths of the zones of inhibition
were measured. The average maximum diameters for each of three
standard samples were plotted against the amount of antibiotic on
semilogarithmic paper. As the relative specific activities for
the various compounds were known, it was possible to determine
the amount of cephalosporin C, the deacetylated metabolite, and

the lactone simultaneously in an unknown sample from the single
standard curve for the parent compound. Reaction mixtures containing cephalosporin C and 7-aminocephalosporanic acid were analyzed for the acid after separation. The chromatogram was air
dried and dipped in a solution of phenylacetyl chloride to convert
the reactive zones to the corresponding phenylacetate. This chromatogram was then bioautographed using *Sarcina lutea* as the test
organism.

McDonald et al. (31) reported that the critical parameters
that must be controlled for quantitative bioautography include the
preparation of the TLC layer, sample application, mobile phase selection, and plate storage. The microbiological conditions that
must be controlled include microorganism choice, organism storage,
preparation of medium, layer thickness, and dye systems for visualization. With careful control of these factors, bioautography
can be routinely used in the assay of lasalocid in diverse biological samples at concentrations of 0.5 ppm and higher.

Since it should be possible to control the factors of variability, it appears that bioautography may see more use in the
future. The availability of more uniform plates, media, prepack-
aging, and test organisms along with careful control of conditions
may make it possible to improve the methodology in this area of
investigation. This is important when the method is the only
available one, particularly in the search for new antibiotics.

15.5 ENZYME INHIBITION TECHNIQUES FOR USE WITH TLC

Thin layer chromatography has seen wide use in pesticide analysis.
Enzyme inhibition reactions have been applied to paper and thin
layer chromatography to detect some organophosphorus pesticides
(31-35) and carbamates (35). Mendoza et al. (36,37) have refined
the methodology to the point where it is more reproducible and
has become generally accepted for detection of pesticides from
biological samples.

The method is based on the inhibition of the hydrolysis of
substrate esters by the pesticide. The pesticide combines with
the active sites on the enzyme to prevent its reaction with the
substrate ester, and no chromogenic product will be formed. On
the TLC plate, the areas where the pesticides are located will appear as colorless spots against an intensely colored background.
The enzyme solution is sprayed on the TLC plate after development.
After a short time, the substrate is sprayed on the plate. If the
pH indicator is used, the pesticide zones appear as colored spots
against a colorless or lightly colored background.

Geike (38) used the inhibition of trypsin for the determination of chlorinated pesticides. Later (39) he used phosphatase
and naphthylphosphate as a substrate. Other examples of substrate

are 1-naphthyl acetate, which is detected by reaction with a diazonium salt after hydrolysis. Indoxyl acetate gives a fluorescent
reaction product if it is hydrolyzed.

The detection limits for each pesticide or group of pesticides
vary with the type or source of the enzymes. This was shown by
Winterlin et al. (40) and by Mendoza and co-workers (36,37,41).
The method is sensitive to as little as 0.5 ng of pesticide (carba-
ryl-using pig liver enzyme). Beef liver esterase was much more
sensitive to organophosphorus pesticides than pig liver esterase.
The opposite was true for carbamates. A marked difference in specificity was noted for inhibition of beef and pig esterases due to
malathion and parathion (41). In contrast, carbaryl was detected
at 0.1 ng by the frozen extract of pig liver and at 0.5 ng by the
freeze-dried extract of pig livers or frozen extract of steer livers. Researchers using the methodology should check these enzyme
preparations.

In other work by Mendoza et al. (37), the specificity of the
enzyme interactions was best demonstrated by the carbaryl-pig liver esterase, demeton-pig or monkey esterases, and by the demeton
compounds-chicken liver esterase combinations. Carbaryl at the 1
ng level and demeton at the 50 ng level inhibited the pig liver
esterases regardless of the method of enzyme extraction. The esterases from the other species were inhibited only by carbaryl exposed to bromine. Demeton was inhibitory to monkey esterases
extracted with water or Tris buffer.

The effect of dilution of the enzyme on the sensitivity of
detection of some carbamates and organophosphorus pesticides is
pronounced. Diluted enzyme sprays gave more sensitive detection
than did concentrated sprays. The difference can be as much as
fivefold. This is because in a concentrated enzyme solution the
number of active sites of the enzyme is greater than the pesticide
could block. The effect of pH of the sprays must also be controlled. No organophosphorus pesticides were detected at pH 5.3
with beef liver esterases and 5-bromoindoxy acetate. A pH of
8.3-9.1 appeared maximal (36).

The method has been used in a number of forensic applications.
Heyndricks et al. (42) reviewed cases of parathion poisioning in
humans. They determined the distributions of parathion in various
organs using fresh horse plasma and 1-naphthyl acetate after separating the pesticide by TLC. The detection limits varied from
0.05 μg for fenthion.

Some general details of the method as outlined by Mendoza and
Shields (41,43) are given in the following sections; they must be
modified according to the material from which the pesticide is to
be extracted.

15.5.1 Extraction of Plant Samples

The sample (50 g) and 50 g of anhydrous sodium sulfate were blend-
ed with 150 mL of chloroform and methanol (90:10) for five min in
a Waring blender. This was filtered through Whatman No. 1 paper.
The solids were rinsed twice with the chlorogorm-methanol solution.
Sodium sulfate was added to remove water. Then the filtrate was
diluted to 200 mL in a volumetric flask.
    Appropriate aliquots were concentrated to obtain equivalent
amounts of plant extract. Equivalent amounts of selected pesticide standards were spotted over the plant extracts on TLC plates.
The resulting chromatograms were then compared to determine the
effects of one extract on the enzyme.

15.5.2 Preparation of Enzymes

In a Virtis homogenizer, blend 50 g of fresh livers for 2 min with
180 mL cold distilled water or Tris buffer solution (pH 8.3) containing nicotinamide, each at 0.01 M. Centrifuge the homogenate
at 2000 G at 4° for 5 min. Freeze the supernatant in 13×100 mm
test tubes. The frozen extracts can be used up to several months
after preparation.
    To prepare the spray, dilute 1 part of the thawed extract with
8 parts of 0.05 m Tris buffer solution before use. The dilution
may be varied depending on the intensity of the indicator used on
the TLC plate. For the preparation as described, one tube diluted
8 times is enough to spray 6 plates.

15.5.3 TLC Procedure

The plates used as recommended by Mendoza et al. (43) for most of
their work were silica gel G-HR coated to a thickness of 750 μ.
Silica gel H and G and aluminum oxide DS-5 have also been used.
The particular pesticide to be used will determine which one of
the sorbents may be used. The best one for a particular purpose
must be found by trial. The plates were heated 1 hr at 110° for
activation before use.
    The pesticides and residues were applied on the plates (10
μL). The solutions were made up in alcohol or acetone to contain
from 1 to 100 ng depending on the inhibition expected. Carbaryl
is sensitive at the 1 ng level, while dimethoate requires 10,000
ng for inhibition. The amount applied will depend on the pesticide
under consideration.
    Most pesticides will be resolved after development with a
solution of acetone in hexane (20:80). A mobile phase travel of
15 cm is sufficient. When the plates are dry (20-30 min), the substrate solution is evenly sprayed with the enzyme solution until

the plate is just thoroughly wet. Allow the plate to dry before spraying with the chromogenic solution. Indophenyl acetate (1 mg/mL in acetone) or 5-bromoindoxyl and indoxyl acetate dissolved in ethanol (0.7 mg/mL) and in acetone (5 mg/mL), respectively, can be used as substrates for organophosphorus and carbamate pesticides. If the pesticide has combined with the active sites on the enzyme, clear colorless zones will appear on the dark background of the plate.

Mendoza et al. (43) also investigated the effect of bromine and UV light on the pesticides. Differences in ability to inhibit esterases and in migration rates on TLC were found with the pesticides and their various metabolites. Some of the pesticides showed greater inhibition on exposure to bromine on the TLC plate prior to development. These differences in inhibition among the various pesticides could be used in identification procedures based on the TLC-enzyme inhibition technique.

These TLC methods in combination with enzymic techniques are versatile and sensitive for detecting pesticides. The combined technique is useful in the analysis of pesticide residues in foods and the environment, in metabolic and forensic investigations, and in analysis of samples with interferences that are too great for other methods. Up-to-date reviews of the present status of this technique have been compiled by Mendoza (45,46).

REFERENCES

1. R. A. deZeeuw, Anal. Chem., *40*, 2134 (1968).
2. R. A. deZeeuw, J. Chromatogr., *48*, 27 (1970).
3. R. A. deZeeuw, J. Pharm. Pharmacol., *20*, 54S (1968).
4. F. Geiss and H. Schlitt, Chromatographia, *1*, 392 (1968).
5. S. Sandroni and H. Schlitt, J.Chromatogr., *52*, 169 (1970).
6. A. V. DeThomas and F. Pascual, J. Chem. Ed., *46*, 319 (1969).
7. V. V. Rachinskii, J. Chromatogr., *33*, 234 (1968).
8. S. Turina, Z. Soljic, and V. Marjanovic, J. Chromatogr., *39*, 81 (1969).
9. S. W. Mayer and E. R. Tompkins, J. Amer. Chem. Soc., *69*, 2866 (1947).
10. S. Turina and V. Jamnicki, Anal. Chem., *44*, 1892 (1972).
11. J. A. Perry, K. W. Haag, and L. J. Glunz, J. Chromatogr. Sci., *11*, 447 (1973).
12. T. H. Jupille and J. A. Perry, J. Chromatogr. Sci., *13*, 163 (1975).
13. T. H. Jupille and J. A. Perry, Science, *194*, 288 (1976).
14. R. N. Rogers, Anal. Chem., *39*, 730 (1967).
15. E. Stahl, J. Chromatogr., *37*, 99 (1968).
16. E. Stahl, Analyst, *94*, 723 (1969).

17. E. Stahl and T. Herting, Chromatographia, 7, 637 (1974).
18. R. R. Goodall and A. A. Levi, Nature, 158, 675 (1946).
19. R. Fischer and H. Lantner, Arch. Pharm., 294, 1 (1961).
20. R. J. R. Nicolaus, C. Coronelli, and A. Binaghi, Farmaco (Pavia) Ed. Pract., 16, 349 (1961).
21. R. M. Kline and T. Golab, J. Chromatogr., 18, 409 (1965).
22. H. Bickek, E. Gaumann, R. Hutter, W. Sackman, E. Vischer, W. Vosen, A. Wettstain, and H. Zahner, Helv. Chim. Acta, 1396 (1962).
23. E. Meyers and D. A. Smith, J. Chromatogr., 24, 129 (1967).
24. E. Meyers and R. C. Erickson, J. Chromatogr., 26, 531 (1967).
25. P. B. Hamilton and C. E. Cook, J. Chromatogr., 35, 295 (1968).
26. M. H. Zuidweg, J. G. Oostendorp, and C. J. K. Box., J. Chromatogr., 42, 552 (1969).
27. D. Perlman, W. L. Lummis, and H. J. Griersbach, J. Pharm. Sci., 58, 633 (1969).
28. D. V. Lefemine and W. K. Hausmann, Antimicrob. Ag. Chemother., 134 (1963).
29. A. L. Homans and A. Fuchs, J. Chromatogr., 51, 327 (1970).
30. R. P. Miller, Antibiotic Chemother., 13, 689 (1962).
31. A. McDonald, G. Chen, P. D. Duke, A. Popick, and R. A. Saperstein, Abstracts, 169th National Meeting, American Chemical Society, Philadelphia, Pa., April 7, 1975.
32. W. P. McKinley and S. L. Read, J. Ass. Off. Agric. Chem., 45, 467 (1962).
33. S. P. McKinley and P. S. Johal, J. Ass. Off. Agric. Chem., 46, 840 (1963).
34. R. Ortloff and P. Franz, Z. Chemie, 5, 388 (1965).
35. J. J. Men and J. B. Bain, Nature, 209, 1351 (1966).
36. C. E. Mendoza, P. J. Wales, H. A. McLeod, and W. P. McKinley, Analyst, 93, 34 (1968).
37. C. E. Mendoza, D. L. Grant, B. Braceland, and K. A. McCully, Analyst, 94, 805 (1969).
38. F. Geike, J. Chromatogr., 52, 447 (1970).
39. F. Geike, J. Chromatogr., 61, 279 (1971).
40. W. Winterlin, G. Walker, and H. Frank, J. Agr. Food Chem., 16, 808 (1966).
41. C. E. Mendoza and J. B. Shields, J. Chromatogr., 50, 92 (9970).
42. A. Heyndrickx, A. Vercruysse, and M. Noe, J. Pharm. Belg., 127 (1967).
43. C. E. Mendoza and J. B. Shields, J. Ass. Off. Anal. Chem., 54, 507 (1971).
44. C. E. Mendoza, P. J. Wales, D. L. Grant, and K. A. McCully, J. Agr. Good Chem., 17, 1196 (1969).
45. C. E. Mendoza, J. Chromatogr., 78, 29 (1973).
46. C. E. Mendoza, Residue Rev., 50, 43 (1974).

CHAPTER 16

The Combination of Thin Layer Chromatography with Other Analytical Techniques

16.1 INTRODUCTION

Thin layer chromatography is a separation process that is useful in a number of ways. The R_f values obtained for separated substances and the comparison of these substances with known compounds separated on the same plate serve as useful guides for the identification of an unknown but are not sufficient for absolute identification of the unknown.

The careful combination of TLC with other chromatographic techniques and with appropriate identification techniques will provide sufficient information about a substance to lead to its identification. Column, high pressure liquid, and gas chromatography are the three most widely used chromatographic techniques in conjunction with TLC, with paper chromatography and electrophoresis being used less frequently. Chromatographic procedures are excellent separation methods for complex mixtures but are of little value by themselves in the identification of the structure of the separated substance. Hand in hand with the chromatographic procedures are the identification methods, which are useful for elucidating the structure of a substance, but these methods require relatively pure material, which the chromatographic procedures can provide.

Identification techniques used include:

1. Chemical reaction such as colored product formation, de-

371

rivatization, chemical pyrolysis. See Chapter 7 for such techniques.
     2. Spectrographic procedures such as IR, mass spectroscopy,
and NMR.
     3. Physical methods such as melting point, sublimation, po-
larography.
     4. Biological or physiological methods such as testing an
antibiotic, pesticidal, or bioautography effect. See Chapter 15.

     The chromatographic and spectrographic methods will be dis-
cussed in depth, as they are the most widely used.

16.2 COLUMN CHROMATOGRAPHY

Some of the correlations and uses of the combination of column
chromatography (CC) and TLC have been reviewed by Janak (1,2) and
by Schlitt and Geiss (3). Since both are forms of liquid chromatography (that is, they employ a liquid mobile phase), separation
mechanisms for TLC and column chromatography become the same if
the mobile phase and sorbent are the same. Often silica gel or
alumina is used as the sorbent in column chromatography, for instance.
     Because TLC has many merits and advantages, it can be used as
a pilot technique to select the best combination of mobile phase
and sorbent for a column separation. Boshoff et al. (4) have recently coupled TLC with high performance liquid chromatography
(HPLC). Solutes from the column were transferred by means of a
fine steel capillary tube (0.25 mm I.D.) on to 5×20 cm chromato-
plates that were moving under the column at a speed of 2 cm/min.
The plate was being heated, and a small vacuum line was placed
above the solute zone at the same time, to evaporate the elution
solvent. With steroids, Boshoff et al. were able to detect 60 ng/
spot, and with the chlorinated pesticide methoxychlor, 0.6 ng/spot.
A spectrofluorimeter was used for detection, after the production
of fluorescent derivatives *in situ*. Snyder (5) claims that the
resolving power of columns in modern, rapid CC is greater than that
of a TLC plate, which is limited by its maximum length. The relative merits of TLC compared to CC are:

     1. Most separated substances may be detected visually, often
with a specific color reaction. Many CC detectors are not specific or sensitive enough for many applications.
     2. The amount of time required to change a set of separation
conditions is very short in TLC compared to CC. It usually only
takes about 0.5 hr to change conditions in TLC, compared with several hours in CC. Column reequilibration after changing the com-

position of the mobile phase often takes at least 1 hr.
    3. With TLC it is possible to do many separations quickly.
    4. The major advantage of TLC is its great economy in mate-
rials and time, which can be used to advantage in itself or to determine the optimum set of conditions for a column separation.

    After the separation conditions have been determined and the
column is being developed, the composition of the eluent may be
monitored by applying aliquots taken at given time or volume intervals to a TLC plate and developing and visualizing it. Results
will show which column fractions may be combined, discarded, or
otherwise acted upon.
    The reverse procedures may also be used; that is, an eluted
zone from a TLC plate may be placed in and developed through a
column.

16.3   GAS CHROMATOGRAPHY

There are two major ways in which thin layer chromatography may be
combined with gas chromatography: indirectly and directly. The
majority of TLC/GC work is done indirectly: The sample is developed on a TLC plate, the separated areas may be tentatively identified, and desired zones are eluted from the plate, concentrated,
derivatized, or otherwise modified, and then subjected to separation and identification by gas chromatography. A number of applications employing this general procedure may be found in the
literature; these include analysis of steroids (6-14), cardiac
glycosides (digoxin) (15), amines (16) nitro compounds (17,18),
imidazoles (19), organophosphorus compounds (20), fat-soluble
vitamins (21), water-soluble vitamins (22), insecticides (23),
pesticides (24,25), Cannabis hallucinogens (26), drugs of abuse
(27,28), antioxidants (29), and unsaturated fatty acid methyl
esters (30-33).
    Two examples are chosen to illustrate the indirect combination
of TLC with GC. Working with urinary estrogens, Touchstone et al.
(7) extracted the steroid conjugates from an entire 24 hr urine
sample and hydrolyzed the extract to release the steroids. The
hydrolysate was extracted several times with ether; then the ether
extract was washed and subjected to an extraction to obtain the
phenolic steroids (estrogens) present. The estrogen extract was
applied to the origin of a silica gel G plate, and known estrogen
standards were applied next to the extract. The plate was developed in 15% acetone in isopropyl ether and dried. The sample lanes
were masked with a clean glass plate, and the reference standards
were visualized with ferric chloride-potassium ferricyanide. The
sample zones corresponding to the standard zones were individually

scraped off into micro separatory funnels containing 15 mL of pH 4 water and extracted three times with 15 mL of ether each time. The ether extracts were washed once with 5 mL of 8% NaHCO$_3$ and twice with 5 mL of water. The ether was evaporated and the residue transferred to a conical tipped tube. A known amount of suitable solvent such as methanol or tertiary butanol was added to the residue, and a suitable aliquot was used for quantitative gas chromatography.

The TLC separation serves to isolate sample components of interest from the unwanted sample contaminants and provide a tentative identification for them. The gas chromatography provides further separation, identification, and, in this case, quantitation of the desired estrogens. The two methods complement each other well for optimum results.

As a second example of the utility of the combination, consider the work of Privett et al. (30-33). Working with methyl esters of unsaturated fatty acids, they separated them on silica gel layers impregnated with silver nitrate, which provides a separation according to the degree of unsaturation. The zones were eluted and reduced by ozonolysis, and the resultant compounds were analyzed by gas chromatography to obtain additional information for identification.

The second major way in which TLC is combined with GC is by direct physical coupling of the GC instrument with the TLC plate. This method has been reviewed by Kaiser (34) and by Janak (35,36), one of the pioneers in this area (37-39). Kaiser points out three advantages to this coupling:

1. It permits a double chromatographic separation in two-dimensions. Results are obtained by comparison of the gas chromatogram with the GC/TLC chromatogram.

2. The combination offers the possibility of individual and multiple qualitative identification as well as quantitative and qualitative determination of individual compounds.

3. The method is a most critical control procedure. Of major value is the disclosure of contradictions in the qualitative results comparing the gas chromatogram with the GC/TLC chromatogram. Such results will allow any quantitative results to also be checked. Overall the two systems complement each other, because the separations are dependent on different characteristics of the sample components.

The first workers to couple a TLC plate to a gas chromatograph were Casu and Cavallotti (40). They positioned the plate under the effluent of the chromatograph, driving it past by means of a gear train from a potentiometric strip chart recorder. The plate retained the sample components, and *in situ* chemical de-

rivatization was used for identification. The separation capabilities of the TLC process were not actually utilized, as the plate
served only as a trap.

Janak was the first to use the TLC separation process after
coupling a plate to the gas chromatograph (37-39,41). Tumlinson
et al. (42,43) manually moved the plate past the GC effluent to
collect the sample components, and they formed derivatives *in situ*
before developing the plate. Curtius and Muller used independent
plate movement and the coupling procedure for steroid analysis
(44). Humphrey has examined volatile oils by this procedure (45).

It is desirable to move the plate continuously past the effluent and subsequently to be able to compare the visualized areas
on the TLC plate with the peaks produced by the chromatograph recorder. Casu and Cavallotti (40) used the drive gears for the
paper on the recorder to turn other gears, which moved the plate.
The effluent from the gas chromatograph was placed about 1 mm above
the surface of the plate layer. Humphrey (45) used the recorder
chart paper itself as the driving mechanism for the plate. The
recorder was equipped with a flat shelf extending 40 cm at an
angle 45° from the point of emergence of the chart paper. The
chart was allowed to run over the sloping shelf, and the TLC plate
was fixed to the paper with a piece of adhesive tape. The open
capillary end of the heated effluent stream splitter was positioned
2 mm above the plate surface. If a destructive detector such as
the hydrogen flame detector is being used in the gas chromatograph,
it is necessary to split the effluent emerging from the column before it arrives at the detector. A small portion, in this case 2%,
is fed to the detector, the other 98% being allowed to pass out to
the TLC plate. This is the same principle used during fraction
collection in preparative gas chromatography. The effluent capillary to the plate was kept at a temperature of 150° for most applications, but 180° or 200° was necessary for oils containing
phenols or sesquiterpene alcohols.

Ruseva-Atanasova and Janak (41), working with fatty acid
methyl esters, logarithmically programmed the speed of the plate;
this resulted in linear and additive spacing of neighboring mem-
bers of a homologous series. Their work illustrates the utility
of the TLC/GC coupling. Gas chromatography of methyl esters on a
nonpolar stationary phase (the first-dimension separation) resulted in a nice separation of the fatty acids according to the number
of carbon atoms in the molecule, regardless of the number of double bonds. A second-dimension separation of the esters on a silver
nitrate-impregnated silica gel plate resulted in a separation according to their degree of unsaturation, regardless of the number
of carbon atoms. The logarithmic variations of the speed of the
plate past the effluent of the gas chromatograph ensured that the

identification of the individual compounds on the developed chromatogram became a matter of linear geometric orientation. This method simplifies the comparison of various lipids, and of differences in their composition depending on their biological source.

Ruseva-Atanasova and Janak point out that both the amount of silver nitrate in the silica gel and the polar component of the mobile phase influence the R_f values obtained. The amount of silver nitrate is important for optimum resolution of areas and for well-defined separations without tailing. Their best separations resulted from a minimum content of 1 g $AgNO_3$ per gram of silica gel. An increase in the amount of the polar component in the mobile phase accelerated the migration of the unsaturated acids up to a limit. Thereafter, tailing of the zones resulted. The optimum mobile phase was 70:30 petroleum ether-diethyl ether.

Figure 16.1 shows the chromatograms typically obtained from a coupling of TLC with GC. After the effluent is collected on the plate, developed, and visualized, it is placed alongside the GC recorder tracing (the gas chromatogram) such that the visualized zones correspond to the peaks. Starting points should be carefully marked before separation. Examination of the spots on the plate will indicate whether the GC peaks truly represent single components or whether inadequate separation is occurring in either dimension.

When the temperature of the GC column is being raised by temperature programming, it is advantageous to move the TLC plate past the column effluent at a linear rate. Dilution of the deposited spot is then avoided.

A number of advantages are afforded by coupled or combination TLC/GC analysis that are not totally present with TLC or GC alone. The flexibility of choice of the column packing for the gas chromatography together with the limited choice of the layer material for the TLC plate is a major advantage. Many kinds of GC column packings are available for specific separations, and there is an abundance of literature detailing extractions, purification procedures, and chromatographic conditions for a wide variety of compounds in many fields. The worker desiring to couple TLC with GC or to use them in combination has a wealth of information available for establishing a qualitative and/or quantitative procedure for a single substance or an entire class.

Many commercial suppliers of TLC plates, equipment, reagents, and reference compounds are also suppliers of columns, packings, liquid phases, reagents, equipment, and reference compounds for GC. Such suppliers include Analabs, Inc., Applied Science Laboratories, and Supelco, Inc.

Another nice advantage to coupled analysis is the flexibility of detection afforded by the two chromatographic methods. General detectors such as the hydrogen flame and argon ionization and spe-

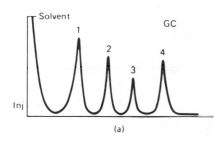

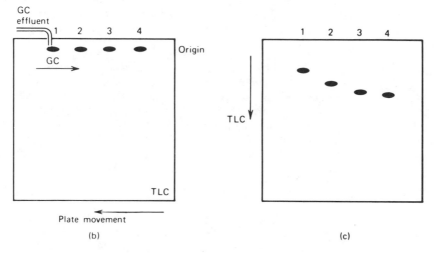

Figure 16.1. Coupling TLC with GC. (a) Gas chromatogram. (b)
Spotted plate. (c) Developed TLC chromatogram.

cific detectors such as the electron capture detector are available
for GC detection according to the type of compounds being separated. This provides some flexibility for the identification of
substances in the first-dimension (GC) separation.

After the separated substances are collected on the TLC plate
and separated in the second dimension, a variety of methods may be
applied for their detection and identification. Many visualization
reagents are very specific in their chemical reaction and provide
a high degree of reliability in their use. In addition to the
qualitative identification afforded by the second-dimension separation and detection procedures, *in situ* quantitative determination
of a wide variety of compound classes is also possible (46). This
is an additional benefit made possible by sophisticated commercial

spectrodensitometers that often saves time, avoids losses, and affords greater sensitivity than elution methods.

Detectors commonly used in gas chromatography may also be used for detection in TLC. This is not the same as coupled GC/TLC, as no separation occurs in a second dimension; only the detector for a gas chromatograph is utilized. Cotgreave and Lynes (47) detected separated substances on a TLC plate by vaporizing them with a furnace and passing them to a suitable detector by means of a gas stream. Padley (48) separated lipids on a thin rod of high-temperature silica glass coated with silica gel G. The rod was then passed through the flame of a flame ionization detector, which combusted and detected the separated lipids. The detector signal was amplified and recorded on a strip chart recorder. Szakasits et al. (49) performed separations of a variety of compounds on 3 mm wide metal strips layered with silica gel. After development, the strip was passed directly between the nozzles of a dual-jet flame ionization detector. The resultant signal was passed to an electrometer, a recorder, and a digital integrator. The strip was passed through at a sufficiently high temperature (300-450°) that all sample components were removed on the first pass. Complete experimental details on the custom fabricated apparatus are given in the paper.

16.4 MASS SPECTROMETRY

The combination of TLC with a mass spectrometer (MS), or with a gas chromatograph coupled to a mass spectrometer (GC/MS), is very useful for the unambiguous identification of separated substances. The major factor limiting the use of such vital combinations for identification has been the high cost of the GC/MS or MS systems. A variety of instrumentation is now available over several price ranges, and this will aid many workers in choosing a system to fit their needs and budget.

Lisboa (50) has shown how the combination of TLC with GC/MS has proved to be indispensable for the positive identification of steroids. Such a combination is valuable for steroids with similar polarities by both TLC and GC, but with different MS fragmentation, and for steroids with similar patterns of mass fragmentation but different chromatographic mobilities. Many examples of the usefulness of this technique in steroid identification are presented.

Majer et al. (51) have combined TLC with high-resolution MS in a highly sensitive, rapid method of analysis for polycyclic aromatic hydrocarbons. Such methods are of major importance in the determination of air pollutants, since many of these particular compounds are carcinogenic and it is therefore necessary to be able to determine very low levels. Majer et al. point out that the sensitivity of the method is several orders of magnitude lower than that of the fluorescence methods that are primarily used for this type of compound.

Details of the quantitative analytical procedure are presented here as a guide for practical application of TLC with MS.

The mass spectra of the individual reference compounds of interest are recorded by evaporating a microgram quantity into the source of the mass spectrometer. On examination of the resulting spectrum, a characteristic ion is selected for the quantitative examination. With the polycyclic compounds this is usually the molecular ion, as it carries a high percentage of the total ion current.

In order to establish calibration curves, known quantities of the compounds of interest are chromatographed in the weight range required. For quantities greater than 10 ng of the polycyclic compounds, the position of the resulting spot may readily be determined by examination under UV light. For smaller quantities, or for compounds not visible under UV, the small quantity must be developed alongside a larger quantity of the same substance that is visualized. The location of the smaller quantity is then found with a straight edge. The spot is marked, scraped off the plate with a spatula, and transferred into a capillary centrifuge tube with the aid of a platinum scoop and a camel hair brush. The sample is then extracted from the layer material by stirring with 20-50 µL of solvent and centrifuging the resulting suspension.

A measured volume of this solution is transferred into the sample probe of the spectrometer. This is introduced into the vacuum system through the insertion lock and held in the cool part of the system until the solvent has evaporated. A reference compound, in this case heptacosafluoro-tri-n-butylamine, is introduced into the source by an alternative inlet for calibration of the mass scale of the instrument. The sample probe is then lowered into the heated part of the source to allow evaporation of the sample while the rise and fall of the ion current is being recorded. The area under this curve is directly proportional to the amount of substance evaporated into the ion source. A calibration curve may be plotted for each known compound.

After an unknown has been separated by TLC, it is eluted and carried through the same procedure. The quantity of the unknown may be obtained by measuring the area under its ion peak and relating this to the calibration curve obtained with the reference compound.

Majer et al. (51) obtained calibration curves that were linear between 10^{-6} and 10^{-10} g. Using the instrument gain and further amplification of the ion current, it is possible to lower the limits of detection still further. They note that the sensitivity may be further increased by decreasing the thickness of the sorbent layer on the TLC plate. A smaller volume of solvent would then be required for extraction.

It is usually necessary to elute the substance to be characterized out of the TLC sorbent before placing it into the spectrometer. If this material were placed directly into the instrument, the instrument would soon become contaminated with sorbent dust, resulting in sporadic operation and eventual shutdown of the instrument for cleaning.

Fetizon (52) isolated several milligrams of substance by extraction from a large number of 250 μm thick plates and used this for mass spectrometry.

Schwartz et al. (53,54), working with metabolites of the drugs diazepam and chlordiazepam, subjected extracts from TLC plates to mass spectrometry. Their spectra had strong background noise, making it difficult to distinguish the signal from the compound being sought. Microfiltration through a membrane filter in a Swinney adapter (see Chapter 12) might have been of value in ridding the extracts of layer sorbent particles.

Boulton and Mayer (55) used TLC/MS for the quantitative analysis of p-tyramines in biological extracts.

Szekely (56) has described a simple technique for preparing an eluted substance for MS without the sorbent layer contamination that leads to high background noise. Shown in Figure 16.2, it consists of a glass capillary 80 mm long and 1 mm inside diameter, such as an open-ended melting point capillary, which is packed with finely powdered potassium bromide sealed in by glass wool. The eluted sample in a small volume (50-100 μL) of volatile solvent is placed into a microcentrifuge tube, reaction vial, or other such vessel. The capillary filled with the potassium bromide is placed into this vessel, and the sample solution is allowed to migrate up. An additional 30-50 μL of solvent is then added to the vessel and allowed to migrate up. This process serves to concentrate the sample substance in the upper portion of the potassium bromide, while the lower portion serves to filter out the sorbent. Most of the lower part of the tube is broken off and discarded. The upper portion of the capillary is introduced through the direct inlet system of the instrument after the sample head has been suitably modified. This procedure requires at least 1 μg of sample.

A microcolumn to hold the TLC sorbent to be eluted may be made from a 230 mm long disposable Pasteur pipet (57), as in Figure 16.3. The sorbent is supported by a small plug of solvent-washed glass wool, and this is eluted with 50-200 μL of solvent. When elution is believed to be complete, the lower portion of the column is broken off and the contained eluate is evaporated on the tip of the solid sample probe of the mass spectrometer. Mass spectra free of impurity peaks have been obtained with as little as 5 μg of sample.

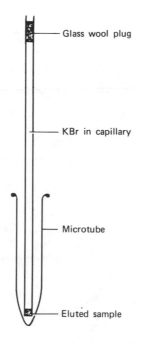

Glass wool plug

KBr in capillary

Microtube

Eluted sample

Figure 16.2. Microcolumn transfer technique (56). See text for explanation.

Throughout any of these elution and preparation procedures, whether for GC, MS, IR, preparative TLC, or other, it is imperative to employ scrupulously clean glassware and apparatus, the finest quality reagents and solvents, and an atmosphere and technique suitable for such detailed micro work. It is desirable, when it is known that isolation and elution are to be carried out, to prewash (predevelop) the TLC plate in a suitable solvent such as methanol. The solvent will carry impurities present in the layer to the top of the plate. If desired, the top 2-3 cm of the layer containing these impurities may then be scraped off and discarded before the plate is used. This procedure may be used for preparative plates also.

Amos (58) has demonstrated the importance of using careful technique and high quality solvents, apparatus, and layer materials. He demonstrated the presence of a readily extractable plasticizer (acetyl tributyl citrate) in the plastic nozzle of a commercial vacuum cleaner type of layer-collection apparatus as described in Chapter 12. The plasticizer was extracted by contacting the dry sorbent being sucked into the extraction thimble.

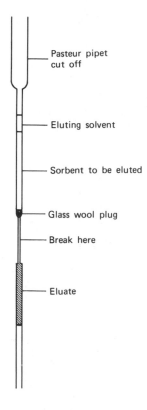

Figure 16.3.   Elution column made from long Pasteur pipet (57).

Removing the plastic nozzle solved the problem, but the glass
plates were scratched. The researcher thought this undesirable and
subsequently used a spatula to scrape off desired zones.

The solvent-extractable impurities in commercial TLC grade
silica gels were measured by Amos. The results are presented in
Table 16.1. These results indicate that Merck silica gel HR contains the least amount of extractable material and that the H
series of silica gel contain fewer impurities than silica gel G or
the other commercial silica gels. The researcher eluting substances from silica gel layers should work with the H and HR series gels whenever possible. Amos notes that the amounts of
ethanol-soluble substances are higher than the acetone-soluble substances because of the increased solubility of inorganic additives
(binders, phosphors, etc.) and silica. Accordingly, polar solvents

TABLE 16.1   SOLVENT-EXTRACTABLE MATERIAL IN TLC GRADE
SILICA GELS (58) (mg/100 g)

	Acetone Solubles	Ethanol Solubles
Merck silica gel		
G	19.2	42.2
H	3.6	28.4
HF 254	4.4	34.0
HF 254 + 366	4.6	31.8
HF 254 silanized	15.4	88.8
PH 254 silanized	7.6	63.0
HR	1.0	28.4
F 254	38.8	92.2
Whatman Chromedia	11.4	23.4
Woelm silica gel	38.2	374.0
B.D.H. silica gel	10.6	73.8
M&B Chromalay	23.2	37.6
Blank on solvent and filter	>0.1	>0.1

such as alcohols are clearly unsuitable for the extraction of TLC
spots for IR or MS examination.

The purity of commercial solvents was also examined by Amos.
He points out that to obtain a spectrum of 10 µg of material
eluted from a TLC spot, no more than 0.5 µg of total impurity can
be tolerated. This equates to not more than 0.5 mL of solvent containing 1 mg/liter of nonvolatile residue being used to elute a
spot (and this does not consider what is eluted as impurity from
the layer material itself). Volumes of a number of commercial solvents were evaporated and the residues weighed to determine if such
criteria could be met. IR spectra were also made of the residues.

It was found that all batches of analytical reagent grade
chloroform were contaminated by phthalate esters. This may be due
to contact with plastic tubing during the bottling process. Ether
residues contained the inhibitor n-propylgallate. Ether is not
recommended for elution; acetone is a suitable substitute. Expensive high purity solvents bottled with plastic or Teflon-lined caps
were found to be generally suitable. When these are not available,
it is suggested that analytical reagent grade solvents be glass
distilled into glass-stoppered bottles of high quality glass.

16.5   INFRARED SPECTROSCOPY

Using the previously described elution techniques and taking suitable precuations, it is possible to isolate quantities of substance suitable for obtaining an infrared spectrum. Additional
elution and purification procedures appropriate for IR work will
be discussed here.

Percival and Griffiths (59) prepared thin layer plates with
a silver chloride layer. Different organic compounds, most of
which were colored dyes, were separated on these plates, and their
infrared spectra were determined *in situ* using an infrared Fourier
transform spectrometer. This method is faster than other methods
but special layers and instruments must be used. This instrument
may also have reduced the detection limits somewhat.

Nash et al. (60) identified the pesticide rotenone in a TLC
chromatogram of technical grade rotenone using potassium bromide
(KBr) microdisks.

McCoy and Fiebig (61) eluted 50-100 µg amounts of separated
substance into the tip of a Pasteur pipet in a manner similar to
the one described in the previous section (Figure 16.3). The
method became cumbersome when the eluting solvent was to be replaced with carbon disulfide or carbon tetrachloride so the spectrum could be run.

Hayden et al. (62) constructed a glass extraction apparatus
similar to a micro Soxhlet extractor for eluting small TLC zones.
The sorbent containing the sample is supported by a small pad of
glass wool rather than an extraction thimble.

A micro collection-elution device also fashioned from a
Pasteur pipet was developed by Clemett (63) and is shown in Figure 16.4. The body of the Pasteur pipet may be cut to any length
to accommodate the desired amount of sorbent material. The tip
end, plugged with glass wool, is connected to a carefully regulated vacuum source, and the desired zone is sucked through the
capillary into the pipet. The plugged pipet may also be placed
into a side-armed test tube, the mouth of which contains a stopper
and the capillary tube going into the pipet body. The vacuum

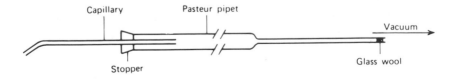

Figure 16.4. Micro vacuum collection device for TLC zones.

source is then connected to the side arm. Vacuum control is easier this way, and the apparatus is not as fragile.

After collection, the pipet becomes a miniature chromatographic column through which the sample is eluted with a small
volume of pure solvent. Clean IR spectra have been obtained with
as little as 20 μg of sample, the eluate being dropped onto a
small amount of KBr powder that was pressed into a microdisk after
the solvent was evaporated. For mass spectrometry, the eluate was
dropped slowly onto the sample probe tip and the solvent was allowed to evaporate. Spectra were obtained from 5 μg of sample.
For gas chromatography, the eluate could be collected in a micro
sample tube, the solvent evaporated under nitrogen, and the residue redissolved in an appropriate amount of suitable solvent for
injection. If evidence of contamination should result from this
method, it may then be desirable to fill the plugged tip of the
pipet with finely powdered KBr to prevent the passage of sorbent
material fines.

Garner and Packer (64) described the Wick-Stick method, kits
for which are available commercially from Harshaw Chemical Co.
The Wick-Stick is a triangle of pressed KBr 25 mm high, 8 mm wide
at the base, and 2 mm thick. For use, the triangle is inserted into
a support holder and placed in a small cylindrical glass vial, all
of which are supplied in the kit. The triangle must stand erect
and not touch the glass. The TLC layer sorbent containing the
sample to be eluted for the IR is placed into the bottom of the
vial, using a funnel so as not to touch the triangle. If a liquid
sample is available, it is carefully run down the side of the vial.
A small volume of solvent for elution is added and the metal dust
cap put on. The elution solvent dissolves the sample out of the
sorbent material and travels up the Wick-Stick by capillary action.
The sample material concentrates at the top of the triangle in the
tip as the solvent evaporates out the hole in the dust cap. Sorbent
material is filtered out by the lower portion of the triangle.
Small samples may be effectively concentrated by adding one or two
more volumes of pure solvent to the vial to repeat the process.
After elution, the triangle is carefully removed with forceps and
placed on a clean inert surface such as a piece of glassine weighing paper. The tip is cut off with a spatula or scalpel and
formed into a KBr pellet in the usual manner. Recovery is usually
on the order of 50-80%, and spectra have been obtained with 10 μg
of separated substance.

Working with amino acid phenylthiohydantoins, Murray and
Smith (65) developed the following simple elution procedure for IR
analysis. The separated area removed from the TLC plate is allowed
to stand overnight with 1 mL of spectral grade 1,2-dichloroethane
in a 100×14 mm tube with a Teflon-lined screw cap. The tube is
centrifuged at about 300 rpm, and 80% of the supernatant is trans-

ferred to a clean microcentrifuge tube with a Pasteur pipet.   Care
is taken not to disturb the sorbent on the bottom of the tube.
The transferred eluate is evaporated under nitrogen until a volume
of 5-25 μL remains.   This is spotted onto a silver chloride plate
or a KBr disk with a microsyringe.   The solvent is carefully allowed to evaporate.   Spectra are recorded after horizontal and
vertical adjustment of the sample holder in the beam condenser to
attain maximum absorbance.   The baseline is set by attenuating
the reference beam, and water vapor bands are eliminated by adjustment of the balance control.   All spectra are recorded with
the slit program designated as "normal."

Spectral grade dichloroethane was used because spectral grade
acetone and methanol contained a ketonic residue (0.0003-0.0005%)
that yielded an intense spectrum upon evaporation of 1 mL volumes.
The dichloroethane must not be exposed to air and light for extended periods, or it becomes unsuitable for this procedure, apparently due to oxidation and/or polymerization reactions.

Rice (66) developed a direct-transfer method that reduced
handling of the area to be eluted to an absolute minimum.   This
was done by marking the area around the sample spot in the shape
of a teardrop.   An area around this teardrop was then scraped away
and discarded for the next step.   Powdered KBr (10-20 mg) was
placed in this cleared area at the tip of the sample spot teardrop.
Pure solvent for elution was added dropwise from a syringe onto
the spot to elute the sample into the KBr.   Amos (58) used this
procedure but found "that considerable losses of solute occurred,
particularly with colorless compounds.   Further more, it was found
to be suitable only for well resolved spots."   The latter point is
evident because of the area around the teardrop spot that must be
cleared to allow for the KBr.

A variation of this technique was used by deKlein (67).   The
difference consisted in arranging the KBr powder in a semicircle
in the clear area around the zone to be eluted.   The powder is
placed 1-2 mm away from the sample area and does not touch it as
in the Rice method.   The major disadvantage to this is that no
portion of the KBr acts as a filter for the sorbent material, as
could be the case with the Rice method.   The whole of the KBr must
be used to form the pellet for the IR.   Szekely (68) has noted
that a good recovery is possible with the semicircle arrangement
of KBr, but the IR spectra obtained through such a procedure ex-
hibited severe contamination, particularly when polar eluting solvents were used.

Amos (58) has developed an elution technique that he prefers.
In its entirety it is as follows: 40 μL of a 5% solution of the
sample is applied as a 2 cm band on a 5×20 cm silica gel HR plate
750 μ thick.   Two μL of a 1% solution of the same sample is chromatographed alongside and used as a reference.   The chromatogram

is developed using only very pure solvents in the mobile phase,
and the reference lane is sprayed with iodine solution (a suitable
one is listed in Chapter 7) so that the location of the separated
components from the large sample band can be ascertained. The
zone desired for elution is scraped off with a microspatula and
placed on top of a small amount of powdered KBr that is tamped
down the cone joint of an 18 gauge syringe needle, as in Figure
16.5. The KBr serves to filter out the fines from the sorbent
material. A 1 mL syringe is filled with high purity acetone and
connected to the needle, and the resulting solute is eluted drop
by drop onto a 10 mg pile of powdered KBr. Each drop is allowed
to evaporate before addition of the next drop. About 20 drops is
usually sufficient to transfer the sample out of the sorbent into
the KBr. The eluted sample and the KBr are mixed with a micro-

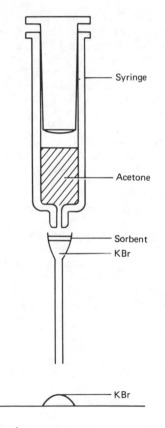

Figure 16.5. Syringe elution apparatus.

spatula and pressed into a 1.5 mm diameter disk. Excellent spectra were obtained on 5-30 µg of solute, using a Perkin-Elmer 521
IR spectrophotometer.

Thin layer separation was combined with GC and IR for the
identification of nitro compounds (69).

16.6 MISCELLANEOUS COMBINATIONS

Irvine and Anderson (70) defined the technique of autotransfer
chromatography (ATC) as chromatography in two or more dimensions,
where the stationary phase is changed and there is facile total
transfer of sample components. Such an example would be the
transfer from TLC to paper chromatography. This technique may
provide additional information about the compounds being separated
by virtue of the direct coupling of the two procedures. Using
this definition, the direct coupling of the effluent of a gas
chromatograph to a TLC plate would be an autotransfer technique.
The TLC/PC coupling provides a good deal of flexibility for the
separation because the nature of the TLC layer and its mobile
phase is variable, the solvent used for the actual transfer from
the TLC layer to the paper is variable, and the mobile phase used
for the development of the paper is still another variable that
may be changed to fit the separation.

Irvine and co-workers (71) used this technique and combined
it with elution from the paper and mass spectrometry for the identification and characterization of pyrroles and indoles from urine
extracts.

Differential thermal analysis (DTA) has been combined with
TLC for the identification of dicarboxylic acids at the 20-60 µg
level (72).

One of the newest techniques used in combination with TLC is
photoacoustic spectrometry. Rosencwaig and Hall (73) note that
the use of conventional spectrometric methods for the location of
compounds on TLC plates has not proved to be reliable, especially
in the UV wavelength region, because the sorbent material is usually too opaque to perform transmission spectrometry and too highly scattering to carry out reflection spectrometry. Photoacoustic
spectrometry is able to locate compounds *in situ* on the plate.

In this technique, a solid sample is placed inside a closed
cell containing air and a sensitive microphone. The sample is
then illuminated with chopped monochromatic light, and any light
absorbed by the solid is converted, in part or in whole, into heat
by nonradiative transition processes. The resultant periodic heat
flow from the solid absorber to the surrounding gas creates pressure fluctuations in the cell that are detected by the microphone.
There is a close correspondence between the amount of light ab-

sorbed by the sample and the magnitude of the acoustic signal. A photoacoustic spectrum thus corresponds, qualitatively at least, to an optical absorption spectrum, provided that the nonradiative processes dominate in the dissipation of the absorbed light energy. Since only the absorbed light is converted into sound, light scattering presents no problems. Rosencwaig has obtained optical data on miscellaneous organics, inorganic materials (74), and hemoproteins (75).

Issaq and Barr (76) combined TLC with flameless atomic absorption spectrometry (FAAS) to identify an inorganic compound in an impure organometallic complex and to determine the recovery and purity of the organometallic samples. Samples of the organometallic tellurium diethyldithiocarbamate (TDDC) always produced two spots upon TLC on silica gel, and it was impossible to determine which spot was the TDDC and which was the impurity. To distinguish between them, both spots were eluted from the plate using an automatic elution system and subjected to atomic absorption spectrometry for tellurium. Only one spot showed the presence of tellurium, and this therefore was the TDDC.

The many examples cited in this chapter show how the separation methods of thin layer chromatography complement the analytical methods necessary for the absolute identification of a substance. TLC provides an excellent purification method for separating a sought substance or substances from other contaminants in the sample. Analytical techniques can then be used to identify the separated substance.

REFERENCES

1. J. Janak, J. Chromatogr., *48*, 279 (1970).
2. J. Janak, in *Pharmaceutical Applications of Thin Layer and Paper Chromatography*, K. Macek, Ed., Elsevier, Amsterdam, 1972.
3. H. Schlitt and F. Geiss, J. Chromatogr., *67*, 261 (1972).
4. P. R. Boshoff, B. J. Hopkins, and V. Pretorius, J. Chromatogr., *126*, 35 (1976).
5. L. R. Snyder, J. Chromatogr. Sci., 7, 360 (1969).
6. J. C. Touchstone and M. F. Dobbins, J. Steroid Biochem., *6*, 1389 (1975).
7. J. C. Touchstone, T. Murawec, O. Brual, and M. Breckwoldt, Steroids, *17*, 385 (1971).
8. M. Breckwoldt, T. Murawec, and J. C. Touchstone, Steroids, *17*, 305 (1971).
9. H. H. Wotiz and S. C. Chattoraj, Anal. Chem., *36*, 1466 (1964).
10. W. P. Collins, and I. F. Sommerville, Nature, *208*, 836 (1964).
11. A. Pinelli and M. L. Formento, J. Chromatogr., *68*, 67 (1972).

12. A. Vermeulen, Clin. Chem. Acta, *34*, 223 (1971).
13. F. L. Berthou, L. G. Bardou, and H. H. Floch, J. Steroid Biochem., *3*, 819 (1972).
14. M. Luisi, C. Fassora, C. Levanti, and F. Franchi, J. Chromatogr., *58*, 213 (1971).
15. E. Watson and S. M. Kalman, J. Chromatogr., *56*, 209 (1971).
16. T. A. Smith, Anal. Biochem., *33*, 10 (1970).
17. J. C. Hoffsommer, J. Chromatogr., *51*, 243 (1970).
18. A. Copin, M. Severin, and J. Evrard, J. Chromatogr., *68*, 89 (1972).
19. C. G. Begg and M. R. Grimmett, J. Chromatogr., *73*, 238 (1972).
20. P. J. Bloom, J. Chromatogr., *75*, 261 (1973).
21. J. R. Evand, Clin. Chim. Acta, *42*, 343 (1972).
22. R. T. Nuttall and B. Bush, Analyst (London), *96*, 875 (1971).
23. P. W. Albro, L. Fishbein, and J. Fawkes, J. Chromatogr., *65*, 521 (1972).
24. R. M. Pfister, P. R. Dugan, and J. I. Frea, Science, *166*, 878 (1969).
25. J. L. Radomski and A. Rey, J. Chromatogr. Sci., *8*, 108 (1970).
26. H. V. Street, J. Chromatogr., *48*, 291 (1970).
27. S. J. Mule, J. Chromatogr., *55*, 255 (1971).
28. D. Sohn and J. Simon, Clin. Chem., *18*, 405 (1972).
29. E. E. Stoddard, J. Assoc. Offic. Anal. Chem., *55*, 1081 (1972).
30. O. S. Privett and E. C. Nickell, J. Am. Oil Chemists' Soc., *39*, 414 (1962).
31. O. S. Privett and E. C. Nickell, J. Lipid Res., *4*, 208 (1963).
32. O. S. Privett, M. L. Blank, and O. Pomanus, J. Lipid Res., *4*, 260 (1963).
33. O. S. Privett and E. C. Nickell, J. Am. Oil Chemists' Soc., *41*, 72 (1964).
34. R. Kaiser, in *Thin-Layer Chromatography, A Laboratory Handbook*, E. Stahl, Ed., Springer-Verlag, New York, 1969.
35. J. Janak, J. Chromatogr., *48*, 279 (1970).
36. J. Janak, in *Pharmaceutical Applications of Thin-Layer and Paper Chromatography*, K. Macek, Ed., Elsevier, Amsterdam, 1972.
37. J. Janak, Nature, *195*, 696 (1962).
38. J. Janak, J. Gas Chromatogr., *1*, 20 (1963).
39. J. Janak, J. Chromatogr., *15*, 15 (1964).
40. B. Casu and L. Cavallotti, Anal. Chem., *34*, 1514 (1962).
41. N. Ruseva-Atanasova and J. Janak, J. Chromatogr., *21*, 207 (1966).
42. J. H. Tumlinson, J. P. Minyard, P. A. Hedin, and A. C. Thompson, J. Chromatogr., *29*, 80 (1967).
43. J. P. Minyard, J. H. Tumlinson, A. C. Thompson, and P. A. Hedin, J. Chromatogr., *29*, 88 (1967).
44. H. C. Curtius and M. Muller, J. Chromatogr., *32*, 222 (1968).

45. A. M. Humphrey, J. Chromatogr., *53*, 375 (1970).
46. J. C. Touchstone, Ed., *Quantitative Thin Layer Chromatography*, Wiley, New York, 1973.
47. T. Cotgreave and A. Lynes, J. Chromatogr., *30*, 117 (1967).
48. F. B. Padley, J. Chromatogr., *39*, 37 (1969).
49. J. J. Szakasits, P. V. Peurifoy, and L. A. Woods, Anal. Chem., *42*, 351 (1970).
50. B. P. Lisboa, J. Chromatogr., *48*, 364 (1970).
51. J. R. Majer, R. Perry, and M. J. Reade, J. Chromatogr., *48*, 328 (1970).
52. M. Fetizon, in *Thin Layer Chromatography*, G. B. Marini-Bettolo, Ed., Elsevier, Amsterdam, 1964.
53. M. A. Schwartz, P. Bommer, and F. M. Vane, Arch. Biochem. Biophys., *121*, 508 (1967).
54. M. A. Schwartz, F. M. Vane, and E. Postma, Biochem. Pharmacol., *17*, 965 (1968).
55. A. A. Boulton and J. R. Mayer, J. Chromatogr., *48*, 322 (1970).
56. G. Szekely, J. Chromatogr., *48*, 313 (1970).
57. M. J. Rix, B. R. Webster, and I. C. Wright, Chem. Ind. (London), 452 (1969).
58. R. Amos, J. Chromatogr., *48*, 343 (1970).
59. C. J. Percival and P. R. Griffiths, Anal. Chem., *47*, 154 (1975).
60. N. Nash, P. Allen, A. Bevenue, and H. Beckman, J. Chromatogr., *12*, 421 (1963).
61. R. N. McCoy and E. C. Fiebig, Anal. Chem., *37*, 593 (1965).
62. A. L. Hayden, W. L. Brannon, and N. R. Craig, J. Pharm. Sci., *57*, 858 (1968).
63. C. J. Clemett, Anal. Chem., *43*, 490 (1971).
64. H. R. Garner and H. Packer, Appl. Spectr, *22*, 122 (1967).
65. M. Murray and G. F. Smith, Anal. Chem., *40*, 440 (1968).
66. D. D. Rice, Anal. Chem., *39*, 1906 (1967).
67. W. J. DeKlein, Anal. Chem., *41*, 667 (1969).
68. S. Szekely, in *Pharmaceutical Applications of Thin Layer and Paper Chromatography*, K. Macek, Ed., Elsevier, Amsterdam, 1972, p. 104.
69. A. Copin, M. Severin, and J. Evrard, J. Chromatogr., *68*, 89 (1972).
70. D. G. Irvine and M. E. Anderson, J. Chromatogr., *20*, 541 (1965).
71. D. G. Irvine, W. Bayne, and J. R. Majer, J. Chromatogr., *48*, 334 (1970).
72. R. V. Mangravite, Anal. Chem., *40*, 250 (1968).
73. A. Rosencwaig and S. S. Hall, Anal. Chem., *47*, 548 (1975).
74. A. Rosencwaig, Opt. Commun., *7*, 305 (1973).
75. A. Rosencwaig, Science, *181*, 657 (1973).
76. H. J. Issaq and E. W. Barr, Anal. Chem., *49*, 189 (1977).

Appendix of Suppliers

A. H. Thomas Co.   General supplies
Vine Street at Third
Philadelphia, PA   19105

Alltech Associates   General supplies
2051 Waukegan Road
Deerfield, IL   60015

Analabs, Inc.   Apparatus; plates
Division of Foxboro Analytical
80 Republic Drive
North Haven, CT   06473

Analtech, Inc.   General supplies
75 Blue Hen Drive
Newark, DE   19711

Analytical Instrument Specialties   Spotting apparatus
P. O. Box 596
Libertyville, IL   60048

Applied Analytical Industries, Inc.   Camag apparatus; plates;
4517 Franklin Avenue   sorbents
Wilmington, NC   28403

Applied Science Laboratories   General supplies
P. O. Box 440
State College, PA   16801

Beta Analytical, Inc.   Berthold scanners
309 Newton Square
Coraopolis, PA   15108

Bio-Rad Laboratories   Sorbent material
2200 Wright Avenue
Richmond, CA   94804

Brinkmann Instruments, Inc.   Apparatus; plates; sorbents
Division of Sybron Corporation
Cantiague Road
Westbury, NY   11590

Eastman Kodak Co.   Reagents; Chromagram sheets
343 State Street
Rochester, NY   14650

Gelman Sciences, Inc.   Supplies; plates
600 S. Wagner Road
Ann Arbor, MI   48106

J. T. Baker Chemical Co.   Reagents; plates, so bents
222 Red School Lane
Phillipsburg, NJ   08865

Kontes Glass Co.   Apparatus; densitometer
Spruce Street
Vineland, NJ   08360

MCB Reagents   Reagents; plates; sorbents
2909 Highland Avenue
Cincinnati, OH   45212

New England Nuclear   Radioisotope supplies
540 Albany Street
Boston, MA   02118

Pharmacia Fine Chemicals, Inc.   Sephadex
800 Centennial Avenue
Piscataway, NJ   08854

Pierce Chemical Co.   Derivatization reagents
Box 117
Rockford, IL   61105

Regis Chemical Co.   PMD apparatus
8210 Austin Avenue
Morton Grove, IL 60053

Schleicher and Schuell, Inc.   Plates; sorbents
Keene, NH 03431

Schoeffel Instrument Co.   Plate scorer; densitometer
Kratos
24 Booker Street
Westwood, NJ 07675

Shandon Southern Instruments   General supplies
515 Broad Street
Sewickley, PA 15143

Supelco, Inc.   General supplies
Supelco Park
Bellefonte, PA 16823

Tridom Chemical Co. C_{12} bonded plates
255 Oser Avenue
Hauppauge, NY 11787

Universal Scientific, Inc.   Woelm plates; sorbents
2070 Peachtree Industrial Ct.
Suite 101, Box 80402
Atlanta, GA 30341

Whatman, Inc.   Plates; sorbents
9 Bridgewell Pl.
Clifton, NJ 07014

Index